Lecture Notes in Bioinformatics 5817

Edited by S. Istrail, P. Pevzner, and M. Waterman

Subseries of Lecture Notes in Computer Science

Lecture Notes in Mathematics

Francesca D. Ciccarelli
István Miklós (Eds.)

Comparative Genomics

International Workshop, RECOMB-CG 2009
Budapest, Hungary, September 27-29, 2009
Proceedings

Volume Editors

Francesca D. Ciccarelli
European Institute of Oncology
IFOM-IEO Campus, Via Adamello, 16, 20139 Milan, Italy
E-mail: francesca.ciccarelli@ifom-ieo-campus.it

István Miklós
Hungarian Academy of Sciences
Rényi Institute
Reáltanoda utca 13-15, 1053 Budapest, Hungary
E-mail: myklosi@renyi.hu

Library of Congress Control Number: 2009936053

CR Subject Classification (1998): F.2, G.3, E.1, H.2.8, J.3

LNCS Sublibrary: SL 8 – Bioinformatics

ISSN 0302-9743
ISBN-10 3-642-04743-2 Springer Berlin Heidelberg New York
ISBN-13 978-3-642-04743-5 Springer Berlin Heidelberg New York

springer.com

Printed in Germany

Typesetting: Camera-ready by author, data conversion by Scientific Publishing Services, Chennai, India
Printed on acid-free paper SPIN: 12769131 06/3180 5 4 3 2 1 0

Preface

As genome-sequencing costs continue their downward spiral, sequencing of closely related organisms has become increasingly affordable. The growing amount of genomic data available demands for the constant development of computational tools to be applied in comparative genomics. The RECOMB Workshop on Comparative Genomics (RECOMB-CG) is devoted to bringing together scientists working on all aspects of comparative genomics, from computer scientists, mathematicians and statisticians working on novel computational approaches for genome analysis and comparison, to biologists applying these computational tools to study the structure and the evolution of prokaryotic and eukaryotic genomes.

This volume contains the 19 papers presented at the 7th Annual RECOMB-CG workshop held during September 27–29, 2009 at the Renyi Institute, in Budapest, Hungary. The papers published in these proceedings were selected for oral presentation from 31 submissions from scientists around the world. Each paper was reviewed by at least three members of the Program Committee in a stringent and thoughtful peer-review process.

The conference itself was enlivened by invited keynote presentations from Richard Durbin (Welcome Trust Sanger Institute), Jotun Hein (Department of Statistics, University of Oxford, UK), Laurence Hurst (Department of Biology and Biochemistry, University of Bath, UK), Csaba Pál (Evolutionary Systems Biology Group, Biological Research Center, Szeged, Hungary), Lisa Stubbs (School of Molecular and Cellular Biology, University of Illinois at Urbana-Champaign) and Jeroen Raes (EMBL Heidelberg). These talks were supplemented by both the presentations of the papers in this volume and a poster session. Together, talks and papers highlighted the state-of-the-art of comparative genomics tools and applications. From the inference of evolution in genetic regulatory networks, to the divergent fates of gene and genome duplication events, to the importance of new computational approaches to unraveling the structural evolution of genomes, these presentations illustrate the crucial role of comparative genomics in understanding genome function.

RECOMB-CG 2009 would not have been possible without the participation of the many scientists who contributed their time and effort to making the conference a success. We thank the scientists who submitted their work for presentation at the conference, those members of the Program Committee who made every effort to ensure fair and balanced review of the many papers submitted for consideration, the members of the local Organizing Committee for arranging all the myriad details of the organizational aspects of the event, and the members of the Steering Committee for their ongoing dedication and guidance. RECOMB-CG 2009 is also deeply indebted to its sponsors including the National Office for Research and Technology, the Hungarian Academy of Sciences, the Renyi Institute and the Hungarian Society for Bioinformatics.

It is the continued support and dedication of this community that allows RECOMB-CG to bring together comparative genomics researchers from across the globe to exchange ideas and information and focus the force of comparative genomics on improving our understanding of genome evolution and function.

August 2009

Francesca Ciccarelli
Istvan Miklós

Table of Contents

Yeast Ancestral Genome Reconstructions: The Possibilities of Computational Methods

Eric Tannier

INRIA Rhône-Alpes, Université de Lyon, F-69000, Université Lyon 1, CNRS, UMR 5558, Laboratoire de Biométrie et Biologie Évolutive, F-69622, Villeurbanne, France

Abstract. In 2006, a debate has risen on the question of the efficiency of bioinformatics methods to reconstruct mammalian ancestral genomes. Three years later, Gordon *et al.* (PLoS Genetics, 5(5), 2009) chose not to use automatic methods to build up the genome of a 100 million year old *Saccharomyces cerevisiae* ancestor. Their manually constructed ancestor provides a reference genome to test whether automatic methods are indeed unable to approach confident reconstructions. Adapting several methodological frameworks to the same yeast gene order data, I discuss the possibilities, differences and similarities of the available algorithms for ancestral genome reconstructions. The methods can be classified into two types: local and global. Studying the properties of both helps to clarify what we can expect from their usage. Both methods propose contiguous ancestral regions that come very close (> 95% identity) to the manually predicted ancestral yeast chromosomes, with a good coverage of the extant genomes.

1 Introduction

The reconstruction of ancestral karyotypes and gene orders from homologies between extant species is a long-standing problem [7]. It helps to understand the large-scale evolutionary mutations that differentiate the present genomes.

Computational methods to handle gene order and propose ancestral genome architectures have a shorter [17], but prolific [8] history. However, despite the numerous efforts of the computational biology community, two recent rounds of publications have put a doubt on their efficiency. In 2006, comparing ancestral boreoeutherian genome reconstructions made by cytogeneticists on one side and bioinformaticians on the other, Froenicke *et al.* [9] found that the manually constructed and expertized cytogenetics one was not acceptably recovered by computational biology. This provoked many comments [2,15] published the same year. Since then, several bioinformatics teams have tried to approach the manually obtained boreoeutherian ancestor [13,4,1]. In may 2009, Gordon *et al.* [10] published the refined configuration of the genome of an ancestor of *Saccharomyces cerevisiae* (which is approximately as ancient as the boreoeutherian ancestor), and chose not to use any computational framework, arguing that those are still in development and yet cannot handle the available data. It seems to have been confirmed by the publication, a few weeks later, by the Genolevures

F.D. Ciccarelli and I. Miklós (Eds.): RECOMB-CG 2009, LNBI 5817, pp. 1–12, 2009.

consortium [18], of a wide yeast genome comparative study including a reconstruction of an ancestral configuration of some non duplicated species, with a yet unpublished method [11]. Thus the ancestor of duplicated species constructed by Gordon *et al.* [10] is still inaccessible by this approach. Sankoff [16] eventually wrote a comment detailing the deficiencies of computational approaches, yet with an optimistic conclusion for the future.

But right now there still seems to be a gap between formal methods and the application they are made for. This paper intends to fill this gap, and presents efficient implementations of automatic methods on yeast genomes.

In [4], a new computational framework to reconstruct ancestral genomes was presented, generalizing a method by Ma *et al.* [13]. Whereas the earlier methods were all based on a distance minimization principle that is never used in the manual reconstructions, these ones tend to mimic the principles used by cytogeneticists on mammalian data, which appear to be very close the the ones used by Gordon *et al.* [10] on yeast genomic data. This framework can handle the presence of a whole genome duplication in the history of the available species, as it was already remarked in [14], a study using teleost fish genomes.

It was shown in [4] that implementing this method on mammalian genomes resulted in a boreoeutherian ancestor which is very close to the one proposed by cytogeneticists. It shows in particular more similarity than the results of the distance based methods. Here I applied two different frameworks (the one described in [4] and the distance based "Guided Genome Halving" of Zheng *et al.* [20]) on the same yeast data used by Gordon *et al.* [10] and compared the results with the manually constructed ancestral configuration. I used genomic markers that cover a large part of the extant genomes, and which allow to apply both methods on the same instance. Analyzing the differences between the three (one manually and two computationally) constructed ancestors helps to understand the behavior of automatic methods and show what can be expected from them at the present time.

2 The Two Kinds of Methods

In this section, I describe the concepts of the two kinds of methods, and why they often give different results. These reasons, to my mind, are an important part of the terms of the debates that have sometimes opposed biologists and bioinformaticians [9,2,15]. More details on the local method can be found in [4], and on global methods in [8].

2.1 The Local Method

The *local* method consists in three steps:

- identifying *ancestral markers* and their homologs in extant genomes;
- searching for *ancestral syntenies*, *i.e.* sets of ancestral markers that are believed to be contiguous in the ancestral genome, and weighting these ancestral syntenies according to the confidence put in their presence in the ancestral genome, guided by its phylogenetic signal;

- assembling a subset of non conflicting ancestral syntenies into *contiguous ancestral regions.*

It is a general principle implicitly used in cytogenetics studies [9] on mammalian genomes as well as by Gordon *et al.* [10] on yeast. It has been computationally implemented first by Ma *et al.* [13], and then formalized and generalized in [4,14]. It is also used as a first step in [18].

All three steps can be implemented in several ways. The third one benefits from a combinatorial framework based on the consecutive ones property of binary matrices and PQ-trees of weakly partitive set families. A set of ancestral syntenies is said to be conflicting if there is no order of the ancestral markers in which for every ancestral synteny AS, the markers of AS are contiguous. The choice of the subset of non conflicting syntenies is done by combinatorial optimization, maximizing the weight of the subset.

Because the criteria for the presence of an ancestral synteny are stringent — every synteny has to be supported by at least two species which evolutionary path goes though the constructed ancestor, or by its presence in two exemplars in a duplicated species —, this method cannot guarantee that contiguous ancestral regions actually are the chromosomes of the ancestor, but they are chromosome segments. And actually this kind of method often finds slightly more contiguous ancestral regions than the believed number of chromosomes [13,4]. It is also the case for the present study on yeasts (see below). It is a weakness because such methods are often not able to reconstruct full chromosomes, but it is also a strength because every adjacency is well supported, and then is easily examined by manual expertise.

2.2 The Global Method

The first step of the *global* method also consists in identifying *ancestral markers* and their homologs in extant genomes. Then it relies on the definition of a distance that compares two arrangements of the markers. Ancestral genomes are the arrangements that minimize the sum of the distances along a phylogenetic tree.

Like the local method, it has many possible implementations. The principle has first been stated by Sankoff *et al.* [17], and since then many developments have been published, according to the chosen distance and kind of genome [8]. Often the distances are a number of rearrangements that are necessary to transform one genome into another. The classical possible rearrangements are inversions, fusions, fissions, translocation, and block interchanges, sometimes all modeled under the double cut-and-join (DCJ) principle. Such methods have been applied to mammalian and nonduplicated yeast genomes [1,18], and the possibility of whole genome duplications has been added and applied to plant or yeast genomes [20], though all programs handling this possibility are currently limited to the comparison of only two genomes. Statistical models have also been constructed on genome evolution by reversals and applied to the computation of ancestral arrangements of groups of bacteria [5].

2.3 Why the Methods Give Different Results

The debate on boreoeutherian genomes [9,2,15] mainly focused on differences of data acquisition, from homologies detected by chromosome banding or in-situ hybridization to sequence alignment from sequenced and assembled genomes. The similarity of the cytogenetics ancestors [9] and the ones derived from the local method on genomic data [4] eventually drives the debate towards a difference of methodological principle more than differences between disciplines.

Gordon *et al.* [10] and Sankoff [16] give several reasons why automatic methods (and they only think about global methods there) are not yet giving as good results as manual ones on yeast genomes. Some of them call some comments, and an important addition on the differences of objectives between two different methodologies.

- *The ancestral markers should have exactly two homologs in the species affected by the whole genome duplication, while most genes have lost a copy after begin duplicated.* It is true, but ancestral markers are not necessarily genes, and I will show in this paper that the global method can be applied on markers covering almost all extant genes using the double synteny principle.
- *Automatic methods are currently limited to three species if they have to take into account a whole genome duplication event.* This is true for global methods, but I will show that even the results using only two species compare well with manual results on 11 species. The local method can take all 11 species into account, even if the implementation I provide uses two reference species and only the comparisons of the others to these two.
- *There is a large number of optimal solutions.* This is indeed a major problem of both the local (which also has an optimization step) and global methods. It is solved by statistical methods [5] presenting a set of solution according to a probability distribution. But no such method is yet implemented for eukaryote chromosomes.
- *The rearrangement models are too simple* [16]. This is avoided by the local method, which is model-free. It will also be solved on global method by introducing probabilistic models as in [5].

In addition to these good reasons, an important difference between global and local methods (whether they are computationally or manually implemented) is that by definition they use different objective functions. In global methods, the objective is to minimize a distance, while in local methods, it is to maximize a number of syntenies. The following example illustrated by Figure 1 will show that these functions are not equivalent.

Indeed, suppose one genome $G1$ has two chromosomes, organized in an arrangement of six markers: (A, B) and (C, D, E, F). Suppose another genome $G2$ has evolved from a common ancestor, but underwent a branch specific whole genome duplication, followed by some rearrangements, and finally has five chromosomes: $(A, D, -B, -E)$, (F, C), (A), (F) and $(E, B, -D, -C)$ (a minus sign indicates that the marker is read in the reverse direction).

The *Guided Genome Halving* problem asks for an ancestral configuration A_{GGH} which minimizes the distance between A_{GGH} and $G1$ plus the distance

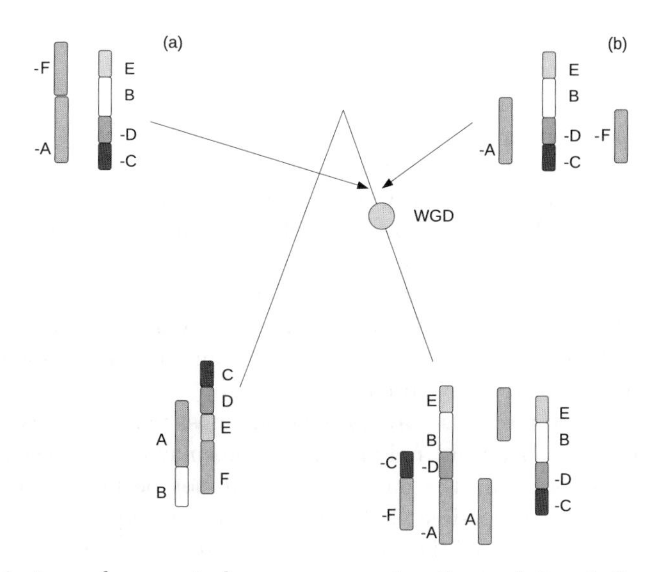

Fig. 1. An instance for ancestral genome reconstruction and its solution using (a) a global method and (b) a local method. All chromosomes are read from top to down for the direction of their markers (a minus sign indicates the reverse direction). The dot indicates the whole genome duplication.

between $A_{GGH} \oplus A_{GGH}$, a duplicated ancestral genome, and $G2$ (for formal definitions, see [19]). Here, taking the DCJ or reversal/translocation distance, there is a unique solution, which has chromosomes (A, F) and $(E, B, -D, -C)$. Its score is 4, since it is at distance 2 from $G1$ (one reversal and one reciprocal translocation), and $A_{GGH} \oplus A_{GGH}$ is at distance two from $G2$ (one fission and one reciprocal translocation). So the argument of the multiple number of solutions does not apply here to explain the presence in the ancestor of this adjacency between markers A and F, which are never seen on the same chromosome on the observed genomes. The scenario has several reciprocal translocations, but these seem to be quite frequent in yeast genome evolution [10], so such a case may not be unlikely. It is also shown by Alekseyev and Pevzner [1] whose Figure 8 shows the reconstruction, with DCJ distance principles, of an ancestral boreoeutherian adjacency which is not an adjacency in any of the observed genomes.

No local method can fuse A and F together. Applying a local method on the same two genomes, ancestral syntenies will fuse C and D, D and B, B and E (all are seen in $G1$ and $G2$, or twice in $G2$). The best ancestral arrangement then consists in the contiguous ancestral regions (A), (F), and $(E, B, -D, -C)$. This genome is at distance 2 from $G2$ (one fusion and one reciprocal translocation), and at distance 3 from $G1$ (one reversal, one fission and one reciprocal translocation), so does not have the best score according to the global method criterion.

Adjacencies like the fusion of A and F in the global method, which are not (or weakly) supported by the data, are difficult to expertize, contrary to every adjacency found by the local method.

On the opposite, there is some proved convergence between the two principles. For example, if an adjacency between two markers is seen at least twice in the data (along an evolutionary path containing the wished ancestor), then the local method will propose it and it will be in the result if it is not in conflict with another synteny (for only two species, there is no possible conflict). And as proved for example in [21], there is always an optimal solution to the global method which contains this adjacency. It may also be true for syntenies that are not adjacencies. The relation between common intervals and adequate subgraphs [21] is still to be explored.

In summary, local methods may not give the most parsimonious solution according to the global criterion, and global methods may infer ancestral syntenies that are not supported by the data, so they cannot be inferred by any local method. Whereas it is possible that the methods converge on major parts of the reconstructions, most of their divergences, observed in [9] on mammals for example, can be explained that way.

On yeast data, the landmark study of Gordon *et al.* [10] allows to perform the comparisons between their manual local approach, and automatic implementations of both principles. I now report the results I got by trying two methods on the same dataset, in order to compare it with the reference ancestor of Gordon *et al.* [10].

3 A Local Method on Yeast Data

3.1 Identification of the Markers

All data I used are from the Yeast Gene Order Browser [3]. For rearrangement studies, the identification of the markers is always the first step. It consists in finding orthologous regions in at least two species. In a whole genome duplication context, it requires that each marker is present in two copies in genomes that derive from this duplication. As the majority of duplicated genes have lost one copy, it is impossible to retain genes as markers and have a good coverage of the genomes as noted in [10]. In this case the markers are obtained with the principle of *double conserved synteny* or DCS [12,6]. It consists in using one duplicated genome and one non duplicated genome, and identify on the first one segments that show homology with two segments of the second one. The principle is well known but rarely formalized. It is for example used by Gordon *et al.* [10] with a human eye to detect the double synteny signal.

I used the genomes of *Lachancea kluyveri* (*Lk*) and *Saccharomyces cerevisiae* (*Sc*), and scanned the genomes looking for a set S of gene families, which have one gene in the first genome, and at least one in the second, and verify

1. the genes of S are contiguous in Lk;
2. the genes of S form two contiguous segments A_{Sc} and B_{Sc} in Sc, of at least 2 genes each;

3. the two sets A_{Lk} and B_{Lk} of genes of Lk which have homologs respectively in A_{Sc} and B_{Sc} form two intersecting segments in Lk;
4. At least one extremity of A_{Sc} (resp. B_{Sc}) is homologous to an extremity of A_{Lk} (resp. B_{Lk});
5. S is maximal for these properties.

The first two conditions impose the presence of one segment in Lk and two orthologous segments in Sc, with a minimum size. It is the basis of the double synteny signal. The presence of at least two genes avoids the possible fortuitous presence of one transposed or misannotated gene. The third condition avoids the ambiguous signal of two successive single syntenies. The fourth condition, which is not necessary for the local method, is used to orient the markers. Indeed, if at least one extremity of two orthologous segments coincide, it is possible to decide a relative orientation of the two. As the global methods work with oriented markers, this condition is useful to run the two methods on the same set of initial markers. This gives a set of 212 makers, which sizes range from 1495bp to 95355bp on the *Lachancea kluyveri* genome, with a median size of 18965bp.

Note that Gordon *et al.* [10] exclude subtelomeric regions from their study because of a lack of synteny conservation. Here these regions are automatically excluded because no DCS — if there are ancestral genes — is found.

Using only these two genomes, 4552 gene families have one gene in Lk and at least one in Sc, so this is a bit less than the 4703 ancestral genes of Gordon *et al.* [10], who used all 11 species. The 212 DCS cover more than 96% of these genes, so both methods will end up with ancestral genomes that have a good coverage of the extant genomes, and take most of the ancestral genes of [10] into account ($> 93\%$). All these DCS are common with the study of Gordon *et al.* [10], who find 3 DCS more, because of a signal I lost by not considering all ancestral genes. The definition I used is then probably very close to the principles applied in [10] for their manual investigation. The number of DCS of Gordon *et al.* [10] is computed in a different way, that is why they claim 182 DCS while they find more DCS in our sense.

3.2 Adjacencies and Common Intervals

As it is done in [4], I computed *ancestral syntenies* by compiling:

- *adjacencies*, *i.e.* pairs of markers that are consecutive in both species or twice in the duplicated species (this is close to the principle described by Figure 1 in [10]);
- *maximal common intervals*, *i.e.* subsets of markers that are contiguous in both species or twice in the duplicated species (Gordon *et al.* [10] also use this kind of relaxed synteny in the case of a synteny signal which is blured by translocations).

This retrieves the information for two species. I then used all the other species mentioned in [10] (6 non duplicated and 5 duplicated species) by computing DCS for *Lachancea kluyveri* and all duplicated genomes (resp. all non duplicated genomes and *Saccharomyces cerevisiae*). For each such pair of species, I computed the maximal common intervals of DCS, mapped the coordinates of their extremities on *Lk* (resp. *Sc*). Then for each common interval, I considered as an ancestral synteny the subset of markers which intersect these coordinates.

This yields a set of ancestral syntenies. Then the assembly of ancestral syntenies is performed as in [4], by finding a maximum weight subset of ancestral syntenies which incidence matrix has the consecutive one property. In general it is an NP-hard problem, but here a simple branch and bound program finds an optimal solution, which is a sign that the data is not very conflictual. And indeed it removes only 32 ancestral syntenies out of 2172 found. But it does not guarantee that this optimal solution is unique.

3.3 Results

The result of the assembly is a set of *contiguous ancestral regions* (this terminology is borrowed from Ma *et al.* [13]). This vocabulary is preferred to "ancestral chromosomes", because as already said, we cannot guarantee that adjacencies that do not fit the ancestral synteny definition cannot exist. Thus it is not possible to assess that we have the ancestral chromosomes as a result of this method only.

And indeed the process ends up with 13 contiguous ancestral regions, while Gordon *et al.* [10] found 8 chromosomes. All the contiguous ancestral regions have a well-defined marker order: the PQ-tree framework allows to represent all solutions of the consecutive ones property for a binary matrix (see [4]), and here there is only one solution. The compared arrangements of Gordon *et al.* [10] and the present method is illustrated on Figure 2 (a) and (b).

The breakpoints distance between the two (according to the definition in [19], which was also proposed earlier in the supplementary material of [13]) is 9. Which, for 212 markers, makes 95.7% identity. The main differences are that the contiguous ancestral regions are not assembled into chromosomes, and that one contiguous ancestral region fuses segments that are believed to belong to two different chromosomes in [10]. The lack of assembly is due to the fact that the procedure misses some information to link blocks, and this is in part due to the DCS it missed at the marker construction step.

The 95.7% identity still means that a rough computational method can come very close to a manually constructed, curated and expertized result.

4 A Global Method on Yeast Data

I used the same 212 oriented markers and ran the Zheng *et al.* [20] "Guided Genome Halving" program on them. Only two species are taken into account. It finds an ancestral configuration which tends to minimize the DCJ distance to the two genomes. Here no guarantee is given that the solution is optimal, as well as that other solutions may not have the same score. But with this objective function, the solution of Gordon *et al.* [10] is suboptimal, an requires two more rearrangements. The results are drawn on Figure 2 (c), where DCS joined by adjacencies found by all three methods are linked into a single block.

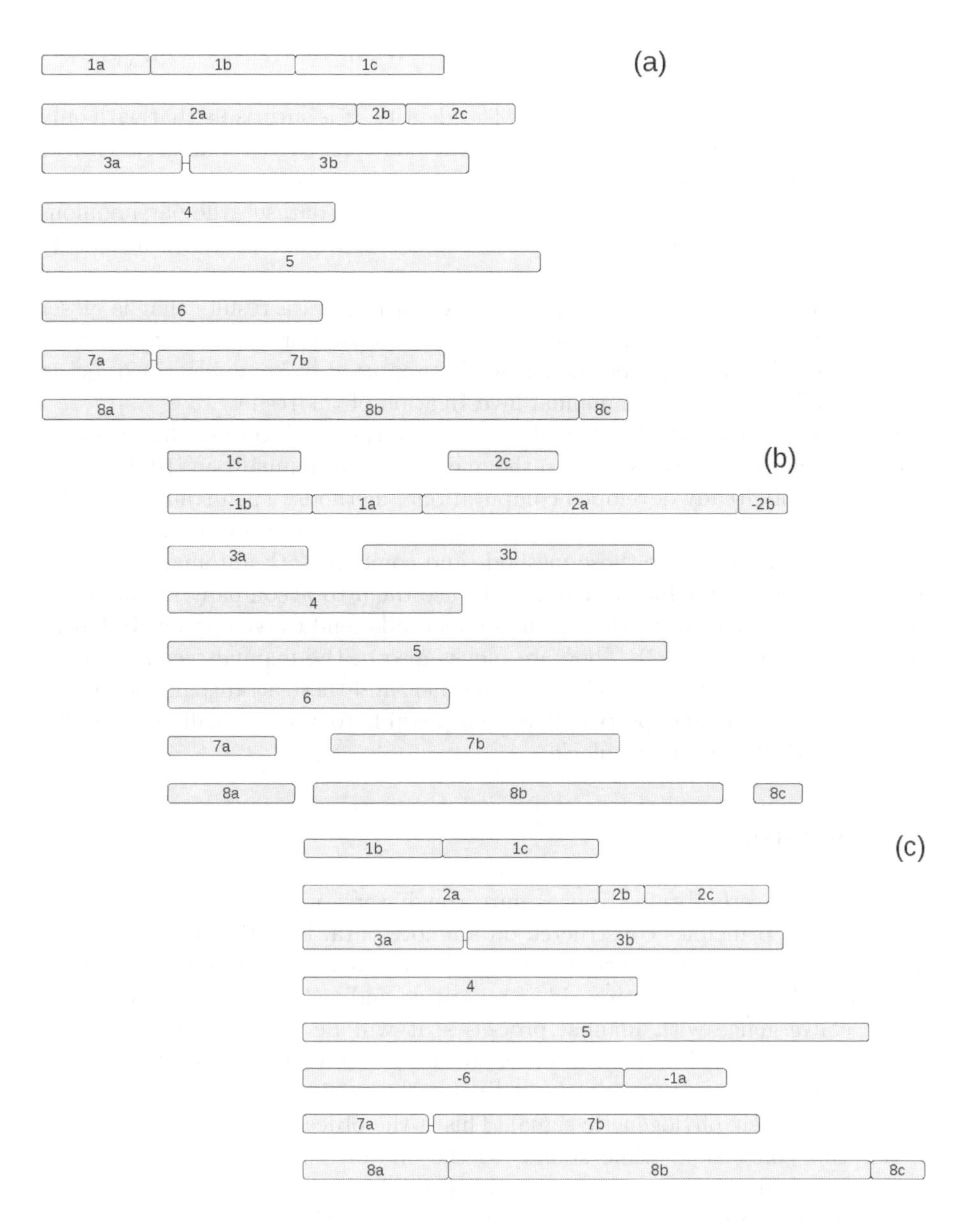

Fig. 2. The arrangements of the yeast ancestral genome obtained (a) by Gordon *et al.* [10], (b) by the local method, and (c) by the global method. Horizontal lines linking two blocks indicate the positions of the DCS that are inferred by Gordon *et al.* [10] but not in this study. Signs on the figures mean the relative orientation of the markers, and numbers refer to the chromosome to which the blocks belong in [10]. Blocks correspond to sets of DCS which are found together by all three methods. The presence of only 16 blocks grouping 212 markers shows the good convergence of the methods.

The constructed ancestor comes very close to the one of Gordon *et al.* [10]. It also has 8 chromosomes, and differs only by a telomeric translocation (where a telomeric segment of a chromosome is translocated to another). The breakpoint distance is 1.5, which gives more than 99% identity. It is surprising that with only two species, the ancestor is so similar to the one constructed manually with 11 species. In this dataset, the possibility to infer less supported syntenies leads to a slightly better result. It is probably because on this dataset, the parsimonious solution in terms of rearrangements is a good approximation of what really happened to *Saccharomyces cerevisiae.*

The drawback of the global method, though it gives a results that is closer to the reference, is that it is difficult to assess the support of each adjacency it infers. In local methods, every adjacency is supported by some identified loci in the data, allowing an easier manual investigation of the result.

Both results, using the local an global frameworks, tend to show that it is possible to obtain very close results to the ones obtained manually in [10] by slight adaptations of already developed computational methods. The manual expertise of Gordon *et al.* [10] will be very precious, at the image of the cytogenetic studies on mammals [9], to refine those methods and come up with automatic methods which are able to convince researchers to use them to accompany a human expertise. For example, here, the automatic methods tend to reconstruct 16 blocks in common (see Figure 2). They are blocks also of the manual reconstruction. The way to assemble those 16 blocks can differ, and these arrangements are still the domain of manual expertise. It is then possible to automatically reduce this work from 4552 genes to 16 blocks.

5 Conclusion

These results were obtained on a quite rough version of automatic methods, adapted from principles constructed on amniote data. But it is sufficient to see that computational frameworks can be used to arrive at very similar results than manual methods while getting rid of repetitive and tedious work. To achieve a better convergence with manual processes, it will be necessary to refine the principles of the local method in order to take all ancestral genes into account, discuss the DCS detection criteria and correctly weight the ancestral syntenies according to their phylogenetic signal. This is the object of a future work.

But this conclusion, while similar to the one of Sankoff [16] since it postpones the total efficiency of computational methods, is a bit more optimistic, since I believe we can already present some results which compare not so badly with the gold standard. With the development of algorithms and methodologies, we have at our disposition all the tools to fulfill the gap between formal algorithmic studies and their application on biological data. My projection of a complete convergence is not later than in an extended version of the present paper.

Acknowledgments

Thanks to Chunfang Zheng for providing her code solving the Guided Halving Problem, to Marie-France Sagot for early discussions on the differences in methodologies, and Ken Wolfe for useful information on the evolution of yeast genomes. E. Tannier is funded by the Agence Nationale pour la Recherche, grant ANR-08-GENM-036-01 "CoGeBi" the Centre National de la Recherche Scientifique (CNRS), and the Ministère des Affaires étrangères et européennes (programme ECO-NET).

References

1. Alekseyev, M.A., Pevzner, P.A.: Breakpoint graphs and ancestral genome reconstructions. Genome Research 19, 943–957 (2009)
2. Bourque, G., Tesler, G., Pevzner, P.A.: The convergence of cytogenetics and rearrangement-based models for ancestral genome reconstruction. Genome Research 16, 311–313 (2006)
3. Byrne, K.P., Wolfe, K.H.: The Yeast Gene Order Browser: combining curated homology and syntenic context reveals gene fate in polyploid species. Genome Research 15(10), 1456–1461 (2005)
4. Chauve, C., Tannier, E.: A Methodological Framework for the Reconstruction of Contiguous Regions of Ancestral Genomes and Its Application to Mammalian Genomes. PLoS Computational Biology 4, 391–410 (2008)
5. Darling, A.E., Miklós, I., Ragan, M.A.: Dynamics of genome rearrangement in bacterial populations. PLoS Genetics 4(7), e1000128 (2008)
6. Dietrich, F.S., Voegeli, S., Brachat, S., Lerch, A., Gates, K., et al.: The *Ashbya gossypii* Genome as a Tool for Mapping the Ancient *Saccharomyces cerevisiae* Genome. Science 304, 304–307 (2004)
7. Dobzhansky, T., Sturtevant, A.H.: Inversions in the chromosomes of Drosophila pseudoobscura. Genetics 23, 28–64 (1938)
8. Fertin, G., Labarre, A., Rusu, I., Tannier, E., Vialette, S.: Combinatorics of Genome Rearrangements. MIT Press, Cambridge (2009)
9. Froenicke, L., Garcia Caldés, M., Graphodatsky, A., Mueller, S., Lyons, L.A., Robinson, T.J., Volleth, M., Yang, F., Wienberg, J.: Are molecular cytogenetics and bioinformatics suggesting diverging models of ancestral mammalian genomes? Genome Research 16, 306–310 (2006)
10. Gordon, J.L., Byrne, K.P., Wolfe, K.H.: Additions, Losses, and Rearrangements on the Evolutionary Route from a Reconstructed Ancestor to the Modern *Saccharomyces cerevisiae* Genome. PLoS Genetics 5(5), e1000485 (2009)
11. Jean, G., Sherman, D.J., Niklski, M.: Mining the semantics of genome superblocks to infer ancestral architectures. Journal of Computational Biology (to appear, 2009)
12. Kellis, M., Birren, B.W., Lander, E.S.: Proof and evolutionary analysis of ancient genome duplication in the yeast *Saccharomyces cerevisiae*. Nature 428, 617–624 (2004)
13. Ma, J., Zhang, L., Suh, B.B., Raney, B.J., Burhans, R.C., Kent, W.J., Blanchette, M., Haussler, D., Miller, W.: Reconstructing contiguous regions of an ancestral genome. Genome Research 16(12), 1557–1565 (2006)

14. Ouangradoua, A., Boyer, F., McPherson, A., Tannier, E., Chauve, C.: Prediction of Contiguous Regions in the Amniote Ancestral Genome. In: Măndoiu, I., Narasimhan, G., Zhang, Y. (eds.) ISBRA 2009. LNCS (LNBI), vol. 5542, pp. 173–185. Springer, Heidelberg (2009)
15. Rocchi, M., Archidiacono, N., Stayon, R.: Ancestral genome reconstruction: An integrated, multi-disciplinary approach is needed. Genome Research 16, 1441–1444 (2006)
16. Sankoff, D.: Reconstructing the History of Yeast Genomes. PLoS Genetics 5(5), e1000483 (2009)
17. Sankoff, D., Sundaram, G., Kececioglu, J.: Steiner points in the space of genome rearrangements. International Journal of the Foundations of Computer Science 7, 1–9 (1996)
18. Souciet, J.L., Dujon, B., Gaillardin, C., et al.: Comparative genomics of protoploid *Saccharomycetaceae*. Genome Research (to appear, 2009)
19. Tannier, E., Zheng, C., Sankoff, D.: Multichromosomal median and halving problems under different genomic distances. BMC Bioinformatics 10(120), 1–15 (2009)
20. Zheng, C., Zhu, Q., Adam, Z., Sankoff, D.: Guided genome halving: hardness, heuristics and the history of the Hemiascomycetes (ISMB). Bioinformatics 24, 96–104 (2008)
21. Xu, A.W.: A Fast and Exact Algorithm for the Median of Three Problem-A Graph Decomposition Approach. In: Nelson, C.E., Vialette, S. (eds.) RECOMB-CG 2008. LNCS (LNBI), vol. 5267, pp. 184–197. Springer, Heidelberg (2008)

Natural Parameter Values for Generalized Gene Adjacency

Zhenyu Yang and David Sankoff

Department of Mathematics and Statistics, University of Ottawa
{sankoff,zyang009}@uottawa.ca

Abstract. Given the gene orders in two modern genomes, it may be difficult to decide if some genes are close enough in both genomes to infer some ancestral proximity or some functional relationship. Current methods all depend on arbitrary parameters. We explore a two-parameter class of gene proximity criteria, and find natural values for these parameters. One has to do with the parameter value where the expected information contained in two genomes about each other is maximized. The other has to do with parameter values beyond which all genes are clustered. We analyse these using combinatorial and probabilistic arguments as well as simulations.

1 Introduction

As genomes of related species diverge through rearrangement mutations, groups of genes once tightly clustered on a chromosome will tend to disperse to remote locations on this chromosome or even onto other chromosomes. Even if most rearrangements are local, e.g., small inversions or transpositions, after a long enough period of time their chromosomal locations may reflect little or none of their original proximity. Given the gene orders in two modern genomes, then, it may be difficult to decide if some set of genes are close enough in both genomes to infer some ancestral proximity or some functional relationship.

There are a number of formal criteria for gene clustering in two or more organisms, giving rise to cluster detection algorithms and statistical tests for the significance of clusters. These methods, comprehensively reviewed by Hoberman and Durand [7], all depend on one or more arbitrary parameters as well as n, the number of genes in common in the two genomes. The various parameters control, in different ways, the proximity of the genes on the chromosome in order to be considered a cluster. Change the parameters and the number of clusters may change, as may the content of each cluster.

In this paper, we define a two-parameter class of gene proximity criteria, where two genes are said to be (i,j)-adjacent if they are separated by $i-1$ genes on a chromosome in either one of the genomes and $j-1$ genes in the other. We define a (θ,ψ) cluster in terms of a graph where the genes are vertices and edges are drawn between those (i,j)-adjacent gene pairs where $\min(i,j) < \min(\theta,\psi)$ and $\max(i,j) < \max(\theta,\psi)$. Then the connected components of the

F.D. Ciccarelli and I. Miklós (Eds.): RECOMB-CG 2009, LNBI 5817, pp. 13–23, 2009.

graph are the (θ, ψ) clusters. These definitions extend our previous notions [10,9] of generalized adjacency, which dealt only with (i, i) adjacency and (θ, θ) clusters (in the present terminology). The virtue of generalized adjacency clusters is that they embody gene order considerations within the cluster. In contrast to r-windows [3] and max-gap clusters [1,8], generalized adjacency cannot have two genes close together in one genome but far apart in the other, although the cluster could be very large.

Of particular interest are $(i, 1)$ adjacencies and $(\theta, 1)$ clusters. These are special in that they pertain to how far apart are genes in one genome that are strictly adjacent in the other genome.

As with other criteria, the quantities θ and ψ would seem arbitrary parameters in our definition of a cluster. The main goal of this paper is to remove some of this arbitrariness, by finding "natural values" for θ and ψ as a function of n, the total number of genes in the genomes. We find two such functions; the first trades off the expected number, across all pairs of genomes, of generalized adjacencies against the parameters θ and ψ, with lower parameter values considered more desirable, i.e., it is good to find a large number of generalized adjacencies, but not at the cost of including unreasonably remote adjacencies.

To do this we first define a wide class of similarities (or equivalently, distances) between two genomes in terms of weights on the (i, j)-adjacencies, namely any system of fixed-sum, symmetric, non-negative weights ω non-increasing in i and j. This is the most general way of representing decreasing weight with increasing separation of the genes on the chromosome. In any pair of genomes, we then wish to maximize the sum of the weights, which essentially maximizes the sensitivity of the criterion. Our main result is a theorem showing that the solution reduces to a uniform weight on gene separations up to a certain value of both θ and ψ, and zero weight on larger separations.

Moreover, the theorem specifies that the optimizing value of $\frac{\theta^2}{2}$ is the record time of a series of $\frac{n^2}{2}$ random variables, where n is the length of the genomes. These are not i.i.d, random variables, however, being highly dependent and, more important, of decreasing mean and variance. We use simulations to investigate the expected value of the record time under a uniform measure on the space of genomes, finding that these increase approximately as $\sqrt{n}$, in contrast with the value $\frac{n^2}{4}$ to be expected if these were i.i.d. variables. Thus it turns out that with genomes lengths of order 10^3 or 10^4, the optimal value of θ is in the range of $8 - 15$.

If we are willing to accept the loss of sensitivity, and prefer to search for clusters more widely dispersed on chromosomes, there is a second set of "natural" parameter values that serve as an upper bound on the meaningful choices θ and ψ. These values are the percolation thresholds of the (θ, ψ) clusters. Beyond these values, tests of significance are no longer meaningful because all clusters rapidly coalesce together. It is no longer surprising, revealing or significant to find large groups of genes clustering together, even in pairs of random genomes.

Percolation has been studied for max-gap clusters [8], but the main analytical results on percolation pertain to completely random (Erdös-Rényi) graphs. The graphs associated with (θ, ψ) clusters manifest delayed percolation, so the use of Erdös-Rényi percolation values would be a "safe" but conservative way of avoiding dangerously high values of the parameters. We show how to translate known results on Erdös-Rényi percolation back to generalized adjacency clusters. We also introduce random bandwidth-limited graphs and use simulations to compare the delays of generalized adjacency and bandwidth-limited percolation with respect to Erdös-Rényi percolation in order to understand what structural properties of generalized adjacency are responsible for the delay.

2 Definitions

Let S be a genome with gene set $V = \{1, \ldots, n\}$. These genes are partitioned among a number of total orders called **chromosomes**. Two genes g and h on the same chromosome are ***i*-adjacent** in S if there are $i-1$ genes between them in S. E.g., 1 and 4 are 2-adjacent on the chromosome 2134.

Let E_S^θ is the set of all i-adjacencies in S, where $1 \leq i \leq \theta$. We define a subset of $C \subseteq V$ to be a **generalized adjacency cluster**, or (θ, ψ) cluster, if it consists of the vertices of a connected component of the graph $G_{ST}^{\theta\psi} = (V, (E_S^\theta \cap E_T^\psi) \cup (E_S^\psi \cap E_T^\theta))$. Fig. 1 illustrates how genomes $S = 123456789$ and $T = 215783649$ determine the $(1,3)$ clusters $\{1,2\}$ and $\{3,4,5,6,7,8\}$.

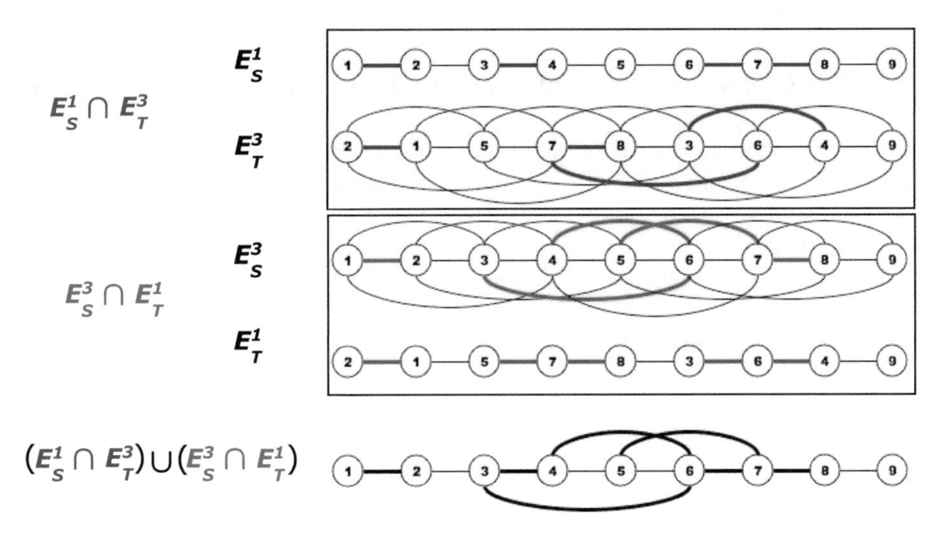

Fig. 1. Determination of $(1,3)$ clusters (or (3,1) clusters)

3 A Class of Genome Distances

Rather than looking directly for natural values of θ and ψ, we first remark that counting all (i,j)-adjacencies (with $i \leq \theta$ and $j \leq \psi$) on the same footing when defining clusters may be giving undue weight to pairs of genes that are remote on one or both genomes, relative to genes that are directly adjacent on both. Thus we are led to consider a general system of (i, j)-dependent weights and to try to optimize this weighting, instead of just the "cut-off" values θ and ψ.

Given two genomes S and T with the same genes. Let ω_{ij} be the **weight** on two genes that are (i, j)-adjacent, i.e., i-adjacent in one of the genomes and j-adjacent in the other, such that

1. $0 \leq \omega_{ij} = \omega_{ji},\ i, j \in \{1, 2, \ldots, n-1\}$
2. $\sum_{i=1}^{n-1} \sum_{j=1}^{n-1} \omega_{ij} = 1$
3. $\omega_{i,j} \geq \omega_{k,l}$ if
   (a) $\max(i, j) < \max(k, l)$
   (b) $\max(i, j) = \max(k, l)$ and
   $\min(i, j) < \min(k, l)$

We define the **distance** between two genomes S and T as

$$d(S, T) = 2(n-1) - \sum_{i=1}^{n-1} \left(n_{ii}\omega_{ii} + \sum_{j=1}^{n-1} n_{ij}\omega_{ij} \right). \tag{1}$$

where n_{ij} is the total number of gene pairs (x, y) that are i-adjacent in S and j-adjacent in T.

4 The Optimum Weight System Is Uniform on $\{i, j\} \leq \theta$

We wish to find a system of weights ω that tends to allocate, inasmuch as possible, higher weight to (i, j)-adjacencies with small i and j, thus emphasizing the local similarities of the two genomes, but not excluding moderate values of i and j, to allow some degree of genome shuffling.

Our strategy is to examine individual pairs of genomes (S, T) first, and that is the topic of this section. In the next section, we will study the consequences of introducing a uniform measure on the space genomes of length n.

Before stating our main results on (i, j)-adjacencies, we prove a special case. We require that every adjacency with non-zero weight be a $(1, j)$-adjacency. Then the definitions of weight and genome distance in Section 2 can be rephrased as following: Let the **weight** ω_i be any non-negative, non-increasing, function on the positive integers such that $\sum_{i=1}^{n-1} \omega_i = 1$. The weight ω induces a **distance** between genomes S and T with the same genes as follows:

$$d(S, T) = 2(n-1) - \sum_{i=1}^{n-1} (n_i^S + n_i^T)\omega_i \tag{2}$$

where n_j^X is the number of pairs of genes that are j-adjacent on genome X and 1-adjacent on the other genome.

Theorem 1. *For genomes S and T, the weight $\boldsymbol{\omega}$ that minimizes the distance (2) has*

$$\omega_i = \begin{cases} \frac{1}{k^*}, & \text{if } 1 \le i \le k^* \\ 0, & \text{otherwise,} \end{cases} \tag{3}$$

where k^* maximizes the function

$$f(k) = \frac{\sum_{i=1}^{k}(n_i^S + n_i^T)}{k} \tag{4}$$

Proof. Let $n_i = n_i^S + n_i^T$. Based on Equation (2), minimizing $d(S,T)$ is equivalent to maximizing the summation

$$R = \sum_{i=1}^{n-1} n_i \omega_i \tag{5}$$

We first note that a uniform upper bound on ω_i is $\frac{1}{i}$. i.e. $0 \le \omega_i \le \frac{1}{i}$. This follows from the non-increasing condition on ω and its sum over all i being 1. Moreover, if $\omega_i = \frac{1}{i}$ for some value of i, then $\omega_1 = \omega_2 = \cdots = \omega_{i-1} = \omega_i = \frac{1}{i}$ and $\omega_{i+1} = \cdots = \omega_{n-1} = 0$, by the same argument.

Now we show that for any solution, i.e., an $\boldsymbol{\omega} = (\omega_1, \omega_2, \ldots, \omega_{n-1})$ that maximizes equation (5), there must be one weight in $\boldsymbol{\omega}$ which attains this upper bound.

To prove this, let weights $\omega_1, \omega_2, \ldots, \omega_{n-1}$ maximize the equation (5) for given values of $n_1, n_2, \ldots, n_{n-1}$, such that $\zeta = \max R$. If all the n_i's are equal or all the ω_i are equal, the theorem holds trivially.

For all other cases, assume that there is no weight in $\boldsymbol{\omega}$ that attains its upper bound. We define the set $\mathcal{C} = \{\, i \mid \omega_i > \omega_{i+1},\ 1 \le i \le n-2\} \ne \varnothing$. Let $\xi = \min_{\mathcal{C}}(\min(\omega_i - \omega_{i+1}, \frac{1}{i} - \omega_i)) > 0$, by assumption. We select two weights ω_i and ω_j where $n_i \ne n_j$. Without loss of generality, we fix $i < j$ and $n_i < n_j$. Then we define

$$\zeta' = \sum_{k=1}^{i-1} n_k\omega_k + n_i(\omega_i - \xi) + \sum_{k=i+1}^{j-1} n_k\omega_k + n_j(\omega_j + \xi) + \sum_{k=j+1}^{n-1} n_k\omega_k \tag{6}$$

$$= \zeta + (n_j - n_i)\xi \tag{7}$$

$$> \zeta. \tag{8}$$

Then ζ is not the maximal value, contradicting the assumption about $\boldsymbol{\omega}$. Hence, there must exist a weight ω_i in $\boldsymbol{\omega}$ attaining its upper-bound $\frac{1}{i}$. Then the optimal weight is $\omega_1 = \omega_2 = \cdots = \omega_{i-1} = \omega_i = \frac{1}{i}$ and $\omega_{i+1} = \cdots = \omega_{n-1} = 0$.

Substituting this $\boldsymbol{\omega}$ in (5), produces the expression of form (4). So maximizing (5) is the same as maximizing (4).

Thus if we set $\theta = k^*$, we should find a large number of generalized adjacencies, but not at the cost of unreasonably increasing the number of potential adjacencies. The cut-off k^* differs widely of course according to the pair of genomes S

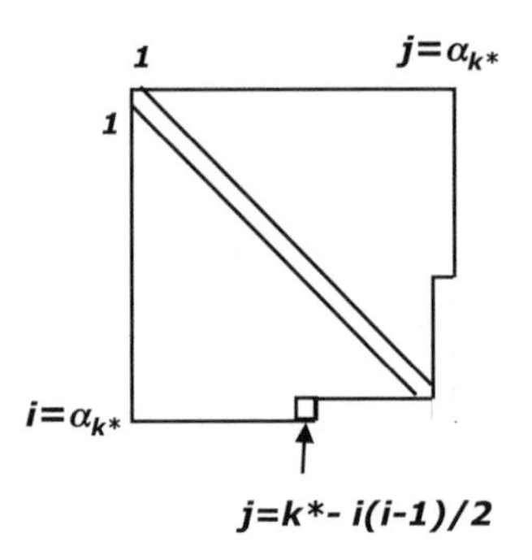

Fig. 2. k is augmented from left to right, starting at the top row, in the lower triangle including the diagonal. Values of ω_{ij} in the upper triangle determined by symmetry.

and T being compared, and this variation increases with n. However, under the uniform measure on the set of permutations, $f(k)$ does not vary much, in the statistical sense, at least as n gets large. Thus we use $\mathrm{E}[k^*]$, as function of n, to find the natural value for the cut-off parameters in the uniform weight-based distance.

Having proved the special case of $(1, j)$-adjacencies, we state the general result for (i, j)-adjacencies. The proof follows the same line as Theorem 1, but its presentation will be postponed to the full version of this paper.

Theorem 2. *Let* $\alpha_k = \lfloor \frac{\sqrt{1+8(k-1)}+1}{2} \rfloor$. *The weight* $\boldsymbol{\omega}$ *that minimizes* $d(S, T)$ *has*

$$\omega_{ij} = \begin{cases} \frac{1}{k^*}, \text{ if } i < \alpha_{k^*},\ j \leq i, \\ \qquad \text{or } i = \alpha_{k^*},\ j \leq k^* - \frac{i(i-1)}{2} \\ \\ 0, \quad \text{otherwise} \end{cases} \tag{9}$$

where k^* *is a natural number and maximizes the function*

$$f(k) = \frac{1}{k} \left[\sum_{i=1}^{\alpha_k - 1} \sum_{j=1}^{i} (n_{ij} + n_{ji}) + \sum_{j=1}^{k - \frac{1}{2}\alpha_k(\alpha_k - 1)} (n_{\alpha_k j} + n_{j \alpha_k}) \right], \tag{10}$$

where n_{ij} *is the number of gene pairs* i*-adjacent on* S *and* j*-adjacent on* T*. (See Fig. 2 for 2-dimensional area measured by* k^**.)*

In analogy to the $(1, \theta)$ clusters mentioned above, we can set $\theta = \psi = \lfloor \frac{\sqrt{1+8(k^*-1)}+1}{2} \rfloor \approx \sqrt{2k^*}$ and use $\mathrm{E}[k^*]$, as function of n, to find the natural value for the cut-off parameters in the uniform weight-based distance.

5 Finding $\mathrm{E}[k^*]$

We will retain the mean of the weight systems over all random pairs of genomes (permutations of length n) as the most natural to use for studying pairs (S, T)

of experimental genomes. The justification of this approach is that it will pick up the maximum resemblance between any two genomes, even random ones, i.e., it maximizes sensitivity. Then the contribution of false positives in reducing $d(S,T)$ for real S and T may be controlled by comparing and "subtracting" the patterns derived from random genomes.

The search for k^* in Theorem 2 over all pairs of random genomes, or even a sample, may seem awkward, but it can be substantially simplified. First we can show the n_{ij} are Poisson distributed, with easily calculated parameters (proof omitted).

Theorem 3. *For two random genomes S and T with n genes the number of pairs of genes n_{ij} that are i-adjacent in S and j-adjacent in T converges to a Poisson distribution with parameter*

$$E(n_{ij}) = \frac{2(n-i)(n-j)}{n(n-1)} \tag{11}$$

Moreover, the number of gene pairs, $N(n,\theta,\psi)$, where the distances are no larger than θ in either of the genomes and no larger than ψ in the other is

$$\begin{aligned} E[N(n,\theta,\psi)] &= \sum_{i=1}^{\theta}\sum_{j=1}^{\psi} E(n_{ij}+n_{ji}) - \sum_{j=2}^{min(\theta,\psi)}\sum_{i=1}^{j-1} E(n_{ij}+n_{ji}) - \sum_{i=1}^{min(\theta,\psi)} E(n_{ii}) \\ &= (4\psi\theta - 2\theta^2) - \frac{2\theta(\psi^2-\theta^2+\psi\theta)}{n} + \frac{\theta(\theta-1)(2\psi^2-2\psi-\theta^2+\theta)}{2n(n-1)}, \end{aligned} \tag{12}$$

where $\theta \leq \psi$.

The $n_{ij}+n_{ji}$ $(1 \leq j \leq i \leq n-1)$ are asymptotically independent, so that we may profit from:

Theorem 4. *Let n_{ij} be the number of gene pairs i-adjacent on S and j-adjacent on T, then the $f(k)$ in Theorem 2 satisfies*

$$\begin{aligned} \mathrm{E}[f(k)] &\rightarrow \left(2-\frac{\alpha}{n}\right)^2 \\ \mathrm{Var}[f(k)] &\rightarrow \frac{8}{\alpha^2}(1+\frac{2}{\alpha}-\frac{2}{\alpha^2}) \end{aligned} \tag{13}$$

as $n \rightarrow \infty$, where $\alpha = \lfloor\frac{\sqrt{1+8(k-1)}+1}{2}\rfloor$.

Proof. Using the Poisson distributions in Theorem 3 in Equation 10 leads to the desired result.

Because it determines a maximum, looking for k^* is similar to the *upper record problem*, i.e., for a series of random variables $X_1, X_2, \ldots$, we consider the new sequence $L(m)$, $(m = 1, 2, \ldots)$, defined in the following manner:

$$L(1) = 1;\ L(m) = \min\{j : X_j > X_{L(m-1)}\}\ (m \geq 2) \tag{14}$$

where $L(m)$ is the index of the mth upper record (or m^{th} record time), while the corresponding r.v. $X_{L(m)}$ is the value of the mth record (or m^{th} record value).

Well-known properties of record times for i.i.d. random variables are:

- The probability that the i-th random variable attains a record is $\frac{1}{i}$.
- The expected number of records up to the i-th random variable is $\log i$.
- the average time at the record for n random variables is $\frac{n}{2}$.

The quantity k^* in Theorem 2 is a record time over $\frac{n(n-1)}{2}$ values of $f(k)$, though these are clearly neither identical nor independent random variables. That both the mean and variance of $f(k)$ are decreasing functions of n means that records become increasingly harder to attain.

This is illustrated in Fig. 3, which compares the proportion of record values at each (i, j) in 50,000 pairs of random genomes of size $n = 10, 30, 100, 300, 1000,$ and 3000, and the accumulated number of record values up to this point, with the corresponding values of i.i.d. random variables. Note that the horizontal axis is k, which maps to $i = \sqrt{2k}$ as a position on the genome.

More important for our purposes is that the average record time is nowhere near half the number of random variables ($\frac{n^2}{4}$ in our case). Fig. 4 clearly shows that k^* is approximately $\sqrt{n}$, so that the cut-off position on the genome of the maximizing weight system will be $o(\sqrt[4]{n})$, actually about $(\sqrt{2}\sqrt[4]{n}$. For genomes of size $n = 12,000$, the expected value of k^* is around 110, so that the cut-off θ for generalized adjacency need not be greater than 15.

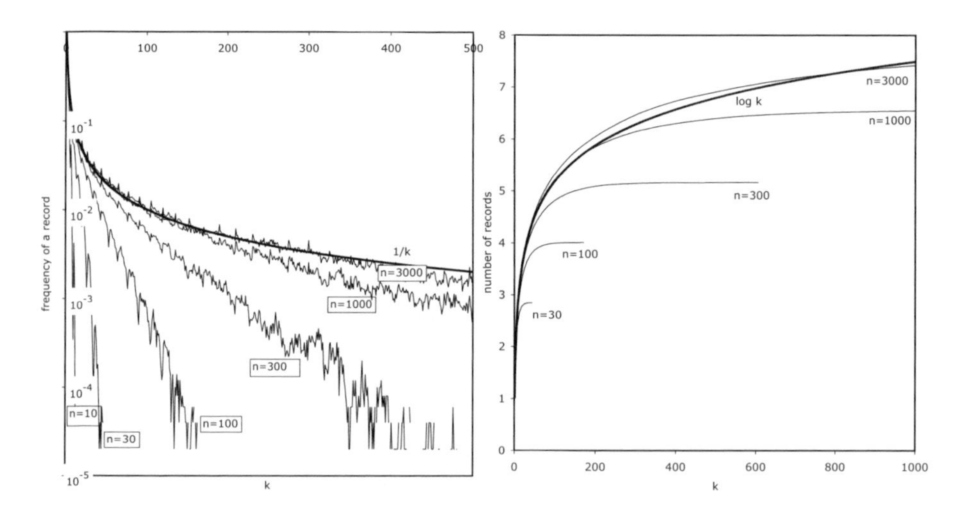

Fig. 3. Comparison of mean optimal k values, over 50,000 pairs of random genomes, with the record behaviour of i.i.d. random variables. Proportion of cases where k is optimal (left) and number of records attained (right), for (i, j) adjacencies as a function of genome size n. As $n \to \infty$, for any k', all curves approach the record time curves for all $k < k'$, but even at $n = 3000$, there is an eventual drop off, due to the declining mean expectations and variances of the n_{ij}.

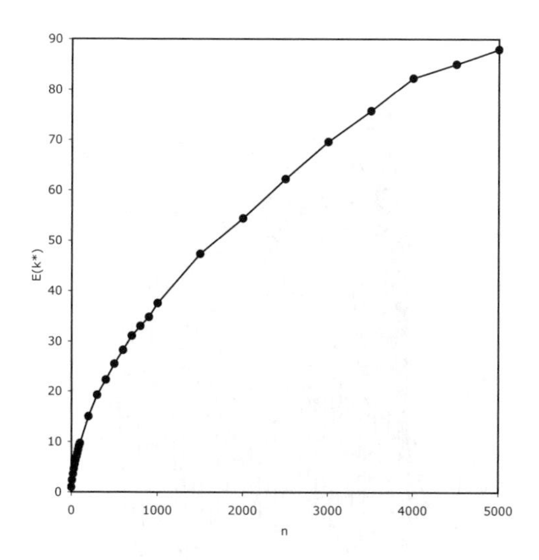

Fig. 4. Average record time as a function of genome length

6 Percolation

Clustering procedures based on parameterized adjacency criteria, e.g., as in Refs. [8,9], can have pathological behaviour as the criteria become less restrictive. At some point, called a *percolation threshold* , instead of large clusters being rare, they suddenly start to predominate and it becomes unusual *not* to find a large cluster. At this point, or earlier, it becomes meaningless to test that the numbers or sizes of clusters exceed those predicted by the null hypothesis of random genomes!

It was established by Erdös and Rényi [4,5,6] that for random graphs where edges are independently present between pairs of the n vertices with probability p, the percolation threshold is $p = \frac{1}{n}$.

When simulating the size of the largest (θ, ψ) cluster as a function of θ and ψ, we obtain graphs like that in Fig. 5.

We note that the percolation of the generalized adjacency graph is delayed considerably compared to unconstrained Erdös-Rényi graphs with the same number of edges, as may be seen in Fig. 6. To understand what aspect of the generalized adjacency graphs is responsible for this delay, we also simulated random graphs of bandwidth $\leq \theta$, since this constraint is an important property of generalized adjacency. It can be seen in Fig. 6, that the limited bandwidth graphs also show delayed percolation, but less than half that of generalized adjacency graphs.

As a control on our simulations, it is known (cf. [2]) that Erdös-Rényi graphs with rn edges, with r somewhat larger than $\frac{1}{2}$ have a cluster of size $(4r-2)n$. Our percolation criterion is that one cluster must have at least $\frac{n}{2}$ vertices. Solving this, we get $r = 0.625$. This means that the $2\theta^2$ edges we use in each of our simulated graphs must be the same as $0.625n$, suggesting that $\theta = 0.56\sqrt{n}$, compared to the $0.61\sqrt{n}$ we found in our limited simulations.

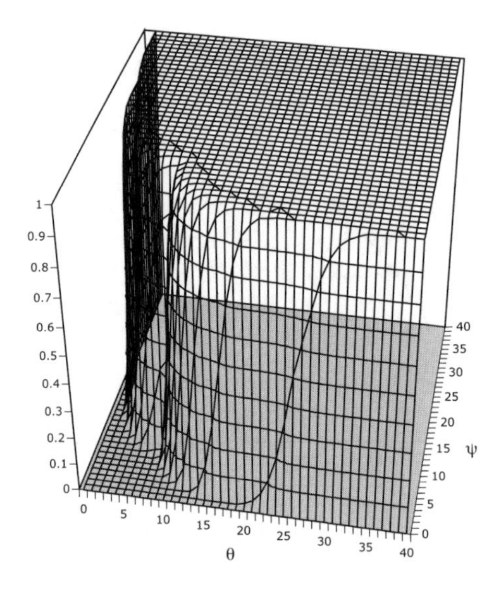

Fig. 5. Proportion of simulations where size of the largest cluster $> \frac{n}{2}$, based on a samples of 50,000 random permutations for $\theta, \psi = 1, 2, \ldots, 99$ and genome size $n = 100$

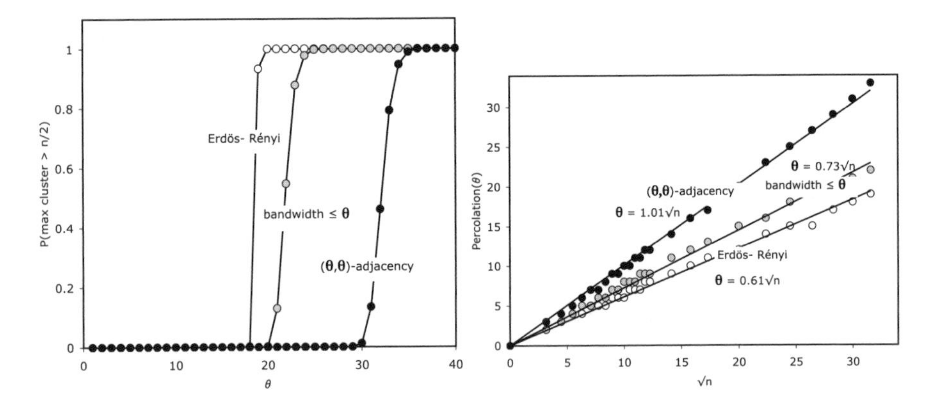

Fig. 6. (left) Simulation with genome length $n = 1000$, with $2\theta^2$ edges in each graph, showing delayed percolation of generalized adjacency graphs with respect to Erdös-Rényi graphs. Bandwidth-limited graphs are also delayed but much less so. (right) Percolation point as a function of $\sqrt{n}$, again with $2\theta^2$ edges per graph. Delay measured by coefficient of $\sqrt{n}$ in equation for trend line.

7 Conclusions

We have defined and explored the notions of generalized gene adjacency and (θ, ψ)-clusters. We have shown that not only simulations, but analytical results are quite feasible. The asymmetry of the criteria in two genomes allows a flexibility not available in previous models.

Of crucial importance is that the natural values of the cut-off we found in Section 5 are less than the percolation points in Section 6.

Future work may help locate k^* analytically as a function of n. Another direction is to pin down other structural properties, beside bandwidth constraints, responsible for the delayed percolation of generalized adjacency graphs.

Acknowledgments

Research supported in part by grants from the Natural Sciences and Engineering Research Council of Canada (NSERC). DS holds the Canada Research Chair in Mathematical Genomics. We would like to thank Wei Xu, Ximing Xu, Chunfang Zheng and Qian Zhu for their constant support and help in this work.

References

1. Bergeron, A., Corteel, S., Raffinot, M.: The algorithmic of gene teams. In: Guigó, R., Gusfield, D. (eds.) WABI 2002. LNCS, vol. 2452, pp. 464–476. Springer, Heidelberg (2002)
2. D'Souza, R., Achlioptas, D., Spencer, J.: Explosive percolation in random networks. Science 323, 1453–1455 (2009)
3. Durand, D., Sankoff, D.: Tests for gene clusters. Journal of Computational Biology 10, 453–482 (2003)
4. Erdös, P., Rényi, A.: On random graphs. Publicationes Mathematicae 6, 290–297 (1959)
5. Erdös, P., Rényi, A.: On the evolution of random graphs. Publications of the Mathematical Institute of the Hungarian Academy of Sciences 5, 17–61 (1960)
6. Erdös, P., Rényi, A.: On the strength of connectedness of a random graphs. Acta Mathematica Scientia Hungary 12, 261–267 (1961)
7. Hoberman, R., Durand, D.: The incompatible desiderata of gene cluster properties. In: McLysaght, A., Huson, D.H. (eds.) RECOMB 2005. LNCS (LNBI), vol. 3678, pp. 73–87. Springer, Heidelberg (2005)
8. Hoberman, R., Sankoff, D., Durand, D.: The statistical analysis of spatially clustered genes under the maximum gap criterion. Journal of Computational Biology 12, 1081–1100 (2005)
9. Xu, X., Sankoff, D.: Tests for gene clusters satisfying the generalized adjacency criterion. In: Bazzan, A.L.C., Craven, M., Martins, N.F. (eds.) BSB 2008. LNCS (LNBI), vol. 5167, pp. 152–160. Springer, Heidelberg (2008)
10. Zhu, Q., Adam, Z., Choi, V., Sankoff, D.: Generalized gene adjacencies, graph bandwidth, and clusters in yeast evolution. Transactions on Computational Biology and Bioinformatics 6(2), 213–220 (2009)

Parking Functions, Labeled Trees and DCJ Sorting Scenarios

Aïda Ouangraoua[1,2] and Anne Bergeron[2]

[1] Department of Mathematics, Simon Fraser University, Burnaby (BC), Canada
aouangra@sfu.ca
[2] Lacim, Université du Québec à Montréal, Montréal (QC), Canada
bergeron.anne@uqam.ca

Abstract. In genome rearrangement theory, one of the elusive questions raised in recent years is the enumeration of rearrangement scenarios between two genomes. This problem is related to the uniform generation of rearrangement scenarios, and the derivation of tests of statistical significance of the properties of these scenarios. Here we give an exact formula for the number of double-cut-and-join (DCJ) rearrangement scenarios of co-tailed genomes. We also construct effective bijections between the set of scenarios that sort a cycle and well studied combinatorial objects such as parking functions and labeled trees.

1 Introduction

Sorting genomes can be succinctly described as finding sequences of rearrangement operations that transform a genome into another. The allowed rearrangement operations are fixed, and the sequences of operations, called *sorting scenarios*, are ideally of minimal length. Given two genomes, the number of different sorting scenarios between them is typically huge – we mean HUGE – and very few analytical tools are available to explore these sets.

In this paper, we give the first exact results on the enumeration and representation of sorting scenarios in terms of well-known combinatorial objects. We prove that sorting scenarios using DCJ operations on *co-tailed* genomes can be represented by *parking functions* and *labeled trees*. This surprising connection yields immediate results on the uniform generation of scenarios [1,11,17], promises tools for sampling processes [6,12] and the development of statistical significant tests [10,15], and offers a wealth of alternate representations to explore the properties of rearrangement scenarios, such as commutation [4,21], structure conservation [3,7], breakpoint reuse [14,16] or cycle length [22].

This research was initiated while we were trying to understand *commuting* operations in a general context. In the case of genomes consisting of single chromosomes, rearrangement operations are often modeled as *inversions*, which can be represented by intervals of the set $\{1, 2, \ldots, n\}$. Commutation properties are described by using overlap relations on the corresponding sets, and a major tool to understand sorting scenarios are *overlap graphs*, whose vertices represent single rearrangement operations, and whose edges model the interactions between

F.D. Ciccarelli and I. Miklós (Eds.): RECOMB-CG 2009, LNBI 5817, pp. 24–35, 2009.

the operations. Unfortunately, overlap graphs do not upgrade easily to genomes with multiple chromosomes, see, for example, [13], where a generalization is given for a restricted set of operations.

We got significant insights when we switched our focus from single rearrangement operations to complete sorting scenarios. This apparently more complex formulation offers the possibility to capture complete scenarios of length d as simple combinatorial objects, such as sequences of integer of length d, or trees with d vertices. It also gives alternate representations of sorting scenarios, using *non-crossing* partitions, that facilitate the study of commuting operations and structure conservation.

In Section 3, we first show that sorting a cycle in the adjacency graph of two genomes with DCJ rearrangement operations is equivalent to refining non-crossing partitions. This observation, together with a result by Richard Stanley [19], gives the existence of bijections between sorting scenarios of a cycle and parking functions or labeled trees. Parking functions and labeled trees are some of a number of combinatorial objects enumerated by the formula $(n+1)^{n-1}$ [8,18,20]. We give explicit bijections with parking functions and labeled trees in Sections 4 and 5. We conclude in Section 6 with remarks on the usefulness of these representations, on the algorithmic complexity of switching between representations, and on generalizations to genomes that are not necessarily co-tailed.

2 Preliminaries

Genomes are compared by identifying homologous segments along their DNA sequences, called *blocks*. These blocks can be relatively small, such as gene coding sequences, or very large fragments of chromosomes. The order and orientation of the blocks may vary in different genomes. Here we assume that the two genomes contain the same set of blocks and consist of either circular chromosomes, or *co-tailed* linear chromosomes. For example, consider the following two genomes, each consisting of two linear chromosomes:

Genome A: $(a\ -f\ -b\ \ e\ -d)\ (-c\ \ g)$
Genome B: $(a\ \ b\ \ c)\ (d\ \ e\ \ f\ \ g)$

The set of *tails* of a linear chromosome $(x_1 \ldots x_m)$ is $\{x_1, -x_m\}$, and two genomes are *co-tailed* if the union of their sets of tails are the same. This is the case for genomes A and B above, since the the union of their sets of tails is $\{a, -c, d, -g\}$.

An *adjacency* in a genome is a sequence of two consecutive blocks. For example, in the above genomes, $(e\ -d)$ is an adjacency of genome A, and $(a\ \ b)$ is an adjacency of genome B. Since a whole chromosome can be flipped, we always have $(x\ y) = (-y\ -x)$.

The *adjacency graph* of two genomes A and B is a graph whose vertices are the adjacencies of A and B, and such that for each block y there is an edge between adjacency $(y\ z)$ in genome A and $(y\ z')$ in genome B, and an edge between $(x\ y)$ in genome A, and $(x'\ y)$ in genome B. See, for example, Figure 1.

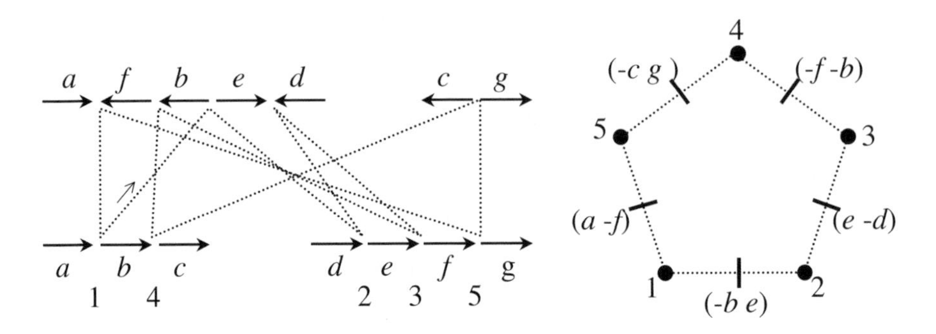

Fig. 1. At the left, the adjacency graph of genome $A = (a\ -f\ -b\ \ e\ -d)\ (-c\ \ g)$ and genome $B = (a\ \ b\ \ c)\ (d\ \ e\ \ f\ \ g)$ is represented by dotted lines. The sign of a block is represented by the orientation of the corresponding arrow. If the – single – cycle is traversed starting with an arbitrary adjacency of genome B, here $(a\ \ b)$, in the direction of the small arrow, then the 5 adjacencies of genome B will be visited in the order indicated by the numbers 1 to 5. At the right, the cycle has been spread out, showing that any DCJ operation acting on two adjacencies of genome A that splits the cycle can be represented by two cuts on the cycle (12345).

Since each vertex has two incident edges, the adjacency graph can be decomposed into connected components that are cycles. The graph of Figure 1 has a single cycle of length 10.

A *double-cut-and-join* (DCJ) rearrangement operation [5,23] on genome A cuts two adjacencies $(x\ y)$ and $(u\ v)$ to produce either $(x\ v)$ and $(u\ y)$, or $(x\ -u)$ and $(-y\ v)$. In simpler words, a DCJ operation cuts the genome at two places, and glues the part in a different order. For example, a DCJ operation on genome $(a\ -f\ -b\ \ e\ -d)\ (-c\ \ g)$ that cuts $(a\ -f)$ and $(-c\ \ g)$ to form $(a\ \ c)$ and $(f\ \ g)$ would produce genome $(a\ \ c)\ (d\ -e\ \ b\ \ f\ \ g)$.

The *distance* between genomes A and B is the minimum number of DCJ operations needed to rearrange – or *sort* – genome A into genome B. The DCJ distance is easily computed from the adjacency graph [5]. For circular chromosomes or co-tailed genomes, the distance is given by:

$$d(A, B) = N - (C + K)$$

where N is the number of blocks, C is the number of cycles of the adjacency graph, and K is the number of linear chromosomes in A. Note that K is a constant for co-tailed genomes. A rearrangement operation is *sorting* if it lowers the distance by 1, and a sequence of sorting operations of length $d(A, B)$ is called a *parsimonious sorting scenario*. It is easy to detect sorting operations since, by the distance formula, a sorting operation must increase by 1 the number of cycles.

A DCJ operation that acts on two cycles of the adjacency graph will merge the two cycles, and can never be sorting. Thus the sorting operations act on a single cycle, and split it into two cycles. The central question of this paper is to enumerate the set of parsimonious sorting scenario. Since each cycle is sorted independently of the others, the problem reduces to enumerating the sorting scenarios of a cycle. Indeed, we have:

Proposition 1. *Given scenarios $S_1, \ldots, S_C$ of lengths $\ell_1, \ldots, \ell_C$ that sort the C cycles of the adjacency graph of genomes A and B, these scenarios can be shuffled into a scenario that sorts genome A to genome B in*

$$\binom{\ell_1 + \ell_2 + \ldots + \ell_C}{\ell_1, \ell_2, \ldots, \ell_C} = \frac{(\ell_1 + \ell_2 + \ldots + \ell_C)!}{\ell_1! \ell_2! \ldots \ell_C!}$$

different ways.

Proof. Since each cycle is sorted independently, the number of global scenarios is enumerated by counting the number of sequences that contains ℓ_m occurrences of the symbol S_m, for $1 \leq m \leq C$, which is counted by a classical formula. For each such sequence, we obtain a scenario by replacing each symbol S_m by the appropriate operation on cycle number m.

3 Representation of Scenarios as Sequences of Fissions

A cycle of length $2n$ of the adjacency graph alternates between adjacencies of genome A and genome B. Given a cycle, suppose that the adjacencies of genome B are labeled by integers from 1 to n in the order they appear along the cycle, starting with an arbitrary adjacency (see Fig. 1). Then any DCJ operation that splits this cycle can be represented by a *fission* of the cycle $(123 \ldots n)$, as

$$(123 \ldots p \| q \ldots t \| u \ldots n)$$

yielding the two non-empty cycles:

$$(123 \ldots pu \ldots n) \text{ and } (q \ldots t).$$

We will always write cycles beginning with their smallest element. Fissions applied to a cycle whose elements are in increasing order always yield cycles whose elements are in increasing order. A fission is characterized by two cuts, each described by the element at the left of the cut. The smallest one, p in the above example, will be called the *base* of the fission, and the largest one, t in the the above example, is called the *top* of the fission. The integer at the right of the first cut, q in the example, is called the *partner* of the base.

In general, after the application of k fissions on $(123 \ldots n)$, the resulting set of cycles will contain $k+1$ elements. The structure of these cycles form a *non-crossing partition* of the initial cycle $(123 \ldots n)$. Namely, we have the following result, which is easily shown by induction on k:

Proposition 2. *Let $k \leq n-1$ fissions be applied on the cycle $(123 \ldots n)$, then the $k+1$ resulting cycles have the following properties:*

1) The elements of each cycle are in increasing order, up to cyclical reordering.

2) [Non-crossing property] If $(c \ldots d)$ *and* $(e \ldots f)$ *are two cycles with* $c < e$*, then either* $d < e$*, or* $c < e \leq f < d$*.*

3) Each successive fission refines the partition of $(123 \ldots n)$ *defined by the cycles.*

A sorting scenario of a cycle of length $2n$ of the adjacency graph can thus be represented by a sequence of $n-1$ fissions on the cycle $(123 \ldots n)$, called a *fission scenario*, and the resulting set of cycles will have the structure $(1)(2)(3) \ldots (n)$. For example, here is a possible fission scenario of (123456789), where the bases of the fissions have been underlined:

$$\begin{aligned}
(123\underline{4}\|5\|6789) &\rightarrow (12346789)(5) \\
(123467\underline{8}\|9\|)(5) &\rightarrow (1234678)(5)(9) \\
(\underline{1}\|234678\|)(5)(9) &\rightarrow (1)(234678)(5)(9) \\
(1)(\underline{2}\|346\|78)(5)(9) &\rightarrow (1)(278)(346)(5)(9) \\
(1)(\underline{2}\|7\|8)(346)(5)(9) &\rightarrow (1)(28)(346)(5)(7)(9) \\
(1)(28)(\underline{3}\|46\|)(5)(7)(9) &\rightarrow (1)(28)(3)(46)(5)(7)(9) \\
(1)(\underline{2}\|8\|)(3)(46)(5)(7)(9) &\rightarrow (1)(2)(3)(46)(5)(7)(8)(9) \\
(1)(2)(3)(\underline{4}\|6\|)(5)(7)(8)(9) &\rightarrow (1)(2)(3)(4)(5)(6)(7)(8)(9)
\end{aligned}$$

Scenarios such as the one above have interesting combinatorial features when all the operations are considered globally, and we will use them extensively in the sequel. A first important remark is that the smallest element of the cycle is always 'linked' to the greatest element through a chain of partners. For example, the last partner of element 1 is element 2, the last partner of element 2 is element 8, and the last partner of element 8 is element 9. We will see that this is always the case, even when the order of the corresponding fissions is arbitrary with respect to the scenario. The following definition captures this idea of chain of partners.

Definition 1. *Consider a scenario* S *of fissions that transform a cycle* $(c \ldots d)$ *into cycles of length 1. For each element* p *in* $(c \ldots d)$*, if* p *is the base of one or more of the fissions of* S*, let* q *be the last partner of* p*, then define recursively*

$$Sup_S(p) = Sup_S(q),$$

otherwise, $Sup_S(p) = p$*.*

In order to see that $Sup(p)$ is well defined, first note that the successive partners of a given base p are always in increasing order, and greater than p. Moreover, the last element of a cycle $(c \ldots d)$ is never the base of a fission. For example, in the above scenario, we would have $Sup_S(1) = Sup_S(2) = Sup_S(8) = Sup_S(9) = 9$.

The following lemma is the key to most of the results that follow:

Lemma 1. *Consider a scenario* S *of fissions that transform a cycle* $(c \ldots d)$ *into cycles of length 1, then* $Sup_S(c) = d$*.*

Proof. If $c = d$, then the result is trivial. Suppose the result is true for cycles of length $\leq n$, and consider a cycle of length $n+1$. The first fission of S will split the cycle $(c \ldots d)$ in two cycles of length $\leq n$. If the two cycles are of the form $(c \ldots d)$ and $(c' \ldots d')$, then c' is, in the worst case, the first partner of c, and cannot be the last since $c \neq d$. Let S' be the subset of S that transform the shorter cycle $(c \ldots d)$ into cycles of length 1. By the induction hypothesis, $Sup_{S'}(c) = d$, but $Sup_S(c) = Sup_{S'}(c)$ since the last partner of c is not in $(c' \ldots d')$.

If the two cycles are of the form $(c \ldots d')$ and $(c' \ldots d)$, consider S_1 the subset of S that transform the cycle $(c \ldots d')$ into cycles of length 1, and S_2 the subset of S that transform the cycle $(c' \ldots d)$ into cycles of length 1. We have, by the induction hypothesis, $Sup_{S_1}(c) = d'$ and $Sup_{S_2}(c') = d$, implying $Sup_S(c) = Sup_S(d')$ and $Sup_S(c') = d$. However, c' is the last partner of d', thus $Sup_S(c) = Sup_S(d') = Sup_S(c') = d$.

4 Fission Scenarios and Parking Functions

In this section, we establish a bijection between fission scenarios and *parking functions* of length $n - 1$. This yields a very compact representation of DCJ sorting scenarios of cycles of length $2n$ as sequences of $n - 1$ integers.

A *parking function* is a sequence of integers $p_1 p_2 \ldots p_{n-1}$ such that if the sequence is sorted in non-decreasing order yielding $p'_1 \leq p'_2 \leq \ldots \leq p'_{n-1}$, then $p'_i \leq i$. These sequences were introduced by Konheim and Weiss [9] in connection with hashing problems. These combinatorial structure are well studied, and the number of different parking functions of length $n - 1$ is known to be n^{n-2}.

Proposition 2 states that a fission scenario is a sequence of successively refined non-crossing partitions of the cycle $(123 \ldots n)$. A result by Stanley [19] has the following immediate consequence:

Theorem 1. *There exists a bijection between fission scenarios of cycles of the form* $(123 \ldots n)$ *and* parking functions *of length* $n - 1$.

Fortunately, in our context, the bijection is very simple: we list the bases of the fissions of the scenario. For example, the parking function associated to the example of Section 3 is 48122324. In general, we have:

Proposition 3. *The sequence of bases of a fission scenario on the cycle* $(123 \ldots n)$ *is a parking function of length* $n - 1$.

Proof. Let $p_1 p_2 \ldots p_{n-1}$ be the sequence of bases of a fission scenario and let $p'_1 p'_2 \ldots p'_{n-1}$ be the corresponding sequence sorted in non-decreasing order. Suppose that there exists a number i such that $p'_i > i$, then there are at least $n - i$ fissions in the scenario with base $p \geq i + 1$. These bases can be associated to at most $n - i - 1$ partners in the set $\{i + 2, i + 3, i + 4, \ldots, n\}$ because a base is always smaller than its partner, but this is impossible because each integer is used at most once as a partner in a fission scenario.

In order to reconstruct a fission scenario from a parking function, we first note that a fission with base p_i and partner q_i creates a cycle whose smallest element is q_i, thus each integer in the set $\{2, 3, \ldots, n\}$ appears exactly once as a partner in a fission scenario.

Given a parking function $p_1p_2 \ldots p_{n-1}$, we must first assign to each base p_i a unique partner q_i in the set $\{2, 3, \ldots, n\}$. By Lemma 1, we can then determine the top t_i of fission i, since the set of fissions from $i+1$ to $n-1$ contains a sorting scenario of the cycle $(q_i \ldots t_i)$. Algorithm 1 details the procedure.

Algorithm 1. [Parking functions to fission scenarios]
Input: a parking function $p_1p_2 \ldots p_{n-1}$.
Output: a fission scenario $(p_1, t_1), \ldots, (p_{n-1}, t_{n-1})$.

$Q \leftarrow \{2, 3, \ldots, n\}$
For p from n − 1 to 1 do:
For each successive occurrence p_i of p in the sequence $p_1p_2 \ldots p_{n-1}$ do:
$q_i \leftarrow$ *The smallest element of Q greater than p_i*
$Q \leftarrow Q \setminus \{q_i\}$
$S \leftarrow \{(p_1, q_1), (p_2, q_2) \ldots, (p_{n-1}, q_{n-1})\}$
For i from 1 to n − 1 do:
$S \leftarrow S \setminus \{(p_i, q_i)\}$
$t_i \leftarrow Sups_S(q_i)$

For example, using the parking function 48122324 and the set of partners $\{2, 3, \ldots, 9\}$, we would get the pairings, starting from base 8 down to base 1:

$$\begin{pmatrix} p_i : 4\,8\,1\,2\,2\,3\,2\,4 \\ q_i : 5\,9\,2\,3\,7\,4\,8\,6 \end{pmatrix}$$

Finally, in order to recover the second cut of each fission, we compute the values t_i:

$$\begin{pmatrix} p_i : 4\,8\,1\,2\,2\,3\,2\,4 \\ t_i : 5\,9\,8\,6\,7\,6\,8\,6 \end{pmatrix}$$

For example, in order to compute t_4, then $S = \{(2, 7), (3, 4), (2, 8), (4, 6)\}$, and $Sups_S(3) = Sups_S(4) = Sups_S(6) = 6$.

Since we know, by Theorem 1, that fissions scenarios are in bijection with parking functions, it is sufficient to show that Algorithm 1 recovers a given scenario in order to prove that it is an effective bijection.

Proposition 4. *Given a fission scenario of a cycle of the form $(123 \ldots n)$, let (p_i, q_i, t_i) be the base, partner and top of fission i. Algorithm 1 recovers uniquely t_i from the parking function $p_1p_2 \ldots p_{n-1}$.*

Proof. By Lemma 1 , we only need to show that Algorithm 1 recovers uniquely the partner q_i of each base p_i. Let p be the largest base, and suppose that p has j partners, then the original cycle must contain at least the elements:

$$(\ldots p\; p+1 \ldots p+j \ldots).$$

We will show that $p+1 \dots p+j$ must be the j partners of p. If it was not the case, at least one of the j adjacencies in the sequence $p\ p+1 \dots p+j$ must be cut in a fission whose base is smaller than p, since p is the largest base, and this would violate the non-crossing property of Proposition 2. Thus Algorithm 1 correctly and uniquely assigns the partners of the largest base. Suppose now that Algorithm 1 has correctly and uniquely assigned the partners of all bases greater than p. The same argument shows that the successive partners of p must be the smallest available partners greater than p.

Summarizing the results so far, we have:

Theorem 2. *If the adjacency graph of two co-tailed genomes has C cycles of length $2(\ell_1+1), \dots, 2(\ell_C+1)$, then the number of sorting scenarios is given by:*

$$\frac{(\ell_1+\ell_2+\dots+\ell_C)!}{\ell_1!\ell_2!\dots\ell_C!} * (\ell_1+1)^{\ell_1-1} * \dots * (\ell_C+1)^{\ell_C-1}.$$

Each sub-scenario that sort a cycle of length $2(\ell_m+1)$ can be represented by a parking function of length ℓ_m.

Proof. Sorting a cycle of length $2(\ell_m+1)$ can be simulated by fissions of the cycle $(12\dots\ell_m+1)$, which can be represented by parking functions of length ℓ_m. The number of different parking functions of length ℓ_m is given by $(\ell_m+1)^{\ell_m-1}$. Applying Proposition 1 yields the enumeration formula.

5 Fission Scenarios and Labeled Trees

Theorem 1 implies that it is possible to construct bijections between fission scenarios and objects that are enumerated by parking functions. This is notably the case of *labeled tree* on n vertices. These are trees with n vertices in which each vertex is given a unique label in the set $\{0,1,\dots,n-1\}$. In this section, we construct an explicit bijection between these trees and fission scenarios of cycles of the form $(123\dots n)$.

Definition 2. *Given a fission scenario S of a cycle of the form $(123\dots n)$, let (p_i, q_i) be the base and partner of fission i.*

The graph T_S is a graph whose nodes are labeled by $\{0,1,\dots,n-1\}$, with an edge between i and j, if $p_i = q_j$, and an edge between 0 and i, if $p_i = 1$.

In the running example, the corresponding graph is depicted in Figure 2 (a). We have:

Proposition 5. *The graph T_S is a labeled tree on n vertices.*

Proof. By construction, the graph has n vertices labeled by $\{0,1,\dots,n-1\}$. In order to show that it is a tree, we will show that the graph has $n-1$ edges and that it is connected. Since each integer in the set $\{2,3,\dots,n\}$ is partner of one and only one fission in S and S contains $n-1$ fissions, T_S has exactly $n-1$ edges. Moreover, by construction, there is a path between each vertex $i \neq 0$ and 0 in T_S, thus T_S is connected.

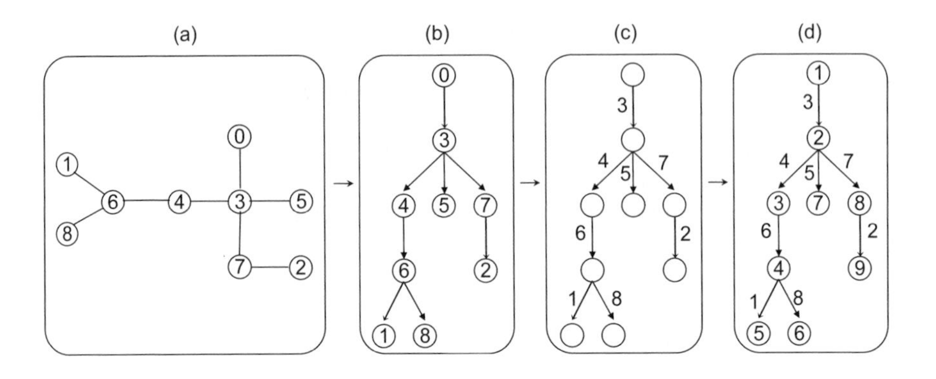

Fig. 2. Construction of a fission scenario. (a) The unrooted tree T_S. (b) The tree is rooted at vertex 0, and the children of each node are ordered. (c) The labels of the nodes are lifted to their incoming edges. (d) The nodes are labeled in prefix order from 1 to 9. The order of the fissions are read on the edges, the source of an edge represent the base p of the fission and its target is the partner. For example, fission #1 has base 4, with partner 5.

Before showing that the construction of T_S yields an effective bijection, we detail how to recover a fission scenario from a tree.

Algorithm 2. [Labeled trees to fission scenarios]
Input: a labeled tree T on n vertices.
Output: a fission scenario $(p_1, t_1), \ldots, (p_{n-1}, t_{n-1})$.

Root the tree at vertex 0.
Put the children of each node in increasing order from left to right.
Label the unique incoming edge of a node with the label of the node.
Relabel the nodes from 1 *to* n *with a prefix traversal of the tree.*
For i *from* 1 *to* $n-1$ *do:*
  $p_i \leftarrow$ *The label of the source* p *of edge* i.
  $t_i \leftarrow$ *The greatest label of the subtree rooted by edge* i.
  Remove edge i *from* T

The following proposition states that the construction of the associated tree T_S is injective, thus providing a bijection between fission scenarios and trees.

Proposition 6. *The trees associated to different fission scenarios are different.*

Proof. Suppose that two different scenarios S_1 and S_2 yield the same tree T. Then, by construction, if T is rooted in 0, for each directed edge from j to i in T, if $j = 0$ then $p_i = 1$ otherwise $p_i = q_j$. Moreover, in a fission scenario, if fission i is the first operation having base p_i, then its partner is $q_i = p_i + 1$, otherwise the non-crossing property of Proposition 2 would be violated. So, using these two properties, the sequences of bases and partners of the fissions in the two scenarios

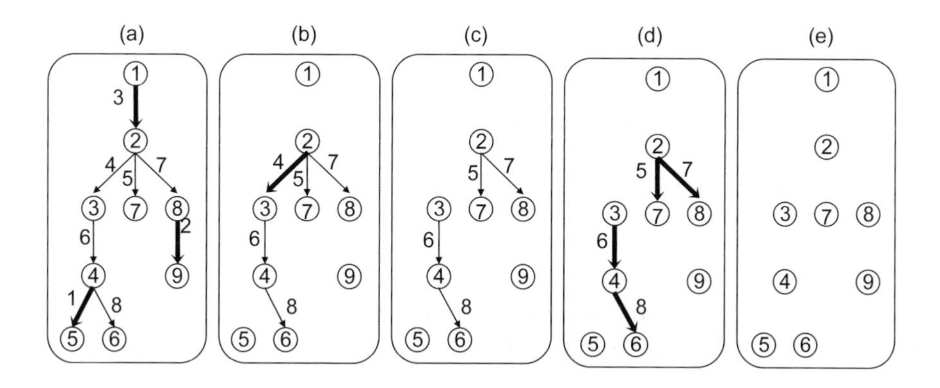

Fig. 3. Sorting directly on a tree: erasing successively the edges 1 to n simulates cycle fissions by creating intermediate forests. In part (b), the fourth fission will split the tree corresponding to cycle (234678) into two trees corresponding to the two cycles (346) and (278).

can be uniquely recovered from T, and thus S_1 and S_2 would correspond to the same parking function.

The tree representation offers another interesting view of the sorting procedure. Indeed, sorting can be done directly on the tree by successively erasing the edges from 1 to $n-1$. This progressively disconnects the tree, and the resulting connected components correspond precisely to the intermediate cycles obtained during the sorting procedure.

For example, Figure 3 gives snap-shots of the sorting procedure. Part (b) shows the forest after the three first operations, the fourth fission splits a tree with six nodes into two trees each with three nodes, corresponding to the cycle splitting of the fourth operation in the running example.

6 Discussion and Conclusions

In this paper, we presented results on the enumeration and representations of sorting scenarios between co-tailed genomes. Since we introduced many combinatorial objects, we bypassed a lot of the usual material presented in rearrangement papers. The following topics will be treated in a future paper.

The first topic is the complexity of the algorithms for switching between representations. Algorithms 1 and 2 are not meant to be efficient, they are rather explicit descriptions of what is being computed. Preliminary work indicates that with suitable data structure, they can be implemented in $O(n)$ running time. Indeed, most of the needed information can be obtained in a single traversal of a tree.

The second obvious extension is to generalize the enumeration formulas and representations to arbitrary genomes. In the general case, when genomes are not necessarily co-tailed, the adjacency graph can be decomposed in cycles and paths, and additional sorting operations must be considered, apart from operations that split cycles [5]. However, these new sorting operations that act on paths create new paths that behave essentially like cycles.

We also had to defer to a further paper the details of the diverse uses of these new representations. One of the main benefits of having a representation of a sorting scenario as a parking function, for example, is that it solves the problem of uniform sampling of sorting scenarios [2]. There is no more bias attached to choosing a first sorting operation, since, when using parking functions, the nature of the first operation depends on the whole scenario. The representation of sorting scenarios as non-crossing partitions refinement also greatly helps in analyzing commutation and conservation properties.

References

1. Ajana, Y., Lefebvre, J.-F., Tillier, E.R.M., El-Mabrouk, N.: Exploring the set of all minimal sequences of reversals - an application to test the replication-directed reversal hypothesis. In: Guigó, R., Gusfield, D. (eds.) WABI 2002. LNCS, vol. 2452, p. 300. Springer, Heidelberg (2002)
2. Barcucci, E., del Lungo, A., Pergola, E.: Random generation of trees and other combinatorial objects. Theoretical Computer Science 218(2), 219–232 (1999)
3. Bérard, S., Bergeron, A., Chauve, C., Paul, C.: Perfect sorting by reversals is not always difficult. IEEE/ACM Transactions on Computational Biology and Bioinformatics 4(1), 4–16 (2007)
4. Bergeron, A., Chauve, C., Hartman, T., St-onge, K.: On the properties of sequences of reversals that sort a signed permutation. In: Proceedings Troisièmes Journées Ouvertes Biologie Informatique Mathématiques, pp. 99–108 (2002)
5. Bergeron, A., Mixtacki, J., Stoye, J.: A unifying view of genome rearrangements. In: Bücher, P., Moret, B.M.E. (eds.) WABI 2006. LNCS (LNBI), vol. 4175, pp. 163–173. Springer, Heidelberg (2006)
6. Braga, M.D.V., Sagot, M.-F., Scornavacca, C., Tannier, E.: Exploring the solution space of sorting by reversals, with experiments and an application to evolution. IEEE/ACM Transactions on Computational Biology and Bioinformatics 5(3), 348–356 (2008)
7. Diekmann, Y., Sagot, M.-F., Tannier, E.: Evolution under reversals: Parsimony and conservation of common intervals. IEEE/ACM Transactions on Computational Biology and Bioinformatics 4(2), 301–309 (2007)
8. Kalikow, L.H.: Enumeration of parking functions, allowable permutation pairs, and labeled trees. PhD thesis, Brandeis University (1999)
9. Konheim, A.G., Weiss, B.: An occupancy discipline and applications. SIAM Journal of Applied Mathematics 14, 1266–1274 (1966)
10. McLysaght, A., Seoighe, C., Wolfe, K.H.: High frequency of inversions during eukaryote gene order evolution. In: Comparative Genomics: Empirical and Analytical Approaches to Gene Order Dynamics, Map Alignment and the Evolution of Gene Families, pp. 47–58. Kluwer Academic Press, Dordrecht (2000)
11. Miklós, I., Darling, A.: Efficient sampling of parsimonious inversion histories with application to genome rearrangement in yersinia. Genome Biology and Evolution 1(1), 153–164 (2009)

12. Miklós, I., Hein, J.: Genome rearrangement in mitochondria and its computational biology. In: Lagergren, J. (ed.) RECOMB-WS 2004. LNCS (LNBI), vol. 3388, pp. 85–96. Springer, Heidelberg (2005)
13. Ozery-flato, M., Shamir, R.: Sorting by translocations via reversals theory. Journal of Computational Biology 14(4), 408–422 (2007)
14. Pevzner, P., Tesler, G.: Human and mouse genomic sequences reveal extensive breakpoint reuse in mammalian evolution. Proceedings of National Academy of Sciences USA 100(13), 7672–7677 (2003)
15. Sankoff, D., Lefebvre, J.-F., Tillier, E.R.M., Maler, A., El-Mabrouk, N.: The distribution of inversion lengths in bacteria. In: Lagergren, J. (ed.) RECOMB-WS 2004. LNCS (LNBI), vol. 3388, pp. 97–108. Springer, Heidelberg (2005)
16. Sankoff, D., Trinh, P.: Chromosomal breakpoint reuse in genome sequence rearrangement. Journal of Computational Biology 12(6), 812–821 (2005)
17. Siepel, A.C.: An algorithm to enumerate all sorting reversals. In: RECOMB 2002: Proceedings of the Sixth annual International Conference on Computational biology, pp. 281–290. ACM, New York (2002)
18. Stanley, R.P.: Enumerative Combinatorics, vol. I. Wadsworth and Brookes/Cole, Monterey, California (1986)
19. Stanley, R.P.: Parking functions and noncrossing partitions. Electronic Journal of Combinatorics 4(2), R20 (1997)
20. Stanley, R.P.: Enumerative Combinatorics, vol. II. Cambridge University Press, Cambridge (1999)
21. Swenson, K.M., Dong, Y., Tang, J., Moret, B.M.E.: Maximum independent sets of commuting and noninterfering inversions. In: 7th Asia-Pacific Bioinformatics Conference (to appear, 2009)
22. Xu, A.W., Zheng, C., Sankoff, D.: Paths and cycles in breakpoint graphs of random multichromosomal genomes. Journal of Computational Biology 14(4), 423–435 (2007)
23. Yancopoulos, S., Attie, O., Friedberg, R.: Efficient sorting of genomic permutations by translocation, inversion and block interchange. Bioinformatics 21(16), 3340–3346 (2005)

Counting All DCJ Sorting Scenarios

Marília D.V. Braga and Jens Stoye

Technische Fakultät, Universität Bielefeld, Germany
mbraga@cebitec.uni-bielefeld.de, stoye@techfak.uni-bielefeld.de

Abstract. In genome rearrangements, the double cut and join (DCJ) operation, introduced by Yancopoulos *et al.*, allows to represent most rearrangement events that could happen in multichromosomal genomes, such as inversions, translocations, fusions and fissions. No restriction on the genome structure considering linear and circular chromosomes is imposed. An advantage of this general model is that it leads to considerable algorithmic simplifications. Recently several works concerning the DCJ operation have been published, and in particular an algorithm was proposed to find an optimal DCJ sequence for sorting one genome into another one. Here we study the solution space of this problem and give an easy to compute formula that corresponds to the exact number of optimal DCJ sorting sequences to a particular subset of instances of the problem. In addition, this formula is also a lower bound to the number of sorting sequences to any instance of the problem.

1 Introduction

Genome rearrangements provide the opportunity for tracking evolutionary events at a structural, whole-genome level. A typical approach is the determination of the minimum number of rearrangement operations that are necessary to transform one genome into another one [7]. The corresponding computational problem is called the *genomic distance problem* [5]. A bit more detailed is the task when in addition to the numeric distance also one or more scenarios of rearrangement operations are to be determined, the so-called *genomic sorting problem.*

Most algorithms that solve the genomic sorting problem will report just one out of a possibly very high number of rearrangement scenarios, and studies of such a particular scenario are not well suited for drawing general conclusions on properties of the relationship between the two genomes under study. Moreover, there are normally too many sorting scenarios in order to enumerate them all [8]. Consequently, people have started characterizing the space of all possible genome rearrangement scenarios without explicit enumeration [2,4]. This space exhibits a nice sub-structure that allows efficient enumeration of substantially different rearrangement scenarios, for example. This may be a good basis for further studies based on statistical approaches or sampling strategies.

Based on the type of genomes and the organism under study, various genome rearrangement operations have been considered. Most results are known for unichromosomal linear genomes, where the only operation is an inversion of

F.D. Ciccarelli and I. Miklós (Eds.): RECOMB-CG 2009, LNBI 5817, pp. 36–47, 2009.

a piece of the chromosome. In this model, the space of all sorting scenarios has been well characterized, allowing to group sorting scenarios into classes of equivalence [2]. The number of classes of equivalence, that can be directly enumerated [4], is much smaller than the total number of scenarios.

In this paper, we study the space of all sorting scenarios under a more general rearrangement operation, called *double-cut and join* (DCJ). This operation was introduced by Yancopoulos *et al.* [9] and further studied in [3]. It acts on multichromosomal linear and/or circular genomes and subsumes all traditionally studied rearrangement operations like inversions, translocations, fusions and fissions. We characterise the sorting sequences and show how to commute operations in order to obtain a sorting sequence from another. In addtion we give a closed formula for the number of DCJ sorting scenarios that is exact for a certain class of instances of the problem, and a lower bound for the general case.

We are aware that recently, and independently of our work, similar results to the ones presented here were obtained by Ouangraoua and Bergeron [6]. However, we believe that our approach is more direct and also more complete, since it is not restricted to *co-tailed* genomes as the approach presented in [6].

2 Genomes, Adjacency Graph and Sorting by DCJ

A multichromosomal genome Π, over a set of markers $\mathcal{G}$, is a collection of linear and/or circular chromosomes in which each marker in $\mathcal{G}$ occurs exactly once in Π. Each marker g in $\mathcal{G}$ is a DNA fragment and has an orientation, therefore, to each $g \in \mathcal{G}$, we define its extremities g^t (tail) and g^h (head) and we can represent an adjacency between two markers a and b in Π by an unordered pair containing the extremities of a and b that are actually adjacent. For example, if the head of marker a is adjacent to the head of marker b in one chromosome of Π, then we have the adjacency $\{a^h, b^h\}$, that we represent as $a^h b^h$ or $b^h a^h$ for simplicity. When the extremity of a marker is adjacent to the telomere of a linear chromosome in Π, we have a singleton instead of a pair. For example, if the extremity b^t is adjacent to a telomere of a linear chromosome, then we have the singleton $\{b^t\}$, that we represent simply as b^t. A genome Π can be represented by the set $V(\Pi)$ containing its adjacencies and singletons [3].

A *double-cut and join* or DCJ operation over a genome Π is the operation that cuts two elements (adjacencies or singletons) of $V(\Pi)$ and joins the separated extremities in a different way, creating two new elements [3]. For example, a DCJ acting on two adjacencies pq and rs would create either the new adjacencies pr and qs, or ps and qr (this could correspond to an inversion, a reciprocal translocation between two linear chromosomes, a fusion of two circular chromosomes, or an excision of a circular chromosome). In the same way, a DCJ acting on one adjacency pq and a singleton r would create either pr and q, or p and qr (in this case, the operation could correspond to an inversion, or a translocation, or a fusion of a circular and a linear chromosome). At last, a DCJ acting on two singletons p and q would create the new adjacency pq (that could represent a circularization of one or a fusion of two linear chromosomes). Conversely, a

DCJ can act on only one adjacency pq and create the two singletons p and q (representing a linearization of a circular, or a fission of a linear chromosome).

Definition 1 ([3]). *Given two genomes Π_1 and Π_2 over the same set of markers $\mathcal{G}$ (with the same content) and without duplications, the* adjacency graph *$AG(\Pi_1, \Pi_2)$ is the graph in which:*

1. *The set of vertices is $V = V(\Pi_1) \cup V(\Pi_2)$.*
2. *For each $g \in \mathcal{G}$, we have an edge connecting the vertices in $V(\Pi_1)$ and $V(\Pi_2)$ that contain g^h and an edge connecting the vertices in $V(\Pi_1)$ and $V(\Pi_2)$ that contain g^t.*

We know that $AG(\Pi_1, \Pi_2)$ is bipartite with maximum degree equal to two (each extremity of a marker appears in at most one adjacency or singleton in Π_1 and in at most one adjacency or singleton in Π_2). Consequently, $AG(\Pi_1, \Pi_2)$ is a collection of cycles of even length and paths, alternating vertices in $V(\Pi_1)$ and $V(\Pi_2)$. A path is said to be *balanced* when it contains the same number of vertices in $V(\Pi_1)$ and in $V(\Pi_2)$, that is, when it contains an even number of vertices (and an odd number of edges). Otherwise the path contains an odd number of vertices (and an even number of edges) and is said to be *unbalanced*.

Observe that both $V(\Pi_1)$ and $V(\Pi_2)$ have an even number of singleton vertices, thus the number of balanced paths in $AG(\Pi_1, \Pi_2)$ is even [3]. An example of an adjacency graph is given in Figure 1.

Given two genomes Π_1 and Π_2 over the same set of markers $\mathcal{G}$ and without duplications, the problem of sorting Π_1 into Π_2 by DCJ operations has been studied by Bergeron *et al.* [3]. The authors proposed a formula for the DCJ distance, based on the adjacency graph $AG(\Pi_1, \Pi_2)$:

$$d(\Pi_1, \Pi_2) = n - c - \frac{b}{2} \tag{1}$$

where n is the number of markers in $\mathcal{G}$, c and b are respectively the number of cycles and balanced paths in $AG(\Pi_1, \Pi_2)$.

Bergeron *et al.* [3] also observed that an optimal DCJ operation either increases the number of cycles by one or the number of balanced paths by two and proposed a simple algorithm to find one optimal sequence of DCJ operations to sort Π_1 into Π_2.

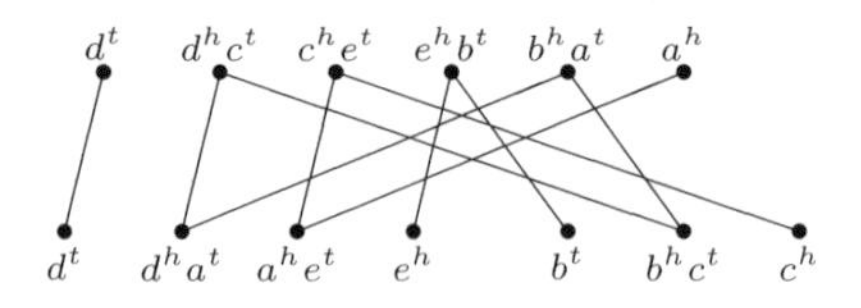

Fig. 1. The adjacency graph for the linear genomes Π_1 and Π_2, defined by the corresponding sets of adjacencies $V(\Pi_1) = \{d^t, d^h c^t, c^h e^t, e^h b^t, b^h a^t, a^h\}$ and $V(\Pi_2) = \{d^t, d^h a^t, a^h e^t, e^h, b^t, b^h c^t, c^h\}$, contains one cycle, one unbalanced, and two balanced paths

3 The Solution Space of Sorting by DCJ

Although an optimal DCJ sequence sorting a genome Π_1 into a genome Π_2 can be easily obtained [3], there are several different optimal sorting sequences, and in this study we approach the problem of characterizing and counting these optimal solutions. We want to analyse the space of solutions sorting Π_1 into Π_2, thus we consider only operations acting on genome Π_1, or, in other words, acting on vertices of $V(\Pi_1)$.

3.1 The Commutation of Two DCJ Operations

Here we represent a DCJ operation ρ by the two pairs of adjacencies concerned, $\rho = (adj_0, adj_1)$, that is, the original pair of adjacencies $adj_0 = \{pq, rs\}$ and the new pair of adjacencies, say $adj_1 = \{pr, qs\}$. (In order to generalize this notation to singletons, any extremity among p, q, r, s could be equal to $\circ$ – a telomere.) We say that the adjacencies adj_0 and adj_1 and the extremities p, q, r and s are *affected* by ρ. Two DCJ operations are said to be *independent* when the set of marker extremities (excluding telomeres) affected by one is independent of the set affected by the other.

Proposition 1. *In any optimal DCJ sorting sequence* $\ldots\rho\theta\ldots$*, two consecutive operations* ρ *and* θ *are independent or share one adjacency, that is, one adjacency created by* ρ *is used (broken) by* θ*.*

Proof. The second operation cannot use both adjacencies created by the first, otherwise they would not be part of an optimal sequence. Thus, the second operation either uses one adjacency created by the first operation or only adjacencies that were not affected by the first. □

Proposition 2. *In any optimal DCJ sorting sequence* $\ldots\rho\theta\ldots$*, the operations* ρ *and* θ *could be commuted to construct another optimal sorting sequence. If* ρ *and* θ *are independent, the commutation is direct, that is, we may simply replace* $\ldots\rho\theta\ldots$ *by* $\ldots\theta\rho\ldots$*. Otherwise, the commutation can be done in two ways, with adjustments as follows: Suppose* $\rho = (\{pv, qr\}, \{pq, rv\})$ *and* $\theta = (\{rv, su\}, \{rs, uv\})$*. Then we could replace* ρ *and* θ *by* $\rho' = (\{pv, su\}, \{ps, uv\})$ *and* $\theta' = (\{ps, qr\}, \{pq, rs\})$ *or alternatively by* $\rho'' = (\{qr, su\}, \{qu, rs\})$ *and* $\theta'' = (\{pv, qu\}, \{pq, uv\})$*.*

Proof. Observe that either $\rho\theta$, or $\rho'\theta'$, or $\rho''\theta''$ transform the adjacencies pv, qr and su into the adjacencies pq, rs and uv, thus the adjacencies that exist before and after $\rho\theta$, $\rho'\theta'$ and $\rho''\theta''$ are the same. Consequentely the remaining operations are still valid. □

The commutation with adjustments is shown in Figure 2.

Theorem 1. *Any optimal DCJ sequence can be obtained from any other optimal DCJ sequence by successive commutations.*

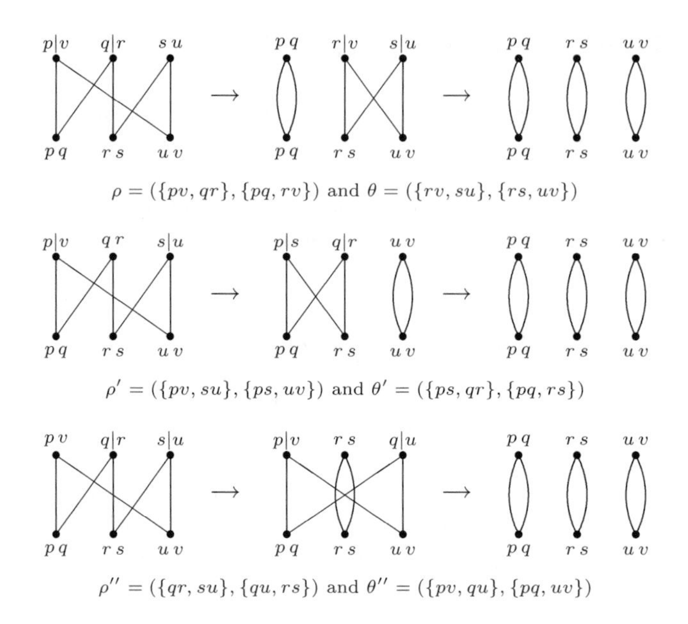

Fig. 2. Example of commutation with adjustments. Observe that the three pairs of consecutive DCJ operations $\rho\theta$, $\rho'\theta'$ and $\rho''\theta''$ transform the adjacencies pv, qr and su into pq, rs and uv.

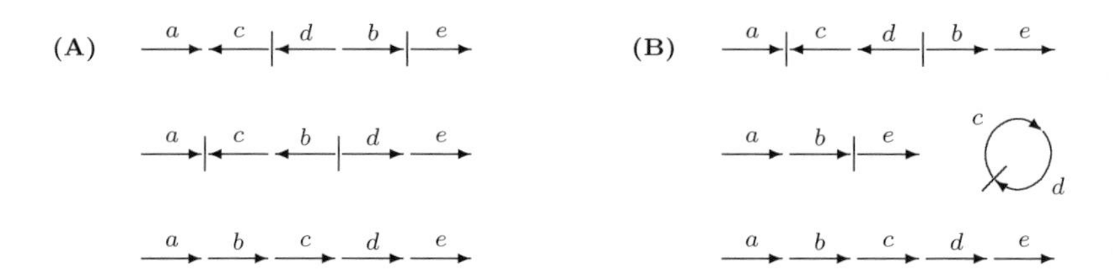

Fig. 3. In this graphic representation of the genomes, each arrow represents a marker (from $\mathcal{G} = \{a, b, c, d, e\}$) with its corresponding orientation. The commutation of two independent DCJ operations sorting $\{a^t, a^h c^h, c^t d^h, d^t b^t, b^h e^t, e^h\}$ into $\{a^t, a^h b^t, b^h c^t, c^h d^t, d^h e^t, e^h\}$ changes the nature of the actual rearrangements. In **(A)** we have $(\{c^t d^h, b^h e^t\}, \{b^h c^t, d^h e^t\})$ followed by $(\{a^h c^h, b^t d^t\}, \{a^h b^t, c^h d^t\})$, a sequence of two inversions, while in **(B)** we have $(\{a^h c^h, b^t d^t\}, \{a^h b^t, c^h d^t\})$ followed by $(\{c^t d^h, b^h e^t\}, \{b^h c^t, d^h e^t\})$, a circular excision followed by a circular integration.

Proof. Any optimal DCJ sequence transforms the same initial set of adjacencies into the same final set of adjacencies. Suppose the final adjacencies are $adj_1, adj_2, \ldots, adj_k$ and consider two optimal DCJ sorting sequences s and t. We can transform s into t with the following procedure: for each adj_i $(1 \leq i \leq k)$, search the operation in s that produces adj_i and move it to the position it occupies in t by commutations. □

One critical aspect of the commutations is that they can change the actual nature of the operations over the genomes, as we can see in Figure 3.

3.2 Counting DCJ Sequences Sorting Components Separately

Proposition 3. *Any DCJ operation acting on vertices of $V(\Pi_1)$ belonging to the same component of $AG(\Pi_1, \Pi_2)$ is optimal.*

Proof. Consider Formula 1 for computing the DCJ distance. We know that an optimal DCJ operation either creates one cycle or two balanced paths [3]. Enumerating the effects of any operation acting on vertices of $V(\Pi_1)$ internal to a component, we have:

- If the vertices are in a cycle, it is split into two cycles, that is, the number of cycles increases by one.
- If the vertices are in a balanced path, it is split into one cycle and one balanced path, that is, the number of cycles increases by one.
- If the vertices are in an unbalanced path, either it is split into one cycle and one unbalanced path, increasing the number of cycles by one; or it is closed into a cycle (when the path has more vertices in $V(\Pi_1)$), also increasing the number of cycles; or it is split into two balanced paths (when the path has more vertices in $V(\Pi_2)$), increasing the number of balanced paths by two.

Thus, any operation acting on vertices of $V(\Pi_1)$ belonging to the same component of $AG(\Pi_1, \Pi_2)$ is optimal. □

Proposition 4. *Given two genomes Π_1 and Π_2, any component of $AG(\Pi_1, \Pi_2)$ can be sorted independently.*

Proof. This is a direct consequence of Proposition 3. □

Let $d(C)$ be the DCJ distance of a component C in $AG(\Pi_1, \Pi_2)$ and $seq(C)$ be the number of DCJ sequences sorting C. Moreover, let EC_{i+1} be an even cycle with $i+1$ edges in $AG(\Pi_1, \Pi_2)$ and let BP_i be a balanced and UP_{i-1} be an unbalanced path with respectively i and $i-1$ edges (observe that i is odd).

Proposition 5. *For any integer $i \in \{3, 5, 7, \ldots\}$, we have $d(UP_{i-1}) = d(BP_i) = d(EC_{i+1}) = \frac{i-1}{2}$ and $seq(UP_{i-1}) = seq(BP_i) = seq(EC_{i+1})$.*

Proof. A balanced path with i edges could be transformed into a cycle with $i+1$ edges by connecting the two ends of the path (this respects the alternation between vertices of $V(\Pi_1)$ and $V(\Pi_2)$). Analogously, an unbalanced path with $i-1$ edges could be transformed into a cycle with $i+1$ edges by the insertion of a double "telomere" vertex on the genome that is under-represented in the path. The two ends of the unbalanced path should then be connected to the new vertex (this also respects the alternation between vertices of $V(\Pi_1)$ and $V(\Pi_2)$) and the new vertex can also be used in optimal DCJ operations within the unbalanced path. An example for the case EC_4, BP_3 and UP_2 is given in

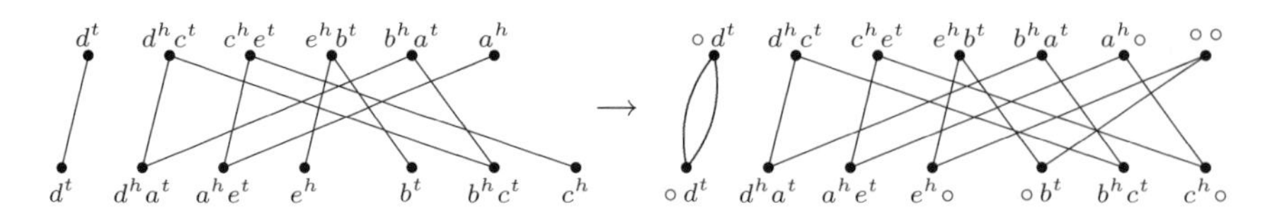

Fig. 4. Adding telomeres to the paths of the adjacency graph. We can see here that the structures of an EC_4, a BP_3 and an UP_2 are identical, therefore these components have the same DCJ distance and the same number of sorting sequences.

Figure 4. It is easy to see that we need $\frac{i-1}{2}$ operations to sort an unbalanced path with $i-1$ edges or a balanced path with i edges or a cycle with $i+1$ edges. □

We denote by $C_i \otimes C_j$ the multiplication of two independent sets of sequences sorting two components C_i and C_j, defined as the set of all sequences that are the result of all possible combinations of each sequence sorting C_i with each sequence sorting C_j. Observe that the operation $\otimes$ is symmetric, that is, $C_i \otimes C_j = C_j \otimes C_i$, and associative, that is, $(C_i \otimes C_j) \otimes C_k = C_i \otimes (C_j \otimes C_k)$. We denote by $||C_i||$ the number of sequences sorting C_i, and by d_i the length of each sequence sorting C_i (the DCJ distance of C_i). Then the number of sequences in $C_i \otimes C_j$ corresponds to $||C_i|| \times ||C_j|| \times M(d_i, d_j)$, where $M(d_i, d_j)$ is the number of possible ways to merge a sequence of length d_i with a sequence of length d_j, such that the merged sequences are subsequences of all resulting sequences.

For example, if $s_1 = \rho_1\rho_2$ and $s_2 = \theta_1\theta_2$, then all possible ways of merging s_1 and s_2 are the 6 sequences $\rho_1\rho_2\theta_1\theta_2$, $\rho_1\theta_1\rho_2\theta_2$, $\rho_1\theta_1\theta_2\rho_2$, $\theta_1\rho_1\rho_2\theta_2$, $\theta_1\rho_1\theta_2\rho_2$, and $\theta_1\theta_2\rho_1\rho_2$, thus $M(2,2) = 6$. The number $M(d_1, d_2, \ldots, d_k)$ corresponds to the multinomial coefficient and can be computed by the following formula:

$$M(d_1, d_2, \ldots, d_k) = \binom{d_1 + d_2 + \ldots + d_k}{d_1, d_2, \ldots, d_k} = \frac{(d_1 + d_2 + \ldots + d_k)!}{d_1! d_2! \ldots d_k!}.$$

Proposition 6. *The number of operations sorting a component C whose distance is d is given by $||C|| = (d+1)^{d-1}$.*

Proof. A cycle with n vertices and n edges has $v = n/2$ vertices in $V(\Pi_1)$, DCJ distance $d = v - 1$, and can be broken with one DCJ operation, resulting in two cycles as follows (each pair of cycles can be obtained in v different ways):

- one of size 2 ($v' = 1; d' = 0$) and one of size $n-2$ ($v' = v-1; d' = d-1$);
- one of size 4 ($v' = 2; d' = 1$) and one of size $n-4$ ($v' = v-2; d' = d-2$);
- one of size 6 ($v' = 3; d' = 2$) and one of size $n-6$ ($v' = v-3; d' = d-3$);
- and so on ...

Thus, the computation of the number of sorting sequences is given by the following recurrence formula on v:

$$T(1) = 1$$

$$T(v) = \frac{v}{2} \sum_{i=1}^{v-1} T(v-i) \otimes T(i).$$

We know that $T(v-i) \otimes T(i) = T(v-i) \times T(i) \times \binom{v-2}{i-1}$. Thus, we have $T(v) = \frac{v}{2} \sum_{i=1}^{v-1} \binom{v-2}{i-1} \times T(v-i) \times T(i)$; or, alternatively $T(v) = \frac{v}{2} \sum_{k=0}^{v-2} \binom{v-2}{k} \times T(v-k-1) \times T(k+1)$, which is also equivalent to

$$T(v) = \sum_{k=0}^{v-2} \binom{v-2}{k} \times (v-k-1) \times T(v-k-1) \times T(k+1).$$

This last recurrence formula is identical to the recurrence formula presented in [10] for counting labeled trees and results in v^{v-2}. Since we have $d = v-1$, we get $T(v) = (d+1)^{d-1}$. □

We call *small components* the paths and cycles of $AG(\Pi_1, \Pi_2)$ with two vertices (BP_1 and EC_2) whose distance is zero; the other components are *big components* (observe that any unbalanced path is a big component). The DCJ distance and number of sequences sorting big components is shown in Table 1.

Table 1. The number of DCJ sequences sorting each type of component independently

unbalanced paths	balanced paths	even cycles	sequence length	number of sequences
UP_2	BP_3	EC_4	1	1
UP_4	BP_5	EC_6	2	3
UP_6	BP_7	EC_8	3	16
UP_8	BP_9	EC_{10}	4	125
UP_{10}	BP_{11}	EC_{12}	5	1296
UP_{12}	BP_{13}	EC_{14}	6	16807
⋮	⋮	⋮	⋮	⋮
UP_{2d}	BP_{2d+1}	EC_{2d+2}	d	$(d+1)^{d-1}$

Proposition 7. *The number of solutions sorting $AG(\Pi_1, \Pi_2)$ obtained by sorting each component independently is*

$$C_1 \otimes C_2 \otimes \ldots \otimes C_k = \frac{(d_1 + d_2 + \ldots + d_k)!}{d_1! d_2! \ldots d_k!} \times \prod_{i=1}^{k} (d_i + 1)^{d_i - 1}$$

where $C_1, C_2, \ldots, C_k$ are the big *components of $AG(\Pi_1, \Pi_2)$ and $d_1, d_2, \ldots, d_k$ are their respective DCJ distances.*

Proof. We know that the number of solutions obtained by merging the sequences sorting the components independently is given by the multinomial coefficient multiplied by the number of sequences sorting each component, the latter being given by the formula in Proposition 6. □

3.3 Towards the General Case

The formula given by Proposition 7 does not correspond to the total number of solutions for a general instance of the problem, due to the recombination of

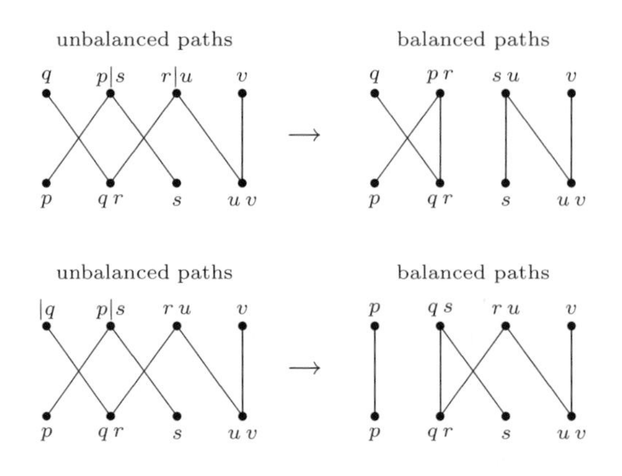

Fig. 5. Here we represent two of many ways of recombining a pair of alternate unbalanced paths into a pair of balanced paths

unbalanced paths. A pair of *alternate unbalanced paths* is composed by one unbalanced path with more vertices in $V(\Pi_1)$ and one unbalanced path with more vertices in $V(\Pi_2)$ and can be recombined at any time and in several different ways into two balanced paths. See an example in Figure 5. This can increase considerably the number of optimal solutions, specially when the number of the two types of unbalanced paths in the graph is big.

Proposition 8. *A DCJ operation acting on vertices of $V(\Pi_1)$ belonging to two different components of $AG(\Pi_1, \Pi_2)$ is optimal, if and only if they are alternate unbalanced paths.*

Proof. Recall that an optimal DCJ operation either increases the number of cycles by one or the number of balanced paths by two [3]. We need to examine all possible DCJ operations acting on two different components of $AG(\Pi_1, \Pi_2)$. If one component is a cycle, and the other component is of type X (X can be either a cycle, or a balanced, or an unbalanced path), then the result is one single component that is also of type X. Thus, the number of cycles in the graph is reduced. If the two components are balanced paths, then the result is either one unbalanced path, or two unbalanced, or two balanced paths. In the first and the second case the number of balanced paths is reduced by two and in the third case it remains unchanged. If one is a balanced and the other is an unbalanced path, then the result is either one balanced path, or one balanced and one unbalanced path. In both cases the number of balanced paths remains unchanged. And the same happens if the two components are unbalanced paths, that do not form an alternate pair. In this case, the result is either one, or two unbalanced paths. Note that all operations enumerated so far can not be optimal. The last possibility is when the components are a pair of alternate unbalanced paths. In this case, any operation acting on two vertices of $V(\Pi_1)$, each one belonging to each one of the unalanced paths, results in a pair of balanced paths (see Figure 5) and is an optimal DCJ operation. □

Proposition 9. *If $AG(\Pi_1, \Pi_2)$ does not contain any pair of alternate unbalanced paths, the components of $AG(\Pi_1, \Pi_2)$ can only be sorted independently.*

Corollary 1. *The formula given in Proposition 7 is a lower bound to the number of DCJ sorting scenarios for any instance of the problem.*

To give an idea of the increase caused by unbalanced paths, we consider the smallest example, that is, $AG(\Pi_1, \Pi_2)$ has only one pair of alternate unbalanced paths with three vertices each. Since the DCJ distance of each one of these unbalanced paths is one, no DCJ operation can be performed on them before the recombination, and they can be recombined into balanced paths in only four different ways, resulting in a small balanced path with two vertices and a big balanced path with four vertices (see Figure 6). Thus, either these two unbalanced paths are sorted independently or they are recombined into balanced paths in one of these four ways. Let $AG(\Pi_1, \Pi_2)$ have k big balanced components (cycles and balanced paths), numbered from C_1 to C_k. We know that the number of solutions sorting the components of $AG(\Pi_1, \Pi_2)$ independently is $\frac{(d_1+d_2+...+d_k+2)!}{d_1!d_2!...d_k!} \times \prod_{i=1}^{k}(d_i+1)^{d_i-1}$. Analogously, the number of solutions sorting the components of $AG(\Pi_1, \Pi_2)$ that contain one of the four ways of recombining the pair of unbalanced paths is $\frac{(d_1+d_2+...+d_k+2)!}{d_1!d_2!...d_k!2!} \times \prod_{i=1}^{k}(d_i+1)^{d_i-1}$. In consequence, the total number of solutions is:

$$\frac{3 \times (d_1 + d_2 + \ldots + d_k + 2)!}{d_1!d_2!\ldots d_k!} \times \prod_{i=1}^{k}(d_i+1)^{d_i-1}.$$

This means that a single shortest pair of alternate unbalanced paths triplicates the number of solutions with respect to the solutions obtained sorting the components individually.

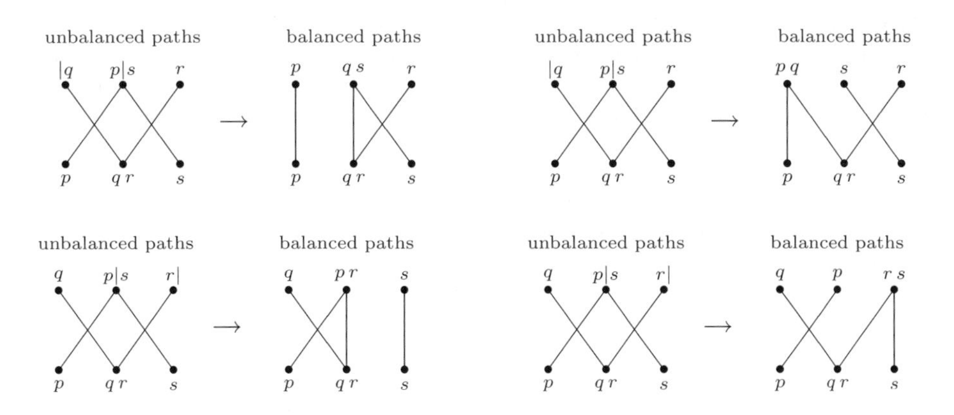

Fig. 6. A pair of shortest alternate unbalanced paths can be recombined into a pair of balanced paths in only four different ways

With this small example, one can figure out the complexity of integrating the recombination of alternate unbalanced paths to the counting formula. The different pairs of alternate unbalanced paths can be recombined separately or simultaneously in many different ways, and not only as the first step sorting the paths to be recombined, but at any time (when these paths have DCJ distance greater than one). However, we conjecture that the formula given in Proposition 7 gives a tight lower bound to the order of magnitude of the number of solutions.

4 Comparing Human, Chimpanzee and Rhesus Monkey

From [1] we obtained a database with the synteny blocks of the genomes of human (*Homo sapiens*), chimpanzee (*Pan troglodytes*) and rhesus monkey (*Macaca mulatta*) and used the formula given in Proposition 7 to compute the number of DCJ scenarios for the pairwise comparison of these genomes. The results are shown in Table 2. We observed that the number of paths, and more particularly the number of unbalanced paths in the corresponding adjacency graphs, is usually small. We know that big paths occur when some extremities of linear chromosomes are different in the two analysed genomes, thus our results suggest that this is unlikely to happen when the genomes are closely related. Comparing the human and chimpanzee genomes, for instance, we have only one unbalanced path, so it is not possible to recombine unbalanced into balanced paths. Consequently we are able to give the exact number of solutions for this instance of the problem.

Table 2. Counting DCJ sorting sequences between human, chimpanzee and rhesus monkey genomes (data obtained from [1]). For each pairwise comparison, the number of sorting sequences is very large and is thus presented approximately (although it can be computed exactly). Observe that the number of paths is usually much smaller than the number of cycles in all pairwise comparisons. Looking at the human *vs.* chimpanzee comparison in particular, we notice that it results in only one unbalanced path, thus none of its sorting sequences can be obtained by recombining unbalanced into balanced paths. This means that the lower bound given by the formula of Proposition 7 is tight in this case.

genomes	**# big cycles**	**# big bal. paths**	**# unbal. paths**	**DCJ distance**	**# DCJ scenarios**
human *vs.* chimpanzee	18	1	$1+0$	22	$\simeq 2.53 \times 10^{21}$
human *vs.* monkey	59	7	$2+4$	106	$\simeq 1.23 \times 10^{177}$
chimpanzee *vs.* monkey	68	8	$1+4$	114	$\simeq 1.53 \times 10^{193}$

5 Final Remarks

In this work we studied the solution space of the sorting by DCJ problem. We were able to characterize the solutions, showing how to transform an optimal sequence into another, and proposed a formula that gives a lower bound to the number of all optimal DCJ sequences sorting one genome into another. This formula can be easily and quickly computed and corresponds to the exact number of sorting sequences for a particular subset of instances of the problem. Although we could identify the structures of the compared genomes that cause the increase of the number of solutions with respect to the given lower bound, finding a general formula to the total number of DCJ sorting scenarios remains an open question.

References

1. Alekseyev, M.A., Pevzner, P.A.: Breakpoint graphs and ancestral genome reconstructions. Genome Res. 19, 943–957 (2009)
2. Bergeron, A., Chauve, C., Hartmann, T., St-Onge, K.: On the properties of sequences of reversals that sort a signed permutation. In: Proceedings of JOBIM 2002, pp. 99–108 (2002)
3. Bergeron, A., Mixtacki, J., Stoye, J.: A unifying view of genome rearrangements. In: Bücher, P., Moret, B.M.E. (eds.) WABI 2006. LNCS (LNBI), vol. 4175, pp. 163–173. Springer, Heidelberg (2006)
4. Braga, M.D.V., Sagot, M.-F., Scornavacca, C., Tannier, E.: Exploring the solution space of sorting by reversals with experiments and an application to evolution. IEEE/ACM Trans. Comput. Biol. Bioinf. 5(3), 348–356 (2008); Preliminary version in Proceedings of ISBRA 2007. LNCS (LNBI), vol. 4463, pp. 293–304 (2007)
5. Hannenhalli, S., Pevzner, P.: Transforming men into mice (polynomial algorithm for genomic distance problem). In: Proceedings of FOCS 1995, pp. 581–592 (1995)
6. Ouangraoua, A., Bergeron, A.: Parking functions, labeled trees and DCJ sorting scenarios. In: Ciccarelli, F.D., Miklós, I. (eds.) RECOMB-CG 2009. LNCS (LNBI), vol. 5817, pp. 24–35. Springer, Heidelberg (2009)
7. Sankoff, D.: Edit Distance for Genome Comparison Based on Non-Local Operations. In: Apostolico, A., Galil, Z., Manber, U., Crochemore, M. (eds.) CPM 1992. LNCS, vol. 644, pp. 121–135. Springer, Heidelberg (1992)
8. Siepel, A.: An algorithm to enumerate sorting reversals for signed permutations. J. Comput. Biol. 10, 575–597 (2003)
9. Yancopoulos, S., Attie, O., Friedberg, R.: Efficient sorting of genomic permutations by translocation, inversion and block interchange. Bioinformatics 21, 3340–3346 (2005)
10. Zeilberger, D.: Yet another proof of Cayley's formula for the number of labelled trees, `http://www.math.rutgers.edu/~zeilberg/mamarim/mamarimPDF/cayley.pdf`

Minimal Conflicting Sets for the Consecutive Ones Property in Ancestral Genome Reconstruction

Cedric Chauve[1], Utz-Uwe Haus[2], Tamon Stephen[1], and Vivija P. You[1]

[1] Department of Mathematics, Simon Fraser University, Burnaby (BC), Canada
{cedric.chauve,tamon,vpy}@sfu.ca

[2] Institute for Mathematical Optimization, University of Magdeburg, Germany
haus@imo.math.uni-magdeburg.de

Abstract. A binary matrix has the Consecutive Ones Property (C1P) if its columns can be ordered in such a way that all 1's on each row are consecutive. A *Minimal Conflicting Set* is a set of rows that does not have the C1P, but every proper subset has the C1P. Such submatrices have been considered in comparative genomics applications, but very little is known about their combinatorial structure and efficient algorithms to compute them. We first describe an algorithm that detects rows that belong to Minimal Conflicting Sets. This algorithm has a polynomial time complexity when the number of 1s in each row of the considered matrix is bounded by a constant. Next, we show that the problem of computing all Minimal Conflicting Sets can be reduced to the joint generation of all minimal true clause and maximal false clauses for some monotone boolean function. We use these methods in preliminary experiments on simulated data related to ancestral genome reconstruction.

1 Introduction

A binary matrix M has the Consecutive Ones Property (C1P) if its columns can be ordered in such a way that all 1's on each row are consecutive. Deciding if a matrix has the C1P can be done in linear-time and space [4,13,15,19,20]. The problem of testing if a matrix has the C1P has been considered in genomics, for problems such as physical mapping [2,6,17] or ancestral genome reconstruction [1,5,18,21].

If a binary matrix M does not have the C1P, a *Minimal Conflicting Set* (MCS) is a submatrix M' of M composed of a subset of the rows of M such that M' does not have the C1P, but every proper subset of rows of M' has the C1P. The Conflicting Index (CI) of a row of M is the number of MCS that contains this row. Minimal Conflicting Sets were introduced in [3]. In the context of ancestral genome reconstruction, MCS were used in [3] and in [23], where the CI was used to rank the rows of M and input into a branch-and-bound algorithm that computes parsimonious evolution scenarios for gene clusters. In both papers, the question of computing the CI of the rows of a non-C1P binary matrix M, or more generally to compute all MCS of M, was raised.

F.D. Ciccarelli and I. Miklós (Eds.): RECOMB-CG 2009, LNBI 5817, pp. 48–58, 2009.

In the present paper, after some preliminaries on the C1P and MCS (Section 2), we attack two problems. First, in Section 3, we consider the problem of deciding if a given row of a binary matrix M belongs to at least one MCS. We show that, when all rows of a matrix are constrained to have a bounded number of 1's, deciding if the CI of a row of a matrix is greater than 0 can be done in polynomial time. Next, in Section 4, we attack the problem of generating all MCS for a binary matrix M. We show that this problem, which is #P-hard, can be approached as a joint generation problem of minimal true clauses and maximal false clauses for monotone boolean functions. This can be done in quasi-polynomial time thanks to an oracle-based algorithm for the dualization of monotone boolean functions [8,9,12]. We implemented this algorithm [14] and applied it on simulated data (Section 5). We conclude by discussing several open problems.

2 Preliminaries

We briefly review here ancestral genome reconstruction and algorithmic results related to Minimal Conflicting Sets. Let M be a binary matrix with m rows and n columns, with e entries 1. We denote by $r_1, \ldots, r_m$ the rows of M and $c_1, \ldots, c_n$ its columns. We assume that M does not have two identical rows, nor two identical columns, nor a row with less than two entries 1 or a column with no entry 1. We denote by $\Delta(M)$ the maximum number of entries 1 found in a single row of M. In the following, we sometimes identify a row of M with the set of columns where it has entries 1.

Minimal Conflicting Sets, Maximal C1P Sets and Conflicting Index. A Minimal Conflicting Set (MCS) is a set R of rows of M that does not have the C1P but such that every proper subset of R has the C1P. The Conflicting Index (CI) of a row r_i of M is the number of MCS that contain r_i. A row r_i of M that belongs to at least one conflicting set is said to be a *conflicting row*.

A Maximal C1P Set (MC1P) is a set R of rows of M that has the C1P and such that adding any additional row from M to it results in a set of rows that does not have the C1P. For a subset $I = \{i_1, \ldots, i_k\}$ of $[n]$, we denote by R_I the set $\{r_{i_1}, \ldots, r_{i_k}\}$ of rows of M. If R_I is an MCS (resp. MC1P), we then say that I is an MCS (resp. MC1P).

Ancestral genome reconstruction. The present work is motivated by the problem of inferring an ancestral genome architecture given a set of extant genomes. An approach to this problem, described in [5], consists in defining an alphabet of genomic markers that are believed to appear uniquely in the extinct ancestral genome. An *ancestral synteny* is a set of markers that are believed to have been consecutive along a chromosome of the ancestor. A set of ancestral syntenies can then be represented by a binary matrix M: columns represent markers and the 1 entries of a given row define an ancestral synteny. If all ancestral syntenies are true positives (i.e. represent sets of markers that were consecutive in the ancestor), then M has the C1P. Otherwise, some ancestral syntenies are false positives that create MCS and the key problem is to remove them. This problem is generally attacked by removing from M the minimum number of rows such that the resulting matrix has the C1P; this optimization problem is NP-hard, even in the simpler case where every row of M contains exactly two entries 1, and is then often attacked using heuristics or branch-and-bound methods [1,5,18,21].

Computing the conflicting index. In the case where each row of M has exactly two entries 1, M naturally defines a graph G_M with vertex set $\{c_1, \ldots, c_n\}$ and where there is an edge between c_i and c_j if and only if there is a row with entries 1 in columns c_i and c_j. A set of rows R of M is an MCS if and only if the subgraph induced by the corresponding edges is a star with four vertices (also called a claw) or a cycle. This implies immediately that the number of MCS can be exponential in n. Also, combined with the fact that counting the number of cycles that contain a given edge in an arbitrary graph is #P-hard [25], this leads to the following result.

Theorem 1. *The problem of computing the Conflicting Index of a row in a binary matrix is #P-hard.*

Testing for Minimal Conflicting Sets. Given a set of p rows R of M, deciding whether these rows form an MCS can be achieved in polynomial time by testing (1) whether they form a matrix that does not have the C1P and, (2) whether every maximal proper subset of R (obtained by removing exactly one row) forms a matrix that has the C1P. This requires only $p + 1$ C1P tests and can then be done in time $O(p(n+m+e))$, using an efficient algorithm for testing the C1P [19].

Generating one Minimal Conflicting Set. It follows from the link between MCS and cycles in graphs that there can be an exponential number of MCS for a given binary matrix M, and that generating all of them is a hard problem. However, generating one MCS is easy and can be achieved in polynomial time by the following simple greedy algorithm:

1. let $R = \{r_1, \ldots, r_m\}$ be the full set of rows of M;
2. for i from 1 to m, if removing r_i from R results in a set of rows that has the C1P then keep r_i in R, otherwise remove r_i from R;
3. the subset of rows R obtained at the end of this loop is then an MCS.

Generating all Minimal Conflicting Sets. Given M and a list $C = \{R_1, \ldots, R_k\}$ of known MCS, the *sequential generation problem* $Gen_{MCS}(M, C)$ is the following: decide if C contains all MCS of M and, if not, compute one MCS that does not belong to C. Using the obvious property that, if R_i and R_j are two MCS then neither $R_i \subset R_j$ nor $R_j \subset R_i$, Stoye and Wittler [23] proposed the following backtracking algorithm for $Gen_{MCS}(M, C)$:

1. Let M' be defined by removing from M at least one row from each R_i in C by recursing on the elements of C.
2. If M' does not have the C1P then compute an MCS of M' and add it to C, else backtrack to step 1 using another set of rows to remove such that each $R_i \in C$ contains at least one of these rows.

This algorithm can require time $\Omega(n^k)$ to terminate, which, as k can be exponential in n, can be superexponential in n. As far as we know, this is the only previously proposed algorithm to compute all MCS.

3 Deciding If a Row Is a Conflicting Row

We now describe our first result, an algorithm to decide if a row of M is a conflicting row (i.e. has a CI greater than 0). Detecting non-conflicting rows is important, for example to speed-up algorithms that compute an optimal C1P subset of rows of M, or in generating all MCS. Our algorithm has a complexity that is exponential in $\Delta(M)$. It is based on a combinatorial characterization of non-C1P matrices due to Tucker [24].

Tucker patterns. The class of C1P matrices is closed under column and row deletion. Hence there exists a characterization of matrices which do not have the C1P by forbidden minors. Tucker [24] characterizes these forbidden submatrices, called M_I, M_{II} M_{III}, M_{IV} and M_V: if M is binary matrix that does not have the C1P, then it contains at least one of these matrices as a submatrix. We call these forbidden matrices the *Tucker patterns*. Patterns M_{IV} and M_V have each 4 rows, while the other three patterns do not have a fixed number of rows or columns; pattern M_I corresponds to the cycle when $\Delta(M) = 2$.

Bounded patterns. Let P be a set of p rows R of M, that defines a $p \times n$ binary matrix. P is said to *contain exactly* a Tucker pattern M_X if a subset of its columns defines a matrix equal to pattern M_X. The following properties are straightforward from the definitions of Tucker patterns and MCS:

Property 1. (1) If a subset of rows of M is an MCS, then the subset contains exactly a Tucker pattern.
(2) If a subset of p rows of M contains exactly a Tucker pattern M_{II} M_{III}, M_{IV} or M_V, then $4 \le p \le \max(4, \Delta(M) + 1)$.

From Property 1.(1), to decide if r_i belongs to at least one MCS, it suffices to decide if it belongs to a set R of p rows of M that contains exactly a Tucker pattern and is an MCS. Moreover, from Property 1.(2), if this Tucker pattern is of type M_{II} to M_V, it can be found by a brute-force approach that considers all sets of at most $\Delta(M)$ rows of R that contain r_i, and test, for each such set of rows, if it is an MCS. This brute-force approach has a worst-case time complexity of $O(\Delta(M)^2 m^{\Delta(M)+1}(n+m+e))$. It also allows to detect if r_i is in an MCS because of a Tucker pattern M_I containing at most $\Delta(M) + 1$ rows, which leaves only the case of patterns M_I (the cycle) with more than $\Delta(M) + 1$ rows.

Arbitrarily long cycles. Let B_M be the bipartite graph defined by M as follows: vertices are rows and columns of M, and every entry 1 in M defines an edge. Pattern M_I with p rows corresponds to a cycle of length $2p$ in B_M. Hence, if R contains M_I with p-row, the subgraph of B_M induced by R contains such a cycle and possibly other edges. Let $C = (r_{i_1}, c_{j_1} \ldots, r_{i_\ell}, c_{j_\ell})$ be a cycle in B_M. We say that a r_{i_q} belonging to C is *blocked in* C if there exists a vertex c_j such $M_{i_q,j} = 1$, $M_{i_{q-1},j} = 1$ ($M_{i_\ell,j} = 1$ if $q = 1$) and $M_{i_{q+1},j} = 1$ ($M_{1,j} = 1$ if $q = \ell$).

Proposition 1. *Let M be a binary matrix that does not have the C1P. Let r_i be a row of M that does not belong to any set R of rows of M that contains exactly a Tucker pattern M_{II} M_{III}, M_{IV} or M_V.*

A subset of p rows of M that contain r_i contains exactly the pattern M_I if and only if r_i belongs to a cycle $C = (r_{i_1}, c_{j_1}, \ldots r_{i_\ell}, c_{j,\ell})$ in B_M and r_i is not blocked in C.

Proof. (Sketch) If r_i belongs to an MCS and is not included in any pattern M_{II} M_{III}, M_{IV} or M_V, then it belongs to a set of rows that contain exactly the pattern M_I where r_i is not blocked (otherwise, this would contradict the fact that these p rows form an MCS).

To show that if r_i is not blocked in a cycle C implies it belongs to an MCS, consider a minimal cycle C where r_i is not blocked. A *chord* in the cycle C is a set of two edges (r, c) and (r', c) such that r and r' belong to C but are not consecutive row vertices in this cycle. If C has no chord, then the vertices it contains define a Tucker pattern M_I and then an MCS. Otherwise, the chord defines two cycles that contain r_i and where r_i is blocked, which contradicts the minimality of C. □

The algorithm. To decide whether r_i belongs to an MCS, we can then (1) check all sets of at least 4 and at most $\Delta(M) + 1$ rows that contain r_i to see if they define an MCS, and, if this is not the case, (2) check whether r_i belongs to an MCS due to pattern M_I. For this second case, we only need to find a cycle where r_i is not blocked. This can be done in polynomial time by considering all pairs of possible rows r_{i_1} and r_{i_2} that each have an entry 1 in a column where r_i has an entry 1 (there are at most $O(m^2)$ such pairs of rows), exclude the cases where the three rows r_i, r_{i_1} and r_{i_2} have an entry 1 in the same column, and then check if there is a path in B_M between r_{i_1} and r_{i_2} that does not visit r_i. This leads to the main result of this section.

Theorem 2. *Let M be an $m \times n$ binary matrix that does not have the C1P, and r_i be a row of M. Deciding if r_i belongs to at least one MCS can be done in $O(\Delta(M)^2 m^{\max(4,\Delta(M)+1)}(n + m + e))$ time.*

4 Generating All MCS with Monotone Boolean Functions

Let $[m] = \{1, 2, \ldots, m\}$. For a set $I = \{i_1, \ldots, i_k\} \subseteq [m]$, we denote by X_I the boolean vector $(x_1, \ldots, x_m)$ such that $x_j = 1$ if and only if $i_j \in I$. A *boolean function* $f : \{0,1\}^m \rightarrow \{0,1\}$ is said to be *monotone* if for every $I, J \subseteq [m]$, $I \subseteq J \Rightarrow f(X_I) \leq f(X_J)$.

Given a boolean function, a boolean vector X is said to be a *Minimal True Clause* (MTC) if $f(X_I) = 1$ and $f(X_J) = 0$ for every $J \subset I$. Symmetrically,

X_I is said to be a *Maximal False Clause* (MFC) if $f(X_I) = 0$ and $f(X_J) = 1$ for every $I \subset J$. We denote by $MTC(f)$ (resp. $MFC(f)$) the set of all MTC (resp. MFC) of f.

For a given $m \times n$ binary matrix M, let $f_M : \{0,1\}^m \rightarrow \{0,1\}$ be the boolean function defined by $f_M(X_I) = 1$ if and only if R_I does not have the C1P, where $I \subseteq [m]$. This boolean function is obviously monotone and the following proposition is immediate.

Proposition 2. *Let $I = \{i_1, \ldots, i_k\} \subseteq [m]$. R_I is an MCS (resp. MC1P) of M if and only if X_I is an MTC (resp. MFC) for f_M.*

Generating Minimal True Clauses for monotone boolean functions. It follows from Proposition 2 that generating all MCS reduces to generating all MTC for a monotone boolean function. This very general problem has been the subject of intense research, and we describe briefly below some important properties.

Theorem 3. [12] *Let $C = \{X_1, \ldots, X_k\}$ be a set of MTC (resp. MFC) of a monotone boolean function f. The problem of deciding if C contains all MTC (resp. MFC) of f is coNP-complete.*

Theorem 4. [11] *The problem of generating all MTC of a monotone boolean function f using an oracle to evaluate this function can require up to $|MTC(f) + MFC(f)|$ calls to this oracle.*

The second property suggests that, in general, to generate all MTC, it is necessary to generate all MFC, and vice-versa. For example, the algorithm of Stoye and Wittler [23] described in Section 2 is a satisfiability oracle based algorithm – it uses a polynomial-time oracle to decide if a given submatrix has the C1P, but it doesn't use this structure any further. Once it has found the complete list C of MCS, it will proceed to check all MC1P sets as candidate conflicting sets before terminating. Since this does not keep the MC1P sets explicitly, but instead uses backtracking, it may generate the same candidates repeatedly resulting in a substantial duplication of effort. In fact, this algorithm can easily be modified to produce *any* monotone boolean function given by a truth oracle.

Joint Generation of MTC and MFC for monotone boolean functions. One of the major results on generating MTC for monotone boolean functions, is due to Fredman and Khachiyan. It states that generating both sets together can be achieved in time quasi-polynomial in the number of MTC plus the number of MFC.

Theorem 5. [9] *Let $f : \{0,1\}^m \rightarrow \{0,1\}$ be a monotone boolean function whose value at any point $x \in \{0,1\}^m$ can be determined in time t, and let C and D be respectively the sets of the MTC and MFC of f. Given two subsets $C' \subseteq C$ and $D' \subseteq D$ of total size $s = |C'| + |D'|$, deciding if $C \cup D = C' \cup D'$, and if $C \cup D \neq C' \cup D'$ finding an element in $(C \backslash C') \cup (D \backslash D')$ can be done in time $O(m(m+t) + s^{o(\log s)})$.*

The key element to achieve this result is an algorithm that tests if two monotone boolean functions are duals of each other (see [8] for a recent survey on this topic). As a consequence, we can then use the algorithm of Fredman and Khachiyan to generate all MCS and MC1P.

5 Experimental Results

We used the `cl-jointgen` implementation of the joint generation method which is publicly available [14] with an oracle to test the C1P property based on the algorithm described in [19]. All datasets and results are available at the following URL: `http://www.cecm.sfu.ca/~cchauve/SUPP/RCG09`.

We generated 10 datasets of adjacencies (the rows of the binary matrices contain each two entries 1) with $n = 40$ and $m = 45$. Each dataset contains exactly 39 true positive (rows $\{i, i+1\}$ for $i = 1, \ldots, 39$) and 6 random false positives (rows $\{i, j\}$ with $j > i+1$). These parameters were chosen to simulate moderately large datasets that resemble real datasets. For a given dataset, the *conflicting ratio* (CR) of a row is the ratio between the CI of this row and the number of MCS. Similarly, the *MC1P ratio* (MR) of a row is the ratio between the number of MC1P that contain the row and the total number of MC1P. The *MCS rank* of a row is its ranking (between 1 and 45) when rows are ranked by increasing CR. The *MC1P rank* of a row is its ranking when rows are ranked by increasing MR. Table 1 shows some statistics on these experiments.

First, we can notice the large difference between the number of MCS and the number of MC1P. This shows that most computation time, in the joint generation, is spent generating MC1P. However if, as expected, false positives have, on average, a higher conflicting ratio than true positives, and conversely a lower MC1P ratio than true positives, it is interesting that the MC1P ratio discriminates much better between false positives and true positives than the

Table 1. Statistics on MCS and MC1P on simulated adjacencies datasets. FP_CR is the Conflicting Ratio for False Positives, TP_CR is for CR the True Positives, FP_MR is the MC1P ratio for False Positives and TP_MR is the MR for True Positives.

Dataset	Number of MCS	Number of MC1P	Average FP_CR	Average TP_CR	Average FP_MR	Average TP_MR	Average FP MCS rank	Average FP MC1P rank
1	55	8379	0.41	0.37	0.34	0.77	18.83	6.83
2	43	4761	0.36	0.32	0.3	0.84	20.33	6
3	38	9917	0.4	0.22	0.34	0.79	33.17	7
4	46	4435	0.5	0.35	0.41	0.8	33.33	9
5	59	6209	0.44	0.3	0.36	0.76	28.33	6
6	45	13791	0,47	0.2	0.39	0.8	32.67	4.67
7	61	2644	0.44	0.31	0.37	0.8	28.83	5.83
8	50	3783	0.43	0.28	0.36	0.81	34.5	6.83
9	57	2575	0.51	0.37	0.43	0.81	32.83	5.17
10	60	3641	0.45	0.31	0.38	0.83	26.33	7.83

Table 2. Distribution of the MCS and MC1P ratios for all rows (ALL), false positives (FP) and true positives (TP). Each cell of the table contains the number of rows whose ratio is in the interval for the column.

Dataset	[0, .1]	(.1, .2]	(.2, .3]	(.3, .4]	(.4, .5]	(.5, .6]	(.6, .7]	(.7, .8]	(.8, .9]	(.9, 1]
MCS ALL	52	16	75	205	87	14	1	0	0	0
MCS FP	0	0	14	31	13	2	0	0	0	0
MCS TP	52	16	61	174	74	12	1	0	0	0
MC1P ALL	10	2	7	18	32	50	73	44	20	194
MC1P FP	0	0	0	6	22	44	64	40	20	194
MC1P TP	10	2	7	12	10	6	9	4	0	0

conflicting ratio. This is seen in the MCS and MC1P ranks: the false positives have an average MCS rank of 28.91, well below the rank that would be expected if they were the rows that have the highest CI (42.17), while they have an average MC1P rank of 6.52, quite close of the 3.5 rank expected if they belonged to the fewest MC1P. To get a better understanding of the usefulness of the MCS ratio and MC1P ratio, Table 2 shows the rough distribution of these ratios.

These result suggest that the increased computation required by generating MC1P brings valuable information in discriminating false positives from true positives, and that the MC1P ratio is a better information to rank rows when trying to compute a maximal MC1P subset of rows.

6 Conclusion and Perspectives

This paper describes preliminary theoretical and experimental results on Minimal Conflicting Sets and Maximal C1P Sets. In particular, we suggested that Tucker patterns are fundamental in understanding the combinatorics of MCS, and that the generation of all MCS is a hard problem, related to monotone boolean functions. From an experimental point of view it appears, at least on datasets of adjacencies, that MC1P offer a better way to detect false positive ancestral syntenies than MCS and the CI. This leaves several open problems to attack.

Detecting non-conflicting rows. The complexity of detecting rows of a matrix that do not belong to any MCS when rows can have an arbitrary number of entries 1 is still open. Solving this problem probably requires a better understanding of the combinatorial structure of MCS and Tucker patterns. Tucker patterns have also be considered in [7, Chapter 3], where polynomial time algorithms are given to compute a Tucker pattern of a given type for a matrix that does not have the C1P. Even if these algorithms can not obviously be modified to decide if a given row belongs to a given Tucker pattern, they provide useful insight on Tucker patterns.

It follows from the dual structure of monotone boolean functions that the question of whether a row belongs to any MCS is equivalent to the question of whether it belongs to any MC1P. Indeed, for an arbitrary oracle-given function, testing if a variable appears in any MTC is as difficult as deciding if a list of MTC is complete. Consider an oracle-given f and a list of its MTC which define a (possibly different) function f'. We can build a new oracle function g with an additional variable x_0, such that $g(x_0, x) = 1$ if and only if $x_0 = 0$ and $f'(x) = 1$ or $x_0 = 1$ and $f(x) = 1$.

Generating all MCS and MC1P. Right now, this can be approached using the joint generation method, but the number of MCS and MC1P makes this approach impractical for large matrices. A natural way to deal with such problem would be to generate at random and uniformly MCS and MC1P. For MCS, this problem is at least as hard as generating random cycles of a graph, which is known to be a hard problem [16]. We are not aware of any work on the random generation of MC1P. An alternative to random generation would be to abort the joint generation after it generates a large number of MCS and MC1P, but the quality of the approximation of the MCS ratio and MC1P ratio so obtained would not be guaranteed. Another approach for the generation of all MCS could be based on the remark that, for adjacencies, it can be reduced to generating all claws and cycles of the graph G_M. Generating all cycles of a graph can be done in time that is polynomial in the number of cycles, using backtracking [22]. It is then tempting to use this approach in conjunction with dynamic partition refinement [13] for example or the graph-theoretical properties of Tucker patterns described in [7].

Combinatorial characterization of false positives ancestral syntenies. It is interesting to remark that, with adjacencies datasets, detecting most false positives can be attacked in a simple way. True positive rows define a set of paths in the graph G_M, representing ancestral genome segments, while false positive rows $\{i, j\}$, unless i or j is an extremity of such a path (in which case it does not exhibit any combinatorial sign of being a false positive), both the vertices i and j belong to a claw in the graph G_M. And it is easy to detect all edges in this graph with both ends belonging to a claw. In order to extend this approach to more general datasets, where $\Delta(M) > 2$, it would be helpful to understand better the impact of adding a false positive row in M. The most promising approach would be to start from the *partition refinement* [13] obtained from all true positive rows and form a better understanding of the combinatorial structure of connected components of the overlap graph that do not have the C1P.

Computation speed. The experiments in this paper took at most a few minutes to complete. We are currently running experiments on larger simulated datasets, as well as real data taken from [21], whose results will be made available on the companion Website of this paper. On larger datasets, especially with matrices with an arbitrary number of entries 1 per row, some connected components of the overlap graph can be very large (see the data in [21] for example). In order to speed up the computations, algorithmic design and engineering developments are required, both in the joint generation algorithm and in the problem of testing the C1P for matrices after rows are added or removed.

Acknowledgments

Cedric Chauve and Tamon Stephen were partially supported by NSERC Discovery Grants. Utz-Uwe Haus was supported by the Magdeburg Center for Systems Biology, funded by a FORSYS grant of the German Ministry of Education and Research.

References

1. Adam, Z., Turmel, M., Lemieux, C., Sankoff, D.: Common intervals and symmetric difference in a model-free phylogenomics, with an application to streptophyte evolution. J. Comput. Biol. 14, 436–445 (2007)
2. Alizadeh, F., Karp, R., Weisser, D., Zweig, G.: Physical mapping of chromosomes using unique probes. J. Comput. Biol. 2, 159–184 (1995)
3. Bergeron, A., Blanchette, M., Chateau, A., Chauve, C.: Reconstructing ancestral gene orders using conserved intervals. In: Jonassen, I., Kim, J. (eds.) WABI 2004. LNCS (LNBI), vol. 3240, pp. 14–25. Springer, Heidelberg (2004)
4. Booth, K.S., Lueker, G.S.: Testing for the consecutive ones property, interval graphs, and graph planarity. J. Comput. Syst. Sci. 13, 335–379 (1976)
5. Chauve, C., Tannier, E.: A methodological framework for the reconstruction of contiguous regions of ancestral genomes and its application to mammalian genome. PLoS Comput. Biol. 4, e1000234 (2008)
6. Christof, T., Jünger, M., Kececioglu, J., Mutzel, P., Reinelt, G.: A branch-and-cut approach to physical mapping of chromosome by unique end-probes. J. Comput. Biol. 4, 433–447 (1997)
7. Dom, M.: Recognition, Generation, and Application of Binary Matrices with the Consecutive-Ones Property. Dissertation, Institut für Informatik, Friedrich-Schiller-Universität, Jena (2008)
8. Eiter, T., Makino, K., Gottlob, G.: Computational aspects of monotone dualization: A brief survey. Disc. Appl. Math. 156, 2035–2049 (2008)
9. Fredman, M.L., Khachiyan, L.: On the complexity of dualization of monotone disjunctive normal forms. J. Algorithms 21, 618–628 (1996)
10. Goldberg, P.W., Golumbic, M.C., Kaplan, H., Shamir, R.: Four strikes againts physical mapping of DNA. J. Comput. Biol. 2, 139–152 (1995)
11. Gunopulos, D., Khardon, R., Mannila, H., Toivonen, H.: Data mining, hypergraph transversals and machine learning. In: PODS 1997, pp. 209–216. ACM, New York (1997)
12. Gurvich, V., Khachiyan, L.: On generating the ireedundant conjunctive and disjunctive normal forms of monotone Boolean functions. Disc. Appl. Math. 96-97, 363–373 (1999)
13. Habib, M., McConnell, R.M., Paul, C., Viennot, L.: Lex-BFS and partition refinement, with applications to transitive orientation, interval graph recognition and consecutive ones testing. Theoret. Comput. Sci. 234, 59–84 (2000)
14. Haus, U.-U., Stephen, T.: CL-JOINTGEN: A Common Lisp Implementation of the Joint Generation Method (2008), `http://primaldual.de/cl-jointgen/`
15. Hsu, W.-L.: A simple test for the Consecutive Ones Porperty. J. Algorithms 43, 1–16 (2002)
16. Jerrum, M.R., Valiant, L.G., Vazirani, V.Y.: Random generation of combinatorial structures from a uniform distribution. Theoret. Comput. Sci. 43, 169–188 (1986)

17. Lu, W.-F., Hsu, W.-L.: A test for the Consecutive Ones Property on noisy data – application to physical mapping and sequence assembly. J. Comp. Biol. 10, 709–735 (2003)
18. Ma, J., et al.: Reconstructing contiguous regions of an ancestral genome. Genome Res. 16, 1557–1565 (2006)
19. McConnell, R.M.: A certifying algorithm for the consecutive-ones property. In: SODA 2004, pp. 761–770. ACM, New York (2004)
20. Meidanis, J., Porto, O., Telle, G.P.: On the consecutive ones property. Discrete Appl. Math. 88, 325–354 (1998)
21. Ouangraoua, A., Boyer, F., McPherson, A., Tannier, E., Chauve, C.: Prediction of contiguous ancestral regions in the amniote ancestral genome. In: Măndoiu, I., Narasimhan, G., Zhang, Y. (eds.) ISBRA 2009. LNCS (LNBI), vol. 5542. Springer, Heidelberg (2009)
22. Read, R.C., Tarjan, R.E.: Bounds on backtrack algorithms for listing cycles, paths, and spanning trees. Networks 5, 237–252 (1975)
23. Stoye, J., Wittler, R.: A unified approach for reconstructing ancient gene clusters. IEEE/ACM Trans. Comput. Biol. Bioinfo (to appear, 2009)
24. Tucker, A.C.: A structure theorem for the consecutive 1's property. J. Combinat. Theory (B) 12, 153–162 (1972)
25. Valiant, L.G.: The complexity of enumeration and reliability problems. SIAM J. Comput. 8, 410–421 (1979)

Finding Nested Common Intervals Efficiently

Guillaume Blin[1] and Jens Stoye[2]

[1] Université Paris-Est, LIGM - UMR CNRS 8049, France
gblin@univ-mlv.fr
[2] Technische Fakultät, Universität Bielefeld, Germany
stoye@techfak.uni-bielefeld.de

Abstract. In this paper, we study the problem of efficiently finding gene clusters formalized by nested common intervals between two genomes represented either as permutations or as sequences. Considering permutations, we give several algorithms whose running time depends on the size of the actual output rather than the output in the worst case. Indeed, we first provide a straightforward $O(n^3)$ time algorithm for finding all nested common intervals. We reduce this complexity by providing an $O(n^2)$ time algorithm computing an irredundant output. Finally, we show, by providing a third algorithm, that finding only the maximal nested common intervals can be done in linear time. Considering sequences, we provide solutions (modifications of previously defined algorithms and a new algorithm) for different variants of the problem, depending on the treatment one wants to apply to duplicated genes.

1 Introduction and Related Work

Computational comparative genomics is a recent and active field of bioinformatics. One of the problems arising in this domain consists in comparing two or more species by seeking for *gene clusters* between their genomes. A gene cluster refers to a set of genes appearing, in spatial proximity along the chromosome, in at least two genomes. Genomes evolved from a common ancestor tend to share the same varieties of gene clusters. Therefore, they may be used for reconstructing recent evolutionary history and inferring putative functional assignments for genes.

The genome evolution process, including – among others – fundamental evolutionary events such as gene duplication and loss [12], has given rise to various genome models and cluster definitions. Indeed, genomes may be either represented as *permutations* (allowing one-to-one correspondence between genes of different genomes) or *sequences* – where the same letter (i.e. gene) may occur more than once (a more realistic model but with higher complexity). In both those models, there may exist, or not, genes not shared between two genomes (often called *gaps*).

Moreover, when modeling genomes for gene order analysis, one may consider either *two* or *multiple* genomes, seeking for *exact* or *approximate* occurrences, finding *all* or just non-extensible (i.e. *maximal*) occurrences.

F.D. Ciccarelli and I. Miklós (Eds.): RECOMB-CG 2009, LNBI 5817, pp. 59–69, 2009.

There are numerous ways of mathematical formalizations of gene clusters. Among others, one can mention *common substrings* (which require a full conservation), *common intervals* [1,7,14,15] (genes must occur consecutively, regardless of their order), *conserved intervals* [4] (common intervals, framed by the same two genes), *gene teams* [16,2,8] (genes in a cluster must not be interrupted by long stretches of genes not belonging to the cluster), and *approximate common intervals* [6,13] (common intervals that may contain few genes from outside the cluster). For more details, please refer to [3].

In this article, we focus on another model – namely the *nested common intervals* – which was mentioned in [9]. In this model, an additional constraint – namely the *nestedness* – (observed in real data [10]) is added to the cluster definition. Hoberman and Durand [9] argued that, depending on the dataset, if the nestedness assumption is not excluding clusters from the data, then it can strengthen the significance of detected clusters since it reduces the probability of observing them by chance.

As far as we know, [9] was the only attempt to take into account the nestedness assumption in a gene cluster model (namely gene teams) and yields to a quadratic-time greedy bottom-up algorithm. In fact, no explicit algorithmic analysis is given in [9], which might be a bit dangerous in view of the fact that genomes may consist of up to $25,000$ and more genes. In the following, we will give some efficient algorithms to find all gene clusters – considering nested common intervals – between two genomes.

Let π_1 and π_2 be our genomes, represented as permutations over $N := \{1,\ldots,n\}$. For any $i \leq j$, $\pi[i,j]$ will refer to the sequence of elements $(\pi[i],\pi[i+1],\ldots,\pi[j])$. Let $\mathcal{CS}(\pi[i,j]) := \{\pi[k] \mid k \in [i,j]\}$ denote the *character set* of the interval $[i,j]$ of π. A subset $C \subseteq N$ is called a *common interval* of π_1 and π_2 if and only if there exist $1 \leq i_1 < j_1 \leq n$ and $1 \leq i_2 < j_2 \leq n$ such that $C = \mathcal{CS}(\pi_1[i_1,j_1]) = \mathcal{CS}(\pi_2[i_2,j_2])$. Note that this definition purposely excludes common intervals of size one since they would not be considered in the more general nested common interval definition. The intervals $[i_1,j_1]$ and $[i_2,j_2]$ are called the *locations* of C in π_1 and π_2, respectively.

Given two common intervals C and C' of π_1 and π_2, C *contains* C' if and only if $C' \subseteq C$. This implies that the location of C' in π_1 (resp. π_2) is included in the location of C in π_1 (resp. π_2). A common interval C is called a *nested common interval* of π_1 and π_2 if either $|C| = 2$, or if $|C| > 2$ and it contains a nested common interval of size $|C|-1$. Note that this recursive definition implies that for any nested common interval C there exists a series of nested common intervals such that $C_2 \subseteq C_3 \subseteq \cdots \subseteq C$ with $|C_i| = i$. A nested common interval of size ℓ is *maximal* if it is not contained in a nested common interval of size $\ell+1$. A maximal nested common interval can however still be contained in a larger nested common interval. For example, considering $\pi_1 := (3,1,2,4,5,6)$ and $\pi_2 := (1,2,3,4,5,6)$, the maximal nested common interval $[4,6]$ in π_1 is contained in $[1,6]$.

The general NESTED COMMON INTERVALS problem may be defined as follows: *Given two genomes, find all their nested common intervals.* One can then consider genomes either as permutations or sequences and might also be interested in finding only the maximal nested common intervals and/or allowing gaps. In the two following sections, we will give efficient algorithms for both permutations and sequences but will leave the case considering gaps as an open problem.

In [1], Bergeron et al. proposed a theoretical framework for computing common intervals based on a linear space basis. Of importance here is the technique proposed in [1] in order to generate the PQ-tree [11] corresponding to a linear space basis for computing all the common intervals of K permutations. Generating this basis can be done in $O(n)$ time for two permutations of size n. Then one can, by a browsing of the tree, generate all the common intervals in $O(n+z)$ time where z is the size of the output. One can adapt this algorithm in order to find nested common intervals in $O(n+z)$ time.

In this work, we did not follow that approach, since (1) PQ-trees are a heavy machinery and quite space consuming in practice, and (2) our aim was to provide easy-to-implement algorithms with small constants in the O-notation. Moreover, algorithms based on PQ-trees do not easily generalize to sequences.

2 Nested Common Intervals on Permutations

As described in [7,14,15], when considering permutations, both common substrings and common intervals can be found in optimal, essentially linear time. As we will show, not surprisingly, finding nested common intervals on permutations can also be done efficiently.

2.1 Finding All Nested Common Intervals

First, one has to notice that the number of nested common intervals can be quadratic in n (e.g. when $\pi_1 = \pi_2$). However, in many practical cases the number of nested common intervals may be much smaller, such that one can still achieve lower time complexity by developing methods whose running time depends on the size of the actual output and not of the output in the worst case.

In the following, we will w.l.o.g. assume that π_1 is the identity permutation and rename π_2 by π for ease of notation. A naive bottom-up algorithm, inspired from the one given in [9] and straightforwardly following the definition of nested common intervals, can be defined as in Algorithm 1.

Clearly such an algorithm requires $O(n+z)$ time to report all nested common intervals where z is the size of the output. However, the output may be highly redundant as several intervals will be identified more than once. The worst case is when one considers $\pi = (1, 2, \ldots, n)$. More precisely, in this case some of the $O(n^2)$ nested common intervals will be reported up to n times, giving a total worst-case runtime of $O(n^3)$.

Therefore, one may be interested in computing an irredundant output. The main improvement we propose consists in a simple preprocessing step that will speed up our algorithm for nested common intervals. Let us define a *run* of two permutations π_1 and π_2 as a pair of intervals $([i_1, j_1], [i_2, j_2])$ such that

Algorithm 1. Find all nested common intervals

```
1: for i ← 1,...,n do
2:   l ← i, r ← i
3:   repeat
4:     l' ← l, r' ← r
5:     if π[l' − 1] = min(CS(π[l', r'])) − 1 or π[r' + 1] = max(CS(π[l', r'])) + 1 then
6:       while π[l − 1] = min(CS(π[l, r'])) − 1 do l−− done
7:       while π[r + 1] = max(CS(π[l', r])) + 1 do r++ done
8:     else
9:       while π[l − 1] = max(CS(π[l, r'])) + 1 do l−− done
10:      while π[r + 1] = min(CS(π[l', r])) − 1 do r++ done
11:    end if
12:    report all intervals [l'', r''] with l ≤ l'' ≤ l' and r' ≤ r'' ≤ r except [l', r']
13:  until l = l' and r = r'
14: end for
```

$\pi_1[i_1, j_1] = \pi_2[i_2, j_2]$ or $\pi_1[i_1, j_1] = \overleftarrow{\pi_2[i_2, j_2]}$ where $\overleftarrow{x} := (x_k, x_{k-1}, \ldots, x_1)$ denotes the reverse of sequence $x = (x_1, x_2, \ldots, x_k)$. A run is *maximal* if it cannot be extended to the left or right. Since a run can also be of size one, two permutations can always be decomposed into their maximal runs with respect to each other. For example, in the following the maximal runs are underlined: $\pi_1 = (\underline{1, 2, 3, 4}, \underline{5}, \underline{6, 7, 8}, \underline{9})$ and $\pi_2 = (\underline{4, 3, 2, 1}, \underline{5}, \underline{9}, \underline{6, 7, 8})$.

Given a decomposition of two permutations into their maximal runs with respect to each other, a *breakpoint* will refer to any pair of neighboring elements that belong to different runs. In the above example, π_1 (resp. π_2) contains three breakpoints $(4, 5)$, $(5, 6)$ and $(8, 9)$ (resp. $(1, 5)$, $(5, 9)$ and $(9, 6)$). By definition, the number of breakpoints is one less than the number of runs. When considering one of π_1 and π_2 as being the identity permutation then a run may be defined as a single interval.

Algorithm 2, hereafter defined, computes irredundant output by making use of the two following simple observations.

Lemma 1. *All subintervals of length at least 2 in a run are nested common intervals.*

Lemma 2. *In the procedure of constructing incrementally a nested common interval by extension, when one reaches up to one end of a run such that the begin/end of this run can be included, then the whole run can be included and all subintervals ending/beginning in this run can be reported as nested common intervals.*

Proof. By definition, the elements of a run $[i, j]$ in π with respect to the identity permutation are consecutive integers strictly increasing or decreasing. Therefore, in an incremental construction by extension of a nested common interval nc that has a run $[i, j]$ as its right (resp. left) neighbor, if one may extend nc by i (resp. j) then, by definition, $\pi[i]$ (resp. $\pi[j]$) is the minimal or maximal element among the elements of the extended interval. Thus, all the elements of the run $[i, j]$ may be added one by one, each leading to a new nested common interval. □

Consequently, by identifying the runs in a preprocessing step (which can be easily done in linear time), whenever during an extension a border of a run is included, the whole run is added at once and all sub-intervals are reported. The details are given in Algorithm 2 which employs a data structure *end* defined as follows: if i is the index of the first or last element of a run in π then $end[i]$ is the index of the other end of that run; $end[i] = 0$ otherwise.

Algorithm 2. Find all nested common intervals, irredundant version

```
1: decompose π into maximal runs w.r.t. id and store them in a data structure end
2: for i ← 1,...,n − 1 do
3:   if end[i] > i then
4:     l ← i, r ← end[i]
5:     report all intervals [l'', r''] with l ≤ l'' < r'' ≤ r
6:     repeat
7:       l' ← l, r' ← r
8:       if π[l'−1] = min(CS(π[l', r'])) − 1 or π[r'+1] = max(CS(π[l', r'])) + 1 then
9:         if π[l − 1] = min(CS(π[l, r'])) − 1 then l ← end[l − 1] end if
10:        if π[r + 1] = max(CS(π[l', r])) + 1 then r ← end[r + 1] end if
11:      else
12:        if π[l − 1] = max(CS(π[l, r'])) + 1 then l ← end[l − 1] end if
13:        if π[r + 1] = min(CS(π[l', r])) − 1 then r ← end[r + 1] end if
14:      end if
15:      report all intervals [l'', r''] with l ≤ l'' ≤ l' and r' ≤ r'' ≤ r except [l', r']
16:    until l = l' and r = r'
17:  end if
18: end for
```

It is easily seen that the time complexity remains in $O(n + z)$ with z being the output size, except that this time the output is irredundant (i.e. each nested common interval is reported exactly once). When applied to $\pi := (1, 2, 3, 6, 4, 5)$ for example, one may check that Algorithm 2 will report locations $[1, 2]$, $[1, 3]$, $[2, 3]$ when $i = 1$, locations $[5, 6]$, $[4, 6]$, $[1, 6]$, $[2, 6]$, $[3, 6]$ when $i = 5$, and nothing for the other values of i. Let us prove this interesting property of Algorithm 2.

Proposition 1. *In Algorithm 2, any breakpoint is considered (i.e. passed through) at most once during the extension procedure described from lines 8 to 14.*

Proof. Assume $bp = (\pi[x], \pi[y])$ is a breakpoint in π with respect to the identity permutation. Then $y = x + 1$ and there are only two possibilities for bp to be passed through twice: either (1) once from the left and once from the right, or (2) twice from the same side.

Let us first consider the case where bp is passed through from both left and right. Therefore assume that

$$\pi = (\ldots, \pi[X], \ldots, \pi[x], \pi[y], \ldots, \pi[Y], \ldots)$$

where $C_1 := \mathcal{CS}(\pi[X, x])$ and $C_2 := \mathcal{CS}(\pi[y, Y])$ are nested common intervals (otherwise *bp* would not be passed through more than once). Further assume w.l.o.g. that *bp* is passed in the extension of interval $[X, x]$ (a similar proof can be easily provided for the extension of $[y, Y]$) whose maximum (or minimum) element is $M := \max(C_1)$ (resp. $m := \min(C_1)$).

Then, in order for $[X, x]$ to be extensible, either $\pi[y] = M+1$ or $\pi[y] = m-1$. Let us assume that $\pi[y] = M + 1$ (the case where $\pi[y] = m - 1$ can be shown similarly). We will show that *bp* cannot be passed through in the other direction, i.e. in an extension of $[y, Y]$. Since $\pi[y] = M + 1$ and each of the two intervals $[X, x]$ and $[y, Y]$ consists of consecutive integers, we have that all elements in C_1 are smaller than any element in C_2. Thus, for an extension in the left direction across *bp*, the largest element of C_1 must be at its right end, i.e. $\pi[x] = M$. However, if this was the case, then $bp = (\pi[x], \pi[y]) = (M, M + 1)$ would not be a breakpoint, a contradiction.

Now let us consider the case where *bp* is passed through twice from the same side, starting with different runs. Therefore assume that

$$\pi = (\ldots, \pi[X'], \ldots, \pi[X], \ldots, \pi[x], \pi[y], \ldots)$$

where $C_1 := \mathcal{CS}(\pi[X, x])$ and $C_2 := \mathcal{CS}(\pi[X', x])$ are nested common intervals derived from different runs (i.e. reported from two different values of i), one in the interval $[X, x]$ and the other in the interval $[X', X - 1]$. Further assume w.l.o.g. that *bp* is passed through in the extension of $[X, x]$ (a similar proof can be easily provided for the extension of $[X', x]$) whose maximum (or minimum) element is $M := \max(C_1)$ (resp. $m := \min(C_1)$).

Then, in order for $[X, x]$ to be extensible, either $\pi[y] = M+1$ or $\pi[y] = m-1$. Let us assume that $\pi[y] = M + 1$ (the case where $\pi[y] = m - 1$ can be shown similarly). We will show that *bp* cannot be passed again in this direction, i.e. in an extension of $[X', x]$. Since $\pi[y] = M + 1$ and, by construction, $C_1 \subset C_2$, we have that all elements in $\mathcal{CS}(\pi[X', X - 1])$ are smaller than any element in C_1. Moreover, since *bp* is a breakpoint, we have that $\pi[x] \neq M$ and, more precisely, $\pi[x] < M$. Then, any extension of $[X', X - 1]$ would not be able to include M since at least one necessary intermediate element (namely $\pi[x]$) would not have been previously included. Thus, all cases are covered and the proposition is proved. □

Irredundancy of the locations of nested common intervals returned by Algorithm 2 follows immediately.

Proposition 2. *In Algorithm 2, two different runs cannot yield the same nested common interval.*

Proof. In order to be possibly reported twice, an interval would have to be a superinterval of two different runs. However, in order to yield the interval, the breakpoint(s) between these two runs would have to be passed through twice, which is not possible by Proposition 1. □

2.2 Finding All Maximal Nested Common Intervals

As previously mentioned, one might also be interested in finding only the maximal nested common intervals in optimal time $O(n + z)$ where z is the number of maximal nested common intervals of π_1 and π_2, since there will be fewer. In fact, we will first prove that the number of maximal nested common intervals is in $O(n)$ leading to an overall linear time algorithm.

Proposition 3. *Every element of π is a member of at most three different maximal nested common intervals.*

Proof. This follows immediately from the correctness of Proposition 1. Indeed, according to Proposition 1 each position can be reached from at most two directions. Thus, the only case where an element of π may be member of exactly three different maximal nested common intervals $nc_1 = [i_1, j_1]$, $nc_2 = [i_2, j_2]$ and $nc_3 = [i_3, j_3]$ is when $i_1 \leq i_2 \leq i_3 \leq j_1 \leq j_2 \leq j_3$. For example, considering $\pi_1 = (2, 1, 3, 4, 6, 5)$ and $\pi_2 = (1, 2, 3, 4, 5, 6)$, the element 3 in π_1 is member of $(2, 1, 3, 4)$, $(3, 4)$ and $(3, 4, 6, 5)$. □

In order to get only the locations of maximal nested common intervals, one has to modify Algorithm 2 such that only the locations at the end of an extension are reported. To do so, one has simply to (1) remove from Algorithm 2 lines 5 and 15 and (2) report the unique interval $[l, r]$ – which is by definition maximal – just after the end of the **repeat ... until** loop (currently line 16).

Clearly, the time complexity of this slightly modified version of Algorithm 2 is unchanged; that is $O(n + z)$ where z is the size of the output. Proposition 3 implies that the number of maximal nested common intervals is in $O(n)$, leading to an overall linear time.

3 Nested Common Intervals on Sequences

In this section, we will give algorithms (mainly ideas due to space constraints) to handle genomes represented as sequences (i.e. genes may be duplicated). In the following, we will assume that our genomes, denoted by S_1 and S_2, are defined over a bounded integer alphabet $\Sigma = \{1, \ldots, \sigma\}$ and have maximal length n. The precise definition of nestedness in sequences is subtle. Therefore, we propose three different variants of the problem, depending on the treatment one wants to apply when, during the extension of an interval, an element that is already inside the interval is met once again.

First, one may just extend the interval "for free", only caring about the "innermost occurrence"; all other occurrences are considered as not contributing to the cluster content. This definition follows the same logic as earlier ones used for common intervals [7,14] and for approximate common intervals [6].

A slight modification of our naive Algorithm 1 leads to an $O(n^3)$ algorithm. Indeed, since the sequences may contain duplicates, one has to start the extension procedure with all possible pairs $(S_1[i], S_2[j])$ where $1 \leq i \leq |S_1|$ and $1 \leq j \leq$

Algorithm 3. Find all maximal nested common intervals in two sequences

```
1: for i ← 1,...,|S1| do
2:   for each occurrence j of S1[i] in S2 do
3:     for each k ← 1,...,σ do c[k] ← (k = S1[i]) done
4:     l1 ← i, r1 ← i
5:     l2 ← j, r2 ← j
6:     repeat
7:       while c[S1[l1 − 1]] = true do l1 ← l1 − 1 done
8:       while c[S1[r1 + 1]] = true do r1 ← r1 + 1 done
9:       while c[S2[l2 − 1]] = true do l2 ← l2 − 1 done
10:      while c[S2[r2 + 1]] = true do r2 ← r2 + 1 done
11:      l1' ← l1, r1' ← r1
12:      if S1[l1 − 1] = S2[l2 − 1] or S1[r1 + 1] = S2[r2 + 1] then
13:        while S1[l1 − 1] = S2[l2 − 1] do
14:          l1−−, l2−−, c[S1[l1]] ← true
15:          while c[S1[l1 − 1]] = true do l1−− done
16:          while c[S2[l2 − 1]] = true do l2−− done
17:        end while
18:        while S1[r1 + 1] = S2[r2 + 1] do
19:          r1++, r2++, c[S1[r1]] ← true
20:          while c[S1[r1 + 1]] = true do r1++ done
21:          while c[S2[r2 + 1]] = true do r2++ done
22:        end while
23:      else
24:        while S1[l1 − 1] = S2[r2 + 1] do
25:          l1−−, r2++, c[S1[l1]] ← true
26:          while c[S1[l1 − 1]] = true do l1−− done
27:          while c[S2[r2 + 1]] = true do r2++ done
28:        end while
29:        while S1[r1 + 1] = S2[l2 − 1] do
30:          r1++, l2−−, c[S1[r1]] ← true
31:          while c[S1[r1 + 1]] = true do r1++ done
32:          while c[S2[l2 − 1]] = true do l2−− done
33:        end while
34:      end if
35:    until l1 = l1' and r1 = r1'
36:    report ([l1, r1], [l2, r2])
37:  end for
38: end for
```

$|S_2|$. Moreover, after each extension step all genes that are already members of the cluster have to be "freely" included. This can be tested efficiently by storing the elements belonging to the current cluster in a bit vector $c[1, \ldots, \sigma]$.

The resulting Algorithm 3 – which clearly runs in $O(n^3)$ time as each pair of index positions (i, j) is considered at most once and for each of them the extension cannot include more than n steps – only reports maximal gene clusters as previously done for permutations.

Second, one may forbid the inclusion of a second copy of a gene in a nested common interval. This problem can also be solved easily, by a quite similar algorithm which stops any extension when a gene already contained in the common interval is encountered.

Finally, one may be interested in finding a bijection (sometimes called *matching* in the computational comparative genomics literature) where, inside a nested common interval, each gene occurrence in S_1 must match a unique gene occurrence in S_2 from the same gene family. Surprisingly, the nestedness constraint leads to a polynomial time algorithm whereas for many other paradigms, considering matching and duplicates leads to hardness [5].

For this last variant, we unfortunately only know a very inefficient algorithm described hereafter. The main idea is to, first, construct a directed acyclic graph G whose vertices correspond to pairs of intervals $([i_1, j_1], [i_2, j_2])$, one from S_1 and one from S_2, that contain the same multiset of characters. In G, an edge is drawn from a vertex $v = ([i_1, j_1], [i_2, j_2])$ to a vertex $v' = ([i'_1, j'_1], [i'_2, j'_2])$ iff the corresponding interval pairs differ by one in length – i.e. $|(j_1 - i_1) - (j'_1 - i'_1)| = 1$ – and the shorter one is contained in the longer one, i.e. $((i_1 = i'_1)$ or $(j_1 = j'_1))$ and $((i_2 = i'_2)$ or $(j_2 = j'_2))$. An illustration is given in Figure 1.

Since, for a given multiset of cardinality ℓ there are at most $(n - \ell + 1)^2$ vertices in the graph, the total number of vertices in G is bounded by $O(n^3)$. Moreover, by definition, each vertex has an output degree of at most four, hence the number of edges is also bounded by $O(n^3)$. Finally, G can clearly be constructed in polynomial time. One can easily see that there is a correspondence between nested gene clusters and directed paths in G starting from vertices corresponding to multisets

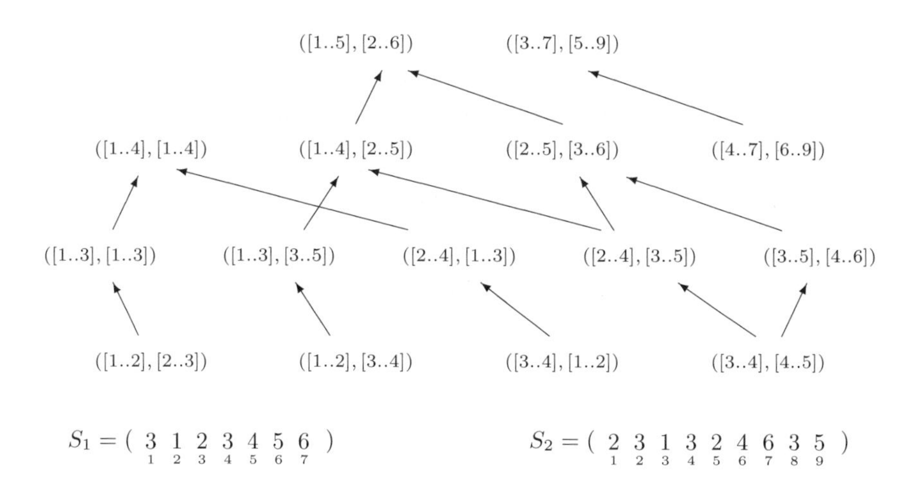

Fig. 1. Graph G for sequences $S_1 = (1, 3, 2, 3, 4, 5, 6)$ and $S_2 = (2, 3, 1, 3, 2, 4, 6, 3, 5)$

of size 2. Indeed, a path $(ci_1, ci_2, \ldots ci_k)$ in this DAG, where $ci_1, ci_2, \ldots ci_k$ are common intervals, induces that $ci_1 \subseteq ci_2 \subseteq \ldots \subseteq ci_k$ and $\forall 1 \leq j < k$ we have $|ci_j| + 1 = |ci_{j+1}|$. Therefore, since any common interval of size 2 is a nested common interval, in any such path, if ci_1 is of size 2 then, by definition, any common interval of this path is a nested common interval.

Thus, the nested gene clusters can be reported in polynomial time. Indeed, building the DAG can be done in $O(n^3)$ time. Then, one has to browse any path starting from a vertex corresponding to a common interval of size 2. Since, there are at most $n-1$ such common intervals, and each vertex has an output degree of at most four, on the whole the number of such paths is bounded by $(n-1) \cdot 4(n-2)$; i.e. $O(n^2)$.

4 Conclusion

In this article, we proposed a set of efficient algorithms considering the nestedness assumption in the common intervals model of gene clusters for genomes represented both as permutations and as sequences. Two main questions remain open: (1) finding a more efficient algorithm for the last variant of nested common intervals on sequences and (2) allowing clusters to be interrupted by up to g consecutive genes from outside the cluster, as in [9].

Acknowledgments

The authors wish to thank Ferdinando Cicalese, Martin Milanič and Mathieu Raffinot for helpful suggestions on the complexity of finding nested common intervals on sequences and on PQ-trees aspects.

References

1. Bergeron, A., Chauve, C., de Montgolfier, F., Raffinot, M.: Computing common intervals of k permutations, with applications to modular decomposition of graphs. SIAM J. Discret. Math. 22(3), 1022–1039 (2008)
2. Bergeron, A., Corteel, S., Raffinot, M.: The algorithmic of gene teams. In: Guigó, R., Gusfield, D. (eds.) WABI 2002. LNCS, vol. 2452, pp. 464–476. Springer, Heidelberg (2002)
3. Bergeron, A., Gingras, Y., Chauve, C.: Formal models of gene clusters. In: Mandoiu, I.I., Zelikovsky, A. (eds.) Bioinformatics Algorithms: Techniques and Applications, ch. 8, pp. 177–202. Wiley, Chichester (2008)
4. Bergeron, A., Stoye, J.: On the similarity of sets of permutations and its applications to genome comparison. J. Comp. Biol. 13(7), 1340–1354 (2006)
5. Blin, G., Chauve, C., Fertin, G., Rizzi, R., Vialette, S.: Comparing genomes with duplications: a computational complexity point of view. ACM/IEEE Trans. Comput. Biol. Bioinf. 14(4), 523–534 (2007)
6. Böcker, S., Jahn, K., Mixtacki, J., Stoye, J.: Computation of median gene clusters. In: Vingron, M., Wong, L. (eds.) RECOMB 2008. LNCS (LNBI), vol. 4955, pp. 331–345. Springer, Heidelberg (2008)

7. Didier, G., Schmidt, T., Stoye, J., Tsur, D.: Character sets of strings. J. Discr. Alg. 5(2), 330–340 (2007)
8. He, X., Goldwasser, M.H.: Identifying conserved gene clusters in the presence of homology families. J. Comp. Biol. 12(6), 638–656 (2005)
9. Hoberman, R., Durand, D.: The incompatible desiderata of gene cluster properties. In: McLysaght, A., Huson, D.H. (eds.) RECOMB 2005. LNCS (LNBI), vol. 3678, pp. 73–87. Springer, Heidelberg (2005)
10. Kurzik-Dumke, U., Zengerle, A.: Identification of a novel *Drosophila melanogaster* gene, *angel*, a member of a nested gene cluster at locus 59F4,5. Biochim. Biophys. Acta 1308(3), 177–181 (1996)
11. Landau, G.M., Parida, L., Weimann, O.: Using pq trees for comparative genomics. In: Apostolico, A., Crochemore, M., Park, K. (eds.) CPM 2005. LNCS, vol. 3537, pp. 128–143. Springer, Heidelberg (2005)
12. Ohno, S.: Evolution by gene duplication. Springer, Heidelberg (1970)
13. Rahmann, S., Klau, G.W.: Integer linear programs for discovering approximate gene clusters. In: Bücher, P., Moret, B.M.E. (eds.) WABI 2006. LNCS (LNBI), vol. 4175, pp. 298–309. Springer, Heidelberg (2006)
14. Schmidt, T., Stoye, J.: Quadratic time algorithms for finding common intervals in two and more sequences. In: Sahinalp, S.C., Muthukrishnan, S.M., Dogrusoz, U. (eds.) CPM 2004. LNCS, vol. 3109, pp. 347–358. Springer, Heidelberg (2004)
15. Uno, T., Yagiura, M.: Fast algorithms to enumerate all common intervals of two permutations. Algorithmica 26(2), 290–309 (2000)
16. Zhang, M., Leong, H.W.: Gene team tree: A compact representation of all gene teams. In: Nelson, C.E., Vialette, S. (eds.) RECOMB-CG 2008. LNCS (LNBI), vol. 5267, pp. 100–112. Springer, Heidelberg (2008)

DCJ Median Problems on Linear Multichromosomal Genomes: Graph Representation and Fast Exact Solutions

Andrew Wei Xu

School of Computer and Communication Sciences
Swiss Federal Institute of Technology (EPFL)
EPFL IC LCBB, Station 14
CH-1015 Lausanne, Switzerland
wei.xu@epfl.ch

Abstract. Given a set of genomes $\mathcal{G}$ and a distance measure d, the genome rearrangement median problem asks for another genome q that minimizes $\sum_{g \in \mathcal{G}} d(q, g)$. This problem lies at the heart of phylogenetic reconstruction from rearrangement data, where solutions to the median problems are iteratively used to update genome assignments to internal nodes for a given tree. The median problem for reversal distance and DCJ distance is known to be NP-hard, regardless of whether genomes contain circular chromosomes or linear chromosomes and of whether extra circular chromosomes is allowed in the median genomes. In this paper, we study the relaxed DCJ median problem on linear multichromosomal genomes where the median genomes may contain extra circular chromosomes; extend our prior results on circular genomes—which allowed us to compute exact medians for genomes of up to 1,000 genes within a few minutes. First we model the DCJ median problem on linear multichromosomal genomes by a *capped multiple breakpoint graph*, a model that avoids another computationally difficult problem—a multi-way capping problem for linear genomes, then establish its corresponding decomposition theory, and finally show its results on genomes with up to several thousand genes.

1 Introduction

Genomes of related species contain large numbers of homologous DNA sequences, including protein-coding genes, noncoding genes, and other conserved genetic units. In the following, we shall use the term *genes* to refer to all of these sequences. These genes typically appear in different orders in different genomes, as a result of evolutionary events that rearranged the gene orders. These gene orders can thus be used to infer phylogenetic relationships; in the process, they may also be used to attempt reconstruction of ancestral gene orders.

We assume that each genome contains the same set of genes and that no gene is duplicated, so that the genes can be represented by integers, each chromosome by a sequence of integers (if a chromosome is circular, this sequence is viewed as circular), and each genome by a collection of such sequences. The integers representing genes have a sign, which denotes the strand on which the genetic information is read; we

F.D. Ciccarelli and I. Miklós (Eds.): RECOMB-CG 2009, LNBI 5817, pp. 70–83, 2009.

assume that the sign of each gene is known. A genome is linear if it only contains linear chromosome, circular otherwise. When a genome contains one chromosome, we call it unichromosomal; otherwise, we call it multichromosomal.

A breakthrough in the study of genome rearrangements was the characterization of the mathematical structure of reversals (usually called inversions by biologists) and the first polynomial-time algorithm to compute a shortest series of reversals to transform one genome into another, whether unichromosomal [5] or multichromosomal [6]. Later work yielded an optimal linear-time algorithm to compute the reversal distance [2] and a $O(n \log n)$ algorithm to sort almost all permutations by reversals [9]. While reversals have been documented by biologists since the 1930s (in the pioneering work of the fly geneticists Sturtevant and Dobzhansky), the most studied operator in the last few years is a mathematical construct that unifies reversals with translocations, fusions, and fissions, and can implement any transposition in two moves, the double-cut-and-join (DCJ) operation [16]. If we let n be the number of genes, χ denote the number of linear chromosomes, and c, p_e, and p_o denote the numbers of cycles, even-sized paths, and odd-sized paths, respectively, in the breakpoint graph between two genomes, the DCJ distance between these genomes is given by the formula:

$$d_{DCJ} = n + \chi - c - p_e - \frac{p_o}{2}. \quad (1)$$

1.1 The DCJ Median Problem

The median problem for genome rearrangements is defined as follows: given a set $\mathcal{G}$ of genomes and a distance measure d, find a genome q that minimizes $\sum_{g \in \mathcal{G}} d(q, g)$. This problem is central to both phylogenetic reconstruction using gene-order data and ancestral reconstruction of gene orders. The median problem is known to be NP-hard under most rearrangement distances proposed to date, such as breakpoint, reversal, and DCJ distances; except when the median genome can contain extra circular chromosomes, the problem becomes polynomial under the breakpoint distance [11].

DCJ operations may create intermediate genomes with one or more extra circular chromosomes in addition to the original collection. We will refer to *relaxed* and *strict* versions of the DCJ median problem depending on whether such extra circular chromosomes are, respectively, allowed or forbidden. Our focus here is on the relaxed version, in part because of its simplicity and in part because, as seen in our experimental results, the average number of extra circular chromosomes does not exceed 0.5, so that solutions for the relaxed version are frequently also solutions for the strict version and those that are not can be transformed (by merging the circular chromosomes) into good approximate solutions to the strict version.

The median problem is NP-hard for the DCJ distance, for both strict and relaxed versions, on both circular and linear genomes [4,11]. There are exact algorithms for the reversal median problem and the strict DCJ median problem [4,8,17], but limited to small sizes; and there are also heuristics [3,1,7,10] of varying speed and accuracy.

In previous work [15,14], we developed an decomposition approach to the relaxed DCJ median problem on circular genomes. We used the *multiple breakpoint graph* (MBG) [4] to model the median problem; and the decomposition relies on the concept of *adequate subgraphs* of the MBG. We proved that such subgraphs enable a divide-and-conquer approach in which the MBG is decomposed, optimal solutions found recursively for its parts, and then optimal solutions for the complete instance created by combining these optimal solutions. We also showed that there are infinitely many types of adequate subgraphs [14], among which those of small sizes immediately tell us which adjacencies should exist in the median genomes. Applying this decomposition method to simulated data, we showed that instances with up to 1,000 genes with moderate numbers of rearrangement events ($\leq 0.9n$) can be solved in a few minutes.

1.2 Contributions and Presentation of Results

In this paper, we consider the relaxed DCJ median problem on linear multichromosomal genomes with equal or unequal numbers of chromosomes. Compared to its circular counterpart [15] this problem introduces a multi-way capping problem. Caps are used to delimit the ends of linear chromosomes; for a genome with χ linear chromosomes, there are $(2\chi)!$ different ways to label its 2χ caps. For a median problem containing three genomes with χ linear chromosomes each, one of these genomes can have fixed consecutive labels from 1 to 2χ, while each of the other two genomes needs to pick a labeling from $(2\chi)!$ possibilities each, for a total of $\left((2\chi)!\right)^2$ choices. For $\chi = 23$ (as in the human genome), this number is 3×10^{115}. The optimal choice is assumed to be one that yields a minimum number of evolutionary events for the median problem.

We introduce the *capped multiple breakpoint graph* (CMBG) to model the DCJ median problem on linear genomes, where a single node is used to represent all caps, following the idea in the *flower graph* [13]. In so doing, we completely avoid the capping problem. When the median genome is obtained, the optimal capping can be determined by solving pairwise capping problems between the median and each of the given genomes—and optimal pairwise cappings can be determined in time linear in the number of genes [16].

The rest of this paper is organized as follows. In Section 2, we briefly review basic concepts and results from [15,14] for the circular case. We defined the capped multiple breakpoint graph in Section 3 and its *regular adequate subgraphs* and *capped adequate subgraphs* in Section 4, where we also listed the most frequent adequate subgraphs. In Section 5 we present algorithm ASMedian-linear (similar to the one for the circular case), which finds exact solutions by iteratively detecting the existence of adequate subgraphs on CMBGs. In section 6, we present the results of tests on simulated data with varying parameters.

2 Basic Definitions and Previous Results on Circular Genomes

For the median problem on circular genomes, the breakpoint graph extends naturally to a *multiple breakpoint graph* (MBG) [4]. We call the number of given genomes $N_{\mathcal{G}}$ the *rank* of an MBG. We label the given genomes and the edges representing their adjacencies by integers from 1 to $N_{\mathcal{G}}$. Finally we define the *size* of an MBG or its subgraph as half the number of vertices it contains.

By a 1-edge, 2-edge or 3-edge, we mean an edge with colour 1, 2 or 3. Since edges of the same colour form a perfect matching of the MBG, we use i-matching to denote

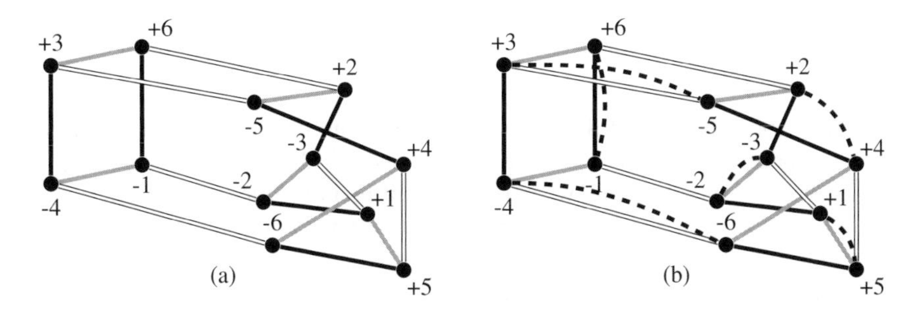

Fig. 1. Multiple breakpoint graph and median graph. Black, gray, double and dashed edges denote edges with colours 1, 2, 3 and 0 correspondingly. (a) A multiple breakpoint graph of three circular genomes, (1 2 3 4 5 6), (1 -5 -2 3 -6 -4) and (1 3 5 -4 6 -2) (b) A median graph with the median genome (1 -5 -3 2 -4 6). (and) are used to indicate the chromosomes are circular.

all edges of colour i, where $1 \leq i \leq 3$. For a candidate median genome, we use a different colour for its adjacency edges, namely colour 0; similarly we have 0-edges and 0-matchings. Adding such a candidate matching to the MBG results in the *median graph*. The set of all possible candidate matchings is denoted by $\mathcal{E}$.

The 0-i *cycles* in a median graph with a 0-matching E, numbering $c_{0,i}$ in all, are the cycles where 0-edges and i-edges alternate. Let $c_E^{\Sigma} = \sum_{1 \leq i \leq 3} c_{0,i}$. Then $c_{\max}^{\Sigma} = \max\{c_E^{\Sigma} : E \in \mathcal{E}\}$ is the maximum number of cycles that can be formed from the MBG. For circular genomes, since the DCJ distance is determined by the number of cycles in the induced breakpoint graph, we have:

Lemma 1. *[15] Minimizing the total DCJ distance in the median problem on circular genomes is equivalent to finding an optimal 0-matching E, i.e., with $c_E^{\Sigma} = c_{\max}^{\Sigma}$.*

A connected MBG subgraph H of size m is an *adequate subgraph* if $c_{\max}^{\Sigma}(H) \geq \frac{1}{2}mN_{\mathcal{G}}$; it is *strongly adequate* if $c_{\max}^{\Sigma}(H) > \frac{1}{2}mN_{\mathcal{G}}$. For the median of three problem where the rank is $N_{\mathcal{G}} = 3$, an adequate subgraph is a subgraph with $c_{\max}^{\Sigma}(H) \geq \frac{3m}{2}$ and a strongly adequate subgraph is one with $c_{\max}^{\Sigma}(H) > \frac{3m}{2}$.

The existence of an adequate subgraphs on an MBG gives a proper decomposition of the MBG into two subproblems, where the optimal solution of the original problem can be found by combing solutions from the two subproblems, as stated in the following theorem.

Theorem 1. *[15][1] The existence of an adequate subgraph gives a proper decomposition from which* an *optimal solution can be found by combining solutions from the two subproblems. The existence of a strongly adequate subgraph gives a proper decomposition from which* all *optimal solutions can be found by combining solutions from the two subproblems.*

Remark 1. Subproblems induced by adequate subgraphs of small sizes are easy to solve. When we only use these small adequate subgraphs (as our algorithms do), we

[1] This theorem has been rephrased, in order to avoid the concept of a *decomposer*, whose definition requires several other concepts.

can encode their solutions into algorithms, so that whenever their existences are detected, adjacencies representing solutions of these subproblems are immediately added into the median genome.

An intuitive understanding of this theorem follows from the definition of an adequate subgraph. For any subgraph H of size m, the theoretical largest value for $c^{\Sigma}_{max}(H)$ is mN_G, which is only possible when edges of different colours coincide. The ability for an adequate subgraph to form half the maximum number or more color alternating cycles, indicates that the decomposition by this subgraph is optimal, as it probably forms more cycles than any other alternative choice.

3 Capped Multiple Breakpoint Graph—A Graph Representation of the Median Problem on Linear Multichromosomal Genomes

In the rest of the paper we study the relaxed DCJ median problem on linear multichromosomal genomes. Here in this section, we introduce its first graph representation. The idea follows a *flower graph* [13], a variant of the breakpoint graph on linear genomes with a single *cap* node delimiting all ends of linear chromosomes. But first let us quickly review the traditional model of using a pair of caps for each linear chromosome [6,12], and show why the problem is simplified when the DCJ distance measure is used.

The process of adding caps is called *capping*. For a pair of genomes with χ linear chromosomes, there are $(2\chi)!$ different ways of capping. Different cappings may lead to different breakpoint graphs, and hence different pairwise distances. The capping problem for two genomes is to identify pairs of telomeres, one telomere from each genome, evolving from the same telomeres in their common ancestor genome. This orthologous relationship is represented, presumably, by an optimal capping, the one giving the smallest distance. Given an optimal capping, since vertices (representing endpoints of genes or telomeres) are incident to two adjacencies, one from each genome, the induced breakpoint graph consists of $c' = c + p_e + \frac{p_o}{2}$ color alternating cycles. As for circular genomes the DCJ distance is determined by its number of cycles c, for linear genomes the DCJ distance is determined by the number of cycles c', in the breakpoint graph induced by an optimal capping. For pairwise genomes under the DCJ distance, optimal cappings can be easily determined [16].

The traditional model of using a pair of caps for each linear chromosome lead to different induced breakpoint graphs, which are all equivalent in the amount of information they carry. We proposed in [13] a succinct graph representation, namely a *flower graph*, in which by allowing cycles and paths to be freely arranged, caps are merged into a single node. Flower graph gives a unique graph representation for pairs of linear genomes, as illustrated by Fig. 2.(a). In determining the pairwise DCJ distance, the significance of introducing the flower graph is limited to giving a mathematically succinct and unique graph representation; an optimal capping is easy to find and its induced breakpoint graph can be thought of as the representative breakpoint graph. However, when there are more than two genomes, such as in the median problem, optimal cappings may be hard to find; any procedure that first requires to find optimal cappings can be computationally very costly. As for the median problem with three given linear multichromosomal genomes, there are $((2\chi)!)^2$ ways of capping; as each capping induces

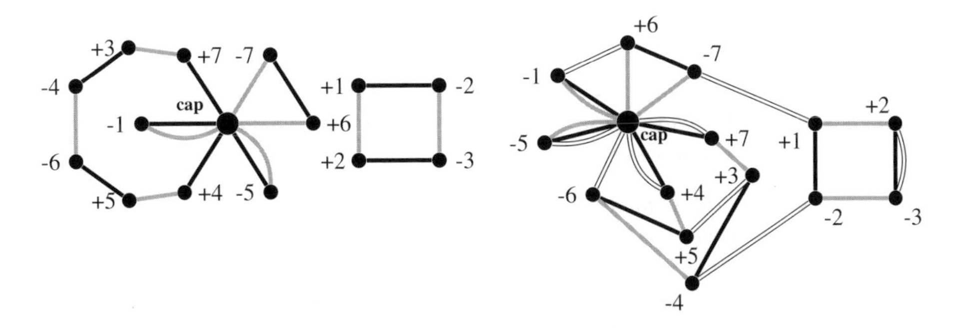

Fig. 2. (a) Flower graph for genomes $\{1\ 2\ 3\ 4;\ 5\ 6\ 7\}$ and $\{1\ \text{-}2\ 3\ \text{-}7;\ 5\ \text{-}4\ 6\}$, where we use black and gray edges to distinguish adjacency edges from different genomes. We have $n = 7, \chi = 2, c = 1, p_o = 2, p_e = 2$, so their DCJ distance is 5. (b) Capped multiple breakpoint graph for genomes $\{1\ 2\ 3\ 4;\ 5\ 6\ 7\}$, $\{1\ \text{-}2\ 3\ \text{-}7;\ 5\ \text{-}4\ 6\}$ and $\{5\ \text{-}3\ \text{-}2\ 4;\ 6\ 1\ 7\}$. Black, gray and double edges represent edges of colours 1 2 and 3 correspondingly. The cap is incident to 6 edges.

a different multiple breakpoint graph and an equivalent instance of the median problem on circular genomes, to find optimal cappings one would need to solve $((2\chi)!)^2$ instances of the median problem on circular genomes.

To model the median problem on linear multichromosomal genomes, following the idea of flower graphs and multiple breakpoint graphs, we propose the *capped multiple breakpoint graph* (CMBG). A CMBG is constructed as follows: each gene is represented by a pair of ordered vertices, a single node named *the cap* is added to delimit ends of all linear chromosomes; adjacencies between genes are represented by coloured edges connecting their corresponding endpoints, and for genes residing at ends of chromosomes, coloured edges are added connecting the cap and their endpoints. Edges representing adjacencies from the same genome are labeled with the same colour. For an instance of the median of three problem, where each genome contains n genes and χ linear chromosomes, the corresponding CMBG has $2n$ regular vertices of degree 3 each, and the cap vertex of degree 6χ. Fig.2.(b) shows a CMBG for the median of three problem with genomes $\{1\ 2\ 3\ 4;\ 5\ 6\ 7\}$, $\{1\ \text{-}2\ 3\ \text{-}7;\ 5\ \text{-}4\ 6\}$ and $\{5\ \text{-}3\ \text{-}2\ 4;\ 6\ 1\ 7\}$.

When the genomes contain different numbers of linear chromosomes, a few null chromosomes are added to equalize the number of chromosomes. A null chromosome consists of two telomeres and no genes; in CMBG, they correspond to edges looping around the cap. When the context is clear, these looping edges can be omitted from the graph.

The definition of the *size* of an MBG as half the number of vertices no longer applies to a CMBG or its subgraphs, as the cap node actually represents a number of telomeres—counting it as one vertex does not reflect the actual number of its adjacent edges. In this case for a CMBG or its subgraph, the *size* is defined as the total number of 0-edges to be added. For a subgraph of a CMBG, it is possible to have different interpretations of its size; however this does not impose much difficulty in defining *adequate subgraphs* for a CMBG—the definition holds if the requirement is satisfied by any interpretation and furthermore such an interpretation is always unique and obvious.

Similar to Lemma 1 we have the following statement for the median problem on linear multichromosomal genomes.

Lemma 2. *Finding the DCJ median on genomes with n genes and χ linear chromosomes is equivalent to finding a set of $n + \chi$ 0-edges for the capped multiple breakpoint graph, satisfying the following properties:*

1. *each regular vertex is incident to E exactly once;*
2. *the cap node is incident to E exactly 2χ times;*
3. *E maximizes the total number of cycles $c'^{\Sigma}_{max} = \max\{\sum_{1\le i\le 3} c'_{0,i}$: for all possible sets of 0-edges$\}$, where $c'_{0,i}$ is equal to the quantity $c + p_e + \frac{p_o}{2}$ between a candidate median genome whose adjacencies represented by a set of 0-edges and the ith given genome.*

Using a single cap node to represent all telomeres for the median problem is more than just for a succinct graph representation; it completely avoids the computationally costly capping problem and allows us to identify the orthologous relationships for telomeres in different genomes in the process of constructing the median genome.

Meanwhile this new representation poses new challenges in finding the median genome; compared to the counterpart problem for circular genomes, the existence of the cap node requires special considerations. The first problem arises in representing the problem when part of the median genome is known; the second problem is about finding the median genome efficiently.

When a partial solution is known for the problem on circular genomes, we can perform shrink operations[4,15]; the resultant multiple breakpoint graph has a smaller size but represents the same problem, which we can further decompose into smaller problems. However on the CMBG, if the 0-edge is incident to the cap node (we call it a capped 0-edge), the shrink operation is not defined as the cap node is incident to many edges and we do not know which edges to choose in order to perform such a shrink operation. One choice, as used in the current paper, is to just keep these capped 0-edges, although this will bring some complications to the graph representation and the implementation of data structures for the algorithms. In the full version of this paper, we will introduce an elegant graph representation with another node called the "cup" which collects the vertices incident to the capped 0-edges; the resultant graphs also consist of regular vertices, one cap node, one cup node and non-0-edges.

To improve the efficiency in finding the median genome, we need to increase the frequency of decompositions carried on the CMBGs. The adequate subgraphs defined in [15] apparently do not apply to any subgraph with the cap node; we need to establish parallel theorems on subgraphs with the cap (capped subgraphs).

4 The Decomposition Theorem and Adequate Subgraphs

For the median problem on circular genomes, [15] shows that the existences of adequate subgraphs allow us to decompose the problem into two subproblems from which the optimal solution of the original problem can be found by combining solutions to the two subproblems. In this section, we will develop the parallel results for the problem

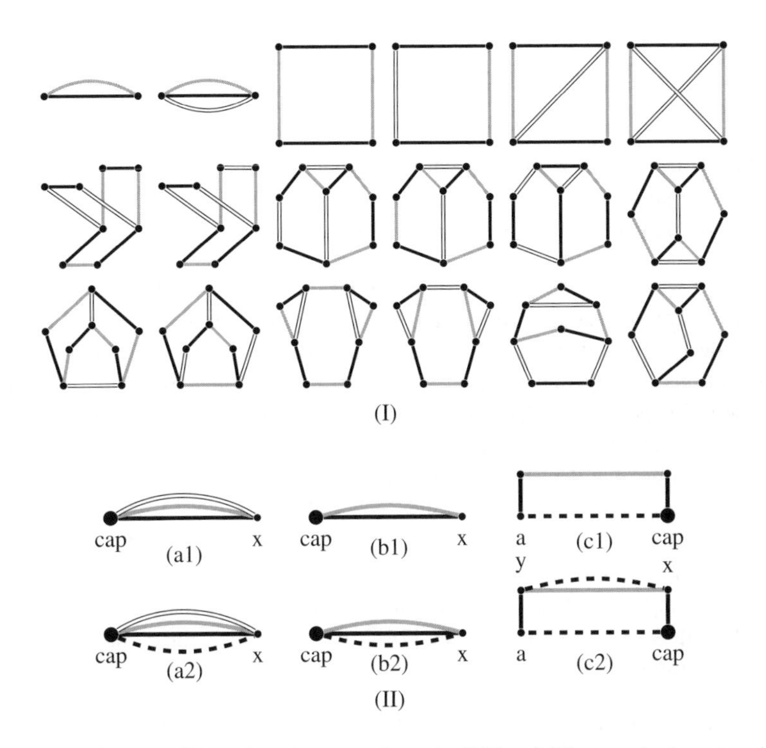

Fig. 3. The most frequent (I) regular adequate subgraphs [15] and (II) capped adequate subgraphs on CMBGs. Black, gray and double edges represent adjacency edges from extant genomes and dashed edges represent adjacency edge in the median genome.

on linear multichromosomal genomes. The cap node in the capped multiple breakpoint graph requires us to distinguish two types of subgraphs: regular subgraphs—the ones not containing the cap node; capped subgraphs—the ones containing the cap node. Parallel to these definitions, we have two types of adequate subgraphs: regular adequate subgraphs and capped adequate subgraphs. The following theorem, with no surprise, states that regular adequate subgraphs defined on CMBGs are identical to adequate subgraphs defined on MBGs. Figure 3.(I) shows the most frequent ones [15].

Theorem 2. *As long as the cap does not involve, regular adequate subgraphs are applicable to capped multiple breakpoint graphs, giving proper decompositions.*

Proof. If we use 2χ caps to delimiter χ linear chromosomes in a traditional way and treat these caps as regular vertices (as these caps are all of degree 2), each CMBG can be transformed into $((2\chi)!)^2$ MBGs. Suppose a regular adequate subgraph exists on the CMBG, then it must also exist in every MBG, since the transformation between the CMBG and the MBGs does not change regular vertices and edges connecting them. This adequate subgraph then decomposes every MBG into two parts, one of which is the adequate subgraph itself. Then the same decomposition induced by this adequate subgraph must happen on the original CMBG.

For a decomposition of a CMBG induced by a capped adequate subgraph, by combining the solutions from the two subproblems, we can find optimal solutions of the following three categories:

1. all optimal solutions with all possible optimal cappings;
2. some optimal solutions with all possible optimal cappings;
3. some optimal solutions with some optimal capping.

In this paper we define capped adequate subgraphs correspond to the first two categories; they are similar to strongly adequate subgraphs and adequate subgraphs on MBGs respectively. Before giving their definitions, we first quickly review some related definitions. Recall that the size of a subgraph (denoted by m) is defined as the number of 0-edges to be added; the rank ($N_{\mathcal{G}}$) of CMBGs is equal to the number of extant genomes involved in the median problem. For a subgraph, we define γ as the number of its edges incident to the cap node. A connected CMBG capped subgraph H of size m is a capped adequate subgraph if $c'^{\Sigma}_{max}(H) \geq \frac{1}{2}(mN_{\mathcal{G}}+\gamma-1)$. It is a capped strongly adequate subgraph if $c'^{\Sigma}_{max}(H) > \frac{1}{2}(mN_{\mathcal{G}}+\gamma-1)$. For the median of three problem, this critical number in above definitions becomes $\frac{1}{2}(3m+\gamma-1)$. It is worth to note that γ for any capped adequate subgraph must be greater than 1, for vertex degrees on any adequate subgraphs are at least 2 [14]. Fig 3.(II) shows the most frequent capped adequate subgraphs.

Theorem 3. *The existence of a capped adequate subgraph on a CMBG gives a proper decomposition from which* an *optimal solution can be found by combining solutions from the two subproblems. The existence of a capped strongly adequate subgraph gives a proper decomposition from which* all *optimal solutions can be found by combining solutions from the two subproblems.*

5 An Exact Algorithm and Lower and Upper Bounds

In this section, we give a high-level description of our algorithm *ASMedian-linear*, which finds exact solutions to the relaxed DCJ median problem on linear multichromosomal genomes. Similar to the algorithm for the circular case, this algorithm iteratively detects existences of regular adequate subgraphs and capped adequate subgraphs. Upon their existences, one or a few adjacencies are added into the median genome. In situations where existences of adequate subgraphs can not be detected (either they do not exist or they have too large sizes to be detected efficiently), this algorithm looks all possible ways of constructing next adjacency edge.

Lower bounds and upper bounds of the total DCJ distance are used to prune obviously bad solutions. The lower bound is derived from the metric property of the DCJ distance (many other distance measures such as reversal distance also have this property),

$$l' = \frac{d_{1,2} + d_{2,3} + d_{1,3}}{2}, \tag{2}$$

where $d_{i,j}$ is the pairwise DCJ distance between genome i and genome j.

Upper bound can be obtained by taking one of the extant genomes which gives the smallest total distance to the remaining genomes as the median genome,

$$u' = d_{1,2} + d_{2,3} + d_{1,3} - \max\{d_{1,2}, d_{2,3}, d_{1,3}\}. \tag{3}$$

When part of the median genome is known, $\tilde{c}$ is used to denote the number of cycles (including paths) formed between existing 0-edges and the given CMBG. Lower/upper bounds of subproblems are denoted by l' and u', whose values are determined by Equations 2 and 3, as each subproblem is viewed as an instance of the median problem. Then the bounds for the original problem are $l = \tilde{c} + l'$ and $u = \tilde{c} + u'$.

It is observed that tightness of lower bound is directly related to algorithms' performance. Any improvement on it shall have a great impact on algorithms' efficiency. An initial value for $d^{\Sigma} = \sum d_{0,i}$ obtained from a fast heuristic algorithm can be used to improve the pruning efficiency in the exact algorithm. At this moment, we use an adequate subgraph based heuristic, which arbitrarily constructs an adjacency edge if adequate subgraphs can not be detected.

Algorithm 1. ASMedian-linear

Input: three genomes with equal or unequal numbers of linear chromosomes
Output: the median genome and the smallest total DCJ distance d^{Σ}

1 run a heuristic algorithm to get an initial value for d^{Σ};
2 construct the capped multiple breakpoint graph, and push it into $\mathcal{L}$, the unexamined list of CMBG or intermediate CMBGs with partial solutions;
3 **while** *$\mathcal{L}$ is not empty and the smallest lower bound l in $\mathcal{L}$ is smaller than d^{Σ}* **do**
4   pop out a (intermediate) CMBG with the smallest lower bound l;
5   **if** *an adequate subgraph (regular or capped) H is detected on this (intermediate) CMBG* **then**
6   add one or a few 0-edges which are guaranteed to exist in an optimal solution, perform shrinking operation for newly added regular 0-edges and push the resultant intermediate CMBG into $\mathcal{L}$;
7   **else**
8   select the vertex v with the smallest label, create a set of intermediate CMBGs by adding one 0-edge incident to v to each of them and shrinking regular 0-edges if there is any, and add them into $\mathcal{L}$;
9   make necessary update for d^{Σ} (the smallest total DCJ distance obtained so far) with upper bounds u derived from newly created intermediate CMBGs ;
10 **return** d^{Σ} as the minimum total distance and the median genome;

6 Performance on Simulated Data

Our algorithm *ASMedian-linear* is implemented in Java, which runs serially on a single CPU. In order to test its performance, we generated sets of simulation data, with varying parameters. In rest of the section, we use n for number of genes in each genome, χ for number of linear chromosomes, r for the total number of reversals used to generate each instance. Three extant genomes in each instance is generated by applying

$r/3$ random reversals of random size on the identity genome (where each chromosome contains roughly the same number of genes, whose labels are consecutive). Each data set contains 100 instances except that the ones in Subsection 6.1 contain 10 instances each.

6.1 Speedups Due to Using Adequate Subgraphs

The program by Zhang et al. [17] is the only published exact solver for the strict DCJ median problem on linear multichromosomal genomes, which exhaustively searches the solution space with a branch-and-bound approach. We compare this program to our algorithm to see gains in speed by using our adequate subgraph based decomposition method. Running times for the program by Zhang et al. are used to to estimate the running times for a relaxed DCJ median solver using a branch-and-bound approach only, as the two problems are closely related with small differences—the relaxed version has a much larger solution space which may require the algorithms to search more solutions for an optimal solution is found; on the other hand the relaxed version may have a smaller optimal total DCJ distances which may let the algorithms terminate earlier compared to the counterpart algorithms for the strict version.

We generated simulations on genomes containing 40 or 50 genes, with varying number of linear chromosomes and varying number of reversals, as shown in Table 1, with 10 instances in each data set. Average running times of the two programs are reported in seconds, together with speedups of our program over Zhang et al's [17] exhaustive search one and average numbers of extra circular chromosomes in the median genomes produced by our algorithm.

The speedups range from 10^1 to 10^8 or even more, increasing along as the numbers of genes and chromosomes increase. Comparisons are carried only on small genomes, otherwise Zhang et al's program can not finish within reasonable time ($\gg$ 400 hours). One can expect much larger speedups on large genomes using our decomposition method. It is safe to say that our adequate subgraph based decomposition method achieves dramatic speedups.

Table 1. Running time comparison between two exact DCJ solvers: our ASMedian-linear for the relaxed version and the one by Zhang et al. for the strict version, on small genomes with varying number of linear chromosomes. For each choice of parameters, results are averaged over 10 simulated instances. Running times for the program by Zhang et al. are used to to estimate the running times for a relaxed DCJ median solver using branch-and-bound approach only, as the two problems are closely related. The table shows our program ASMedian-linear achieves dramatic speedups.

n, r	40, 8			50, 10		
χ	2	4	8	2	4	8
Zhang's	1.9×10^{-2}	1.6×10^{2}	$>1.6\times10^{5}$	2.1×10^{-2}	3.6×10^{3}	$>1.6\times10^{5}$
ASMedian-linear	1.0×10^{-3}	1.0×10^{-3}	1.0×10^{-3}	2.0×10^{-3}	2.0×10^{-3}	2.0×10^{-3}
average # circular chromosomes	0.1	0.0	0.1	0.1	0.0	0.1
speedup	10^1	10^5	$>10^8$	10^2	10^6	$>10^8$

6.2 Performance on Large Genomes

Sets of data on large genomes (with n ranging from 100 to 5000, and χ equal to 2 or 10) are also generated, 100 instances each. The total number of reversals used to generate these data sets is proportional to the number of genes, where this coefficient ranges from 0.3 to 0.9. Average running times are reported in seconds if every instance finishes in 10 minutes; otherwise number of finished instances is reported in parentheses. The last column reports numbers of extra circular chromosomes in the median genomes averaged over all finished instances with the same genome size.

Table 2 and 3 show running times on genomes with 2 or 10 linear chromosomes respectively. All instances with r/n no larger than 0.78 can finished within 1 second for

Table 2. Results for simulated genomes with 2 linear chromosomes. For each data set, 100 instances are simulated and if every instance finishes in 10 minutes, then their average running time is shown in seconds; otherwise number of finished instances is shown with parenthesis. Average numbers of extra circular chromosomes in the median genomes for instances with the same genome size are reported in the last column. As these numbers are no larger than 0.5, our exact solver for the relaxed DCJ median either gives optimal solutions or near-optimal solutions to the strict DCJ median problem on linear multichromosomal genomes.

n	r/n					average # circular chromosomes
	0.3	0.6	0.78	0.84	0.9	
100	$<1\times10^{-3}$	$<1\times10^{-3}$	1×10^{-3}	1×10^{-3}	1×10^{-3}	0.30
200	$<1\times10^{-3}$	1×10^{-3}	4×10^{-3}	7×10^{-3}	1.5×10^{0}	0.40
500	3×10^{-3}	5×10^{-3}	1.0×10^{-2}	1.3×10^{-1}	(98)	0.40
1000	1.4×10^{-2}	1.7×10^{-2}	4.0×10^{-2}	2.5×10^{0}	(80)	0.48
2000	5.7×10^{-2}	6.9×10^{-2}	1.5×10^{-1}	6.9×10^{0}	(21)	0.34
5000	3.7×10^{-1}	4.5×10^{-1}	9.9×10^{-1}	(73)	(0)	0.34

Table 3. Results for simulated genomes with 10 linear chromosomes. For each data set, 100 instances are simulated and if every instance finishes in 10 minutes, then their average running time is shown in seconds; otherwise the number of finished instances is shown with parenthesis. Average numbers of extra circular chromosomes in the median genomes for instances with the same genome size are reported in the last column. As these numbers are no larger than 0.15, our exact solver for the relaxed DCJ median gives optimal solutions in most of the cases, and in the remaining cases it gives near-optimal solutions to the strict DCJ median problem on linear multichromosomal genomes.

n	r/n					average # circular chromosomes
	0.3	0.6	0.78	0.84	0.9	
100	$<1\times10^{-3}$	1×10^{-3}	4×10^{-3}	8×10^{-3}	1.9×10^{-2}	0.08
200	1×10^{-3}	1×10^{-3}	2.8×10^{-2}	4×10^{-3}	(98)	0.13
500	4×10^{-3}	5×10^{-3}	2.3×10^{-2}	2.7×10^{-2}	(73)	0.14
1000	1.4×10^{-2}	1.7×10^{-2}	5.0×10^{-2}	2.7×10^{0}	(52)	0.11
2000	5.5×10^{-2}	6.9×10^{-2}	2.0×10^{-1}	(91)	(20)	0.11
5000	3.7×10^{-1}	4.5×10^{-1}	9.6×10^{-1}	(58)	(0)	0.10

both cases. When r/n is no larger than 0.6, the data sets only differing in χ have almost the same running time.

While as r/n increases, average running time increases quickly and many instances can not finish within 10 minutes. Comparison of Table 2 to Table 3 shows that, instances with 10 linear chromosomes take more time than the ones with 2 linear chromosomes. This is not surprising, because the multichromosomal case is associated with a three way capping problem, whose solution space is $((2\chi)!)^2$, which increases dramatically as χ increases.

Notice that the reported average numbers of extra circular chromosomes in the median genomes are very small (≤ 0.5). This means that on more than half of the instances, our algorithm gives optimal solutions to the problem of the strict version, and on the remaining instances, our algorithm provides near-optimal solutions after merging the extra circular chromosomes.

7 Conclusion

In this paper, in order to solve the relaxed DCJ median problem on linear multichromosomal genomes efficiently, we introduce capped multiple breakpoint graphs and their adequate subgraphs. By applying our adequate subgraph based decomposition method, we design a relative efficient algorithm *ASMedian-linear* which quickly gives exact solutions to most instances with number of genes up to thousands and with moderate number of evolution events as in real biology problems. Although the solutions may contain some extra circular chromosomes, which is generally considered to be undesirable, these numbers are either zero or very small. So we actually obtain optimal or near-optimal solutions for the strict DCJ median problems.

Since the median problem is NP-hard (for DCJ distance with relaxed version or strict version, or for reversal distance) and there is a need to solve instances with tens of thousands or even more genes (plus other conserved genetic units) in mammal genomes, highly efficient and accurate heuristics should be considered.

Acknowledgments

We would like to thank Jijun Tang for providing their solver for the strict DCJ median problem on linear multichromosomal genomes. I also want to thank anonymous referees, Bernard Moret and Vaibhav Rajan for their help and suggestions in writing this paper.

References

1. Adam, Z., Sankoff, D.: The ABCs of MGR with DCJ. Evol. Bioinformatics 4, 69–74 (2008)
2. Bader, D., Moret, B., Yan, M.: A fast linear-time algorithm for inversion distance with an experimental comparison. J. Comput. Biol. 8(5), 483–491 (2001)
3. Bourque, G., Pevzner, P.: Genome-scale evolution: Reconstructing gene orders in the ancestral species. Genome Res. 12, 26–36 (2002)

4. Caprara, A.: The reversal median problem. INFORMS J. Comput. 15, 93–113 (2003)
5. Hannenhalli, S., Pevzner, P.: Transforming cabbage into turnip: Polynomial algorithm for sorting signed permutations by reversals. In: Proc. 27th ACM Symp. on Theory of Computing STOC 1995, pp. 178–189. ACM, New York (1995)
6. Hannenhalli, S., Pevzner, P.: Transforming men into mice (polynomial algorithm for genomic distance problem). In: Proc. 43rd IEEE Symp. on Foudations of Computer Science FOCS 1995, pp. 581–592. IEEE Computer Soc., Los Alamitos (1995)
7. Lenne, R., Solnon, C., Stützle, T., Tannier, E., Birattari, M.: Reactive stochastic local search algorithms for the genomic median problem. In: van Hemert, J., Cotta, C. (eds.) EvoCOP 2008. LNCS, vol. 4972, pp. 266–276. Springer, Heidelberg (2008)
8. Siepel, A., Moret, B.: Finding an optimal inversion median: Experimental results. In: Gascuel, O., Moret, B.M.E. (eds.) WABI 2001. LNCS, vol. 2149, pp. 189–203. Springer, Heidelberg (2001)
9. Swenson, K., Rajan, V., Lin, Y., Moret, B.: Sorting signed permutations by inversions in o(nlogn) time. In: Batzoglou, S. (ed.) RECOMB 2009. LNCS, vol. 5541, pp. 386–399. Springer, Heidelberg (2009)
10. Swenson, K., To, Y., Tang, J., Moret, B.: Maximum independent sets of commuting and noninterfering inversions. In: Proc. 7th Asia-Pacific Bioinformatics Conf. APBC 2009, vol. 10 (suppl. 1), p. S6 (2009)
11. Tannier, E., Zheng, C., Sankoff, D.: Multichromosomal median and halving problems. In: Crandall, K.A., Lagergren, J. (eds.) WABI 2008. LNCS (LNBI), vol. 5251, pp. 1–13. Springer, Heidelberg (2008)
12. Tesler, G.: Efficient algorithms for multichromosomal genome rearrangements. J. Comput. Syst. Sci. 65(3), 587–609 (2002)
13. Xu, A.: The distance between randomly constructed genomes. In: Proc. 5th Asia-Pacific Bioinformatics Conf. APBC 2007. Advances in Bioinformatics and Computational Biology, vol. 5, pp. 227–236. Imperial College Press, London (2007)
14. Xu, A.: A fast and exact algorithm for the median of three problem—A graph decomposition approach. In: Nelson, C.E., Vialette, S. (eds.) RECOMB-CG 2008. LNCS (LNBI), vol. 5267, pp. 184–197. Springer, Heidelberg (2008)
15. Xu, A.W., Sankoff, D.: Decompositions of multiple breakpoint graphs and rapid exact solutions to the median problem. In: Crandall, K.A., Lagergren, J. (eds.) WABI 2008. LNCS (LNBI), vol. 5251, pp. 25–37. Springer, Heidelberg (2008)
16. Yancopoulos, S., Attie, O., Friedberg, R.: Efficient sorting of genomic permutations by translocation, inversion and block interchange. Bioinformatics 21, 3340–3346 (2005)
17. Zhang, M., Arndt, W., Tang, J.: An exact median solver for the DCJ distance. In: Proc. 14th Pacific Symposium on Biocomputing PSB 2009, pp. 138–149. World Scientific, Singapore (2009)

Rearrangement Models and Single-Cut Operations

Paul Medvedev[1] and Jens Stoye[2]

[1] Department of Computer Science, University of Toronto
pashadag@cs.toronto.edu
[2] Technische Fakultät, Universität Bielefeld, Germany
stoye@techfak.uni-bielefeld.de

Abstract. There have been many widely used genome rearrangement models, such as reversals, Hannenhalli-Pevzner, and double-cut and join. Though each one can be precisely defined, the general notion of a model remains undefined. In this paper, we give a formal set-theoretic definition, which allows us to investigate and prove relationships between distances under various existing and new models. We also initiate the formal study of single-cut operations by giving a linear time algorithm for the distance problem under a new single-cut and join model.

1 Introduction

In 1938, Dobzhansky and Sturtevant first noticed that the pattern of large scale rare events, called genome rearrangements, can serve as an indicator of the evolutionary distance between two species [9]. With the pioneering work of Sankoff and colleagues to formulate the question of evolutionary distance in purely combinatorial terms [23,22], the mathematical study of genome rearrangements was initiated. Here, the evolutionary distance is determined as the smallest number of rearrangements needed to transform one genome (abstracted as a gene-order) into another. This has given rise to numerous combinatorial problems, including distance, median, aliquoting, and halving problems, which are used to build phylogenetic trees and infer other kinds of evolutionary properties.

An underlying challenge of such approaches is to define an appropriate model, which specifies the kinds of rearrangements allowed. On one hand, the model should be as accurate as possible, including all the possible underlying biological events and weights which reflect their likelihood. On the other, answering questions like the median or distance can be computationally intractable for many models. The trade-offs between these, as well as other, considerations decide which rearrangement model is best suited for the desired type of analysis.

Though the ideas of genomes and rearrangements are inherently biological, they require precise mathematical definitions for the purposes of combinatorial analysis. Earlier definitions of genomes as signed permutations did not generalize well to genomes with duplicates, but recently a more general set-theoretic definition in terms of adjacencies has become used [5]. However, though particular models, like HP or DCJ, have their own precise definitions, the notion of a model, in general, remains undefined. In this paper, we give such a definition and show how current rearrangement models can

F.D. Ciccarelli and I. Miklós (Eds.): RECOMB-CG 2009, LNBI 5817, pp. 84–97, 2009.

be defined within our framework. This allows us to investigate and prove things about the relationship between sorting distances under different models, which we present in combination with what is already known to give an exposition of current results.

Recently, it was observed that most of the events in a parsimonious evolution scenario between human and mouse were operations which cut the genome in only one place, such as fusions, fissions, semi-translocations, and affix reversals (reversals which include a telomere) [7]. Such scenarios have applicability to the breakpoint reuse debate [2,21,24,7] since they can suggest a low rate of reuse. In this paper, we initiate the formal study of such single-cut operations by giving a linear time algorithm to find the minimum distance under a new single-cut and join (SCJ) model[1] and using it to determine the SCJ distance between the human and several other organisms.

2 Preliminaries

We begin by giving the standard definition of a genome, consistent with [5]. We represent the genes by a finite subset of the natural numbers, $N \subset \mathbb{N}$. For a gene $g \in N$, there is a corresponding *head* g_{h} and *tail* g_{t}, which are together referred to as the *extremities* of g. The set of all extremities of all genes in N is called N_{ext}. The set $\{p, q\}$, where p and q are extremities, is called an *adjacency*. We denote by N_{adj} the set of all possible adjacencies of N. The one-element set $\{p\}$, where p is an extremity, is called a *telomere*. We denote by N_{tel} the set of all possible telomeres of N. Telomeres and adjacencies are collectively referred to as *points*. A *genome* $G \subseteq N_{\mathrm{adj}} \cup N_{\mathrm{tel}}$ is a set of points such that each extremity of a gene appears exactly once[2]:

$$\bigcup_{x \in G} x = N_{\mathrm{ext}} \text{ and for all } x, y \in G, x \cap y = \emptyset$$

For brevity, we will sometimes use signed permutation notation to describe a unichromosomal linear genome, such as $G = (1, -2, -3, 4)$; however, this is just a notation and the underlying representation of the genome is always as a set of points. We denote by $N(G)$ the set of genes underlying the genome G. Finally, we define $\mathcal{G}$ to be the set of all possible genomes over all possible gene sets ($\mathcal{G}$ is a countable infinite set).

Though this definition of a genome does not immediately reflect the notion of chromosomes or gene-orders, these are reflected as properties of the genome graph. Given a genome G, its *genome graph* is an undirected graph whose vertices are exactly the points of G. The edges are exactly the genes of G, where edge g connects the two vertices that contain the extremities of g (this may be a loop). It is easy to show that the genome graph is a collection of cycles and paths [5].

We can now define a *chromosome* as a connected component in the genome graph. We also say that a sequence of extremities $p_1, \ldots, p_m$ is *ordered* if there exists a path

[1] Note that here SCJ refers to single-cut **and** join, as opposed to single-cut **or** join which was recently introduced by Feijão and Meidanis [10].

[2] Though this paper focuses on genomes without duplicate genes, this definition could be extended to the more general case by treating the set of genes and its corresponding derivatives as multisets, including the genome.

which traverses the vertices associated with the extremities in the given order. Note that questions like the number of chromosomes, whether two genes lie on the same chromosome, or whether extremities are ordered can all be answered in linear time by constructing and analyzing the genome graph.

Another useful graph is the *adjacency graph* [5]. For two genomes A and B, $AG(A,B)$ is an undirected, bipartite multi-graph whose vertices are the points of A and B. For each $x \in A$ and $y \in B$ there are $|x \cap y|$ edges between x and y. It is not difficult to show that this graph is a vertex-disjoint collection of paths and even-cycles and can be constructed in linear time [5]. We denote by $C_s(A,B)$, $C_l(A,B)$, and $I(A,B)$ the number of short cycles (length two), long cycles (length greater than 2), and odd paths in $AG(A,B)$, respectively. We use the term A-path to refer to a path that has at least one endpoint in A, and BB-path to one with both endpoints in B.

3 Models, Operations, and Events

We now present a formal treatment of rearrangement models, beginning with the main definition:

Definition 1 (Rearrangement Models, Operations and Events). *A rearrangement operation (also called a model) is a binary relation $\mathcal{R} \subseteq \mathcal{G} \times \mathcal{G}$. A rearrangement event is a pair $R = (G_1, G_2) \in \mathcal{G} \times \mathcal{G}$.*[3] *We say that R is an $\mathcal{R}$ event if $R \in \mathcal{R}$.*

For example, if we have genes $\{a,b,c\}$ and a genome $G = \{\{a_h\},\{a_t,b_h\},\{b_t\},\{c_h\},\{c_t\}\}$, then a possible event is $R = (G, \{\{a_h\},\{a_t,b_h\},\{b_t,c_h\},\{c_t\}\})$. This event has the effect of fusing the two chromosomes of G. On the other hand, a fusion operation $\mathcal{R}$ is given by the set of all pairs (G_1,G_2) such that there exist extremities $x,y \in N(G_1)_{\text{ext}}$ and $G_2 \cup \{\{x\},\{y\}\} = G_1 \cup \{\{x,y\}\}$. It is easy to see that R is an $\mathcal{R}$-event. Thus the operation $\mathcal{R}$ captures the general notion of a fusion as a type of rearrangement, while the event R captures this particular instance of a fusion. Most current literature does not make a formal distinction between types of rearrangements (which we call operations) and their particular instances (which we call events), but this is necessary for defining the notion of a rearrangement model.

Current literature often makes the informal distinction between models, such as DCJ or HP, and operations, such as reversals or fusions. Operations are considered more biologically atomic, with a model being a combination of these atomic operations. Here, we maintain this notational consistency; however, we note that the terms model and operation are mathematically equivalent.

In this paper, we focus on the double-cut and join (DCJ) model and its subsets (referred to as *submodels* in the context of models).

Definition 2 (DCJ)**.** *Let G_1 and G_2 be two genomes with equal gene content ($N(G_1) = N(G_2)$). Then $(G_1,G_2) \in$ DCJ if and only if there exist extremities p, q, r, s such that one of the following holds:*

[3] Alternatively, we can treat R as a function in the standard way of viewing relations as functions. Namely, $R : \mathcal{G} \cup \{\emptyset\} \longrightarrow \mathcal{G} \cup \{\emptyset\}$ where $R(C) = G_2$ if $C = G_1$ and $R(C) = \emptyset$ otherwise.

(a) $G_2 \cup \{\{p,q\},\{r,s\}\} = G_1 \cup \{\{p,r\},\{q,s\}\}$
(b) $G_2 \cup \{\{p,q\},\{r\}\} = G_1 \cup \{\{p,r\},\{q\}\}$
(c) $G_2 \cup \{\{q\},\{r\}\} = G_1 \cup \{\{q,r\}\}$
(d) $G_2 \cup \{\{q,r\}\} = G_1 \cup \{\{q\},\{r\}\}$

This definition is equivalent to the one given in [27,5]. A more intuitive interpretation of, for example, (a), is that the event replaces the adjacencies $\{p,q\}$ and $\{r,s\}$ in G_1 with $\{p,r\}$ and $\{q,s\}$ in G_2.

Note that an event that satisfies one of the conditions (b)-(d) of the DCJ model only cuts the genome in one place. These events define the submodel of DCJ called single-cut and join (SCJ), which we will study in Section 6. Other operations, such as reversals, can be defined in a similar manner, though we do not do it here for conciseness. We will, however, define how to restrict a model so that it only deals with linear and/or uni-chromosomal genomes:

Definition 3. *Given a model M, let*

- $M_{lin} = \{(G_1,G_2) \mid (G_1,G_2) \in M$ *and* G_1 *and* G_2 *are linear*$\}$.
- $M_{uni} = \{(G_1,G_2) \mid (G_1,G_2) \in M$ *and* G_1 *and* G_2 *are uni-chromosomal*$\}$.

There are many questions one can pose within any model, including sorting, distance, median, halving, or aliquoting. We will focus on the sorting and distance problems here:

Definition 4 (Sorting Sequence and Distance). *A sequence of events $R_1, R_2, \ldots, R_m$ sorts G_1 into G_2 if $G_2 = R_m(\ldots(R_1(G_1)))$. The sorting distance between G_1 and G_2 under a model $\mathcal{R}$, denoted by $d_{\mathcal{R}}(G_1,G_2)$, is the length of the shortest sorting sequence such that all R_i are $\mathcal{R}$ events. If such a sequence does not exist then we say $d_{\mathcal{R}}(G_1,G_2) = \infty$.*

Given two genomes, we can either find a shortest sorting sequence or just the sorting distance. Since the number of possible events in each step is polynomial, if we have a polynomial time algorithm for distance, then we also have one for sorting [13]. That is why when discussing poly-time complexity, we will focus only on the distance problem. Note however that the precise complexities may differ, as for example is the case for the reversal model, where the distance can be computed in linear time [4] while the best-known sorting algorithms have worst-case time complexity $O(n^{3/2}\sqrt{\log n})$ [25].

4 Submodels of Double-Cut and Join

Motivated by different biological systems, several distinct rearrangement models have been studied. These models differ in various aspects, spanning a whole space of genome rearrangement models. The three most relevant dimensions of this space are (i) the number of chromosomes a genome may have, (ii) the shape that the chromosomes may have (i.e. linear or circular), and (iii) the maximum number of chromosome cuts (and joins) an operation may perform[4]. This three-dimensional space is visualized in Fig. 1.

[4] In this paper, we focus on double and single-cut operations. However, more general k-cut operations have also been considered [3].

Table 1. A description of some of the models in terms of the elementary operations defined in [5]. A dark bullet means that the operation is fully contained in the model, no bullet means that the operation is disjoint with the model, and an empty bullet means that some but not all of the operation is contained in the model. Furthermore, each model is precisely the union of the operations specified by the bullets.

	Model				
Operation	DCJ	HP	SCJ	SCJ_{lin}	$(\text{SCJ}_{\text{uni}})_{\text{lin}}$
PROPER TRANSLOCATION	•	•			
SEMI TRANSLOCATION	•	•	•	•	
PATH FISSION/FUSION	•	•	•	•	
EXCISION/INTEGRATION	•				
REVERSAL	•	•	○	○	○
CIRCULARIZATION/LINEARIZATION	•		•		
CYCLE FISSION/FUSION	•		○		

Each of the corners of the cube can be formally defined by deriving a submodel from either DCJ or SCJ using the linear and/or uni-chromosomal restrictions. For example, HP $=$ DCJ_{lin}, and the front bottom left corner is $(\text{SCJ}_{\text{uni}})_{\text{lin}}$ One can also think of these models in terms of the operations they allow, which is shown in Table 1.

Of the corners of the cube visualized in Fig. 1, some are of particular interest and thus have been studied more than others. The first model to be studied was the REV model, where the only allowed operation was the reversal. This model can only be used to sort linear uni-chromosomal genomes, since a reversal can never change the number of chromosomes or make them circular. The biological motivation for this model goes back to Nadeau and Taylor [19] and it was first formally modeled by Sankoff [22]. The first polynomial-time algorithm for computing the reversal distance and solving the reversal model was given by Hannenhalli and Pevzner [13] in 1995.

In the same year, Hannenhalli and Pevzner also looked at a model where multiple linear chromosomes are allowed, as is often the case in eukaryotic genomes [12]. After its authors, the resulting model is called the HP model. A combination of work showed that the distance under HP can be computed in poly-time [12,26,20,14,6].

A more recently introduced model is the double-cut and join (DCJ), which encompasses all events that can be achieved by first cutting the genome in up to two places, and then rejoining them in different combinations [27]. Though such a model is less biologically realistic than the HP model, there are fast algorithms for solving it [5] which have made it useful as an efficient approximation for the HP distance [1,15,18]. The DCJ model is a superset of all the other models in the cube, and is thus the most general.

The REV, HP, and DCJ models are all double-cut models, in that they allow for the cutting of the genome in two places. However, one can also consider models where only one cut is allowed. These make up the bottom plane of the cube, with the most general of these being the already defined SCJ model. These models have not yet been studied, since they are quite restrictive. However, we became aware of their relevance when we looked at certain rearrangement scenarios in eukaryotic evolution. In particular, while studying rearrangement scenarios between human and mouse with a minimum number of breakpoint reuses [7] it was observed that most of the events (213 out of 246) were

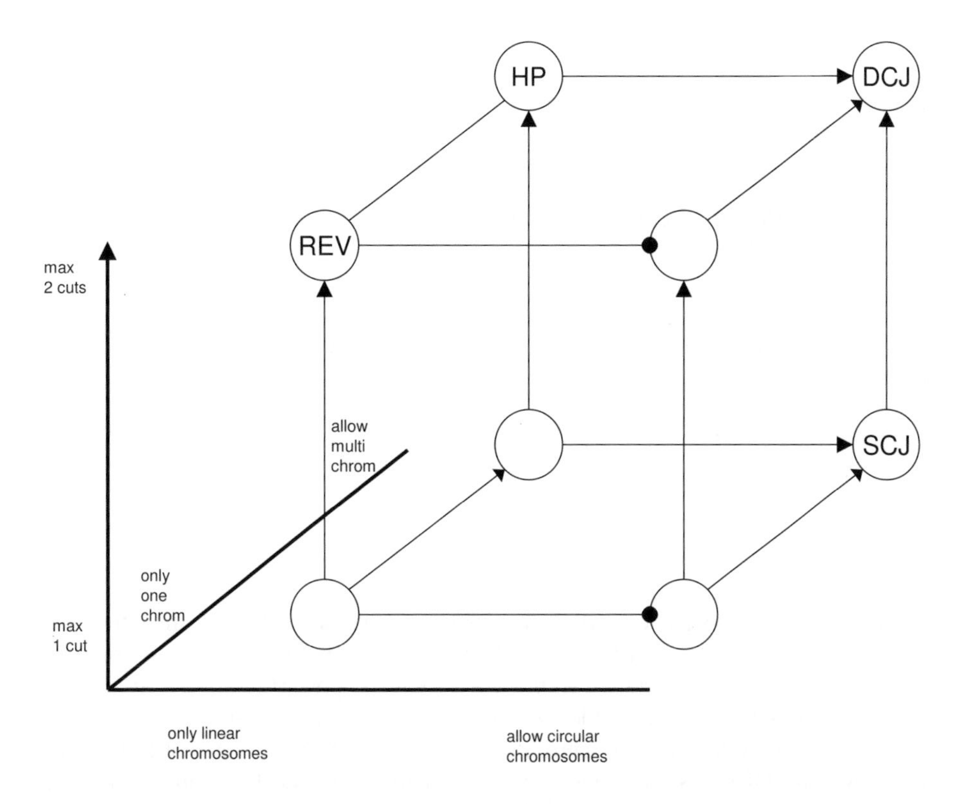

Fig. 1. The space of DCJ genome rearrangement submodels. An arrow on an edge from M to M' indicates that there exists a pair of genomes that are sortable under both models and whose distance under M is strictly more than under M'. An edge between M and M' with a circular ending at M' indicates that if a pair of genomes is sortable under M, then its distances under M and M' are the same.

single-cut (186 semi-translocations and affix reversals, 15 fissions and 12 fusions). This observation raised our interest in studying SCJ and its submodels.

In Section 6, we give a linear time algorithm for the sorting distance in the SCJ model. When restricted to linear chromosomes (SCJ_{lin}), we have the single-cut equivalent of the HP model, allowing fissions, fusions, semi-translocations, and affix reversals. The complexity of this model is unknown. The even more restrictive $(\text{SCJ}_{\text{uni}})_{\text{lin}}$ model consists of only affix reversals, which reverse a prefix/suffix of a chromosome, and is the single-cut equivalent of the REV model. There is a simple 2-approximation algorithm for the sorting distance, which increases the number of short cycles by at least one every two steps. However, the complexity of the problem remains open. It is related to the problem of sorting burnt pancakes [11,8], which is similar except that the chromosome has an orientation and only prefix reversals are allowed. The complexity of this problem is also open.

5 Sorting Distance Relationships

In this section, we study the relationship between sorting distances under the various models represented in the cube of Figure 1. We start with an easy observation, which we sometimes may use without explicitly stating it:

Lemma 1 (Submodel Lemma). *For any two models M and M' with $M' \subseteq M$, and all genomes G_1, G_2,*

$$d_M(G_1, G_2) \leq d_{M'}(G_1, G_2).$$

Proof. Any sorting sequence under M' is by definition also a sorting sequence under M. □

Corollary 1. *For all genomes G_1, G_2 and models M,*

- $d_{\text{DCJ}}(G_1, G_2) \leq d_{\text{SCJ}}(G_1, G_2)$
- $d_M(G_1, G_2) \leq d_{M_{lin}}(G_1, G_2)$
- $d_M(G_1, G_2) \leq d_{M_{uni}}(G_1, G_2)$

In fact, if you compare two models that are connected by an edge of the cube, then the one that is furthest from the origin is at least as powerful as the closer one. An interesting question is when the two models are equally powerful – that is, if a pair of genomes is sortable under both models, then the distances are the same. We first study if the restriction to linear or uni-chromosomal genomes makes the DCJ or SCJ models less powerful. Of course, it is clear from the definitions that, for example, a multi-chromosomal genome can be sorted by SCJ and cannot be sorted by SCJ_{uni}. However, we are interested if there are genomes which can be sorted by both models, but with fewer steps in one than the other.

Lemma 2. *There exist two genomes G_1 and G_2 such that*

- $d_{\text{SCJ}}(G_1, G_2) < d_{\text{SCJ}_{lin}}(G_1, G_2) < \infty$
- $d_{\text{SCJ}}(G_1, G_2) < d_{\text{SCJ}_{uni}}(G_1, G_2) < \infty$
- $d_{\text{DCJ}}(G_1, G_2) < d_{\text{DCJ}_{uni}}(G_1, G_2) < \infty$

Proof. Let $G_1 = (1, 3, 2)$ and $G_2 = (1, 2, 3)$. We can sort G_1 into G_2 using two SCJ events. First, we make an excision by replacing points $\{1_h, 3_t\}$ and $\{2_h\}$ with $\{1_h\}$ and $\{2_h, 3_t\}$. Second, we make an integration by replacing points $\{1_h\}$ and $\{3_h, 2_t\}$ with $\{1_h, 2_t\}$ and $\{3_h\}$. However, there does not exist a sorting sequence of length two under either SCJ_{lin}, SCJ_{uni}, or DCJ_{uni} models, though the genomes are clearly sortable under all these models. □

To complete the picture, it is already known that there are genomes that are sortable in HP but require more steps than in DCJ (see [6] for an example).

We next look if the flexibility of double-cut operations makes the models in the top plane more powerful than their respective counterparts in the bottom plane. There is a simple example that answers this question in the affirmative.

Lemma 3. *There exist two genomes G_1 and G_2 such that*

- $d_{\text{REV}}(G_1, G_2) < d_{(\text{SCJ}_{uni})_{lin}}(G_1, G_2) < \infty$
- $d_{\text{HP}}(G_1, G_2) < d_{\text{SCJ}_{lin}}(G_1, G_2) < \infty$
- $d_{\text{DCJ}_{uni}}(G_1, G_2) < d_{\text{SCJ}_{uni}}(G_1, G_2) < \infty$
- $d_{\text{DCJ}}(G_1, G_2) < d_{\text{SCJ}}(G_1, G_2) < \infty$

Proof. Let $G_1 = (1, -2, 3)$ and $G_2 = (1, 2, 3)$. We can sort G_1 into G_2 using just one event in the REV model, while there is no single SCJ event that does this. However, G_1 is sortable into G_2 using affix reversals. The lemma follows by applying the Submodel Lemma. □

We now compare $(\text{SCJ}_{\text{uni}})_{\text{lin}}$ with SCJ_{lin} to determine if the flexibility to create additional chromosomes in intermediate steps adds power when we are restricted to single-cut operations and linear genomes.

Lemma 4. *There exist two genomes G_1 and G_2 such that*

$$d_{\text{SCJ}_{lin}}(G_1, G_2) < d_{(\text{SCJ}_{uni})_{lin}}(G_1, G_2) < \infty.$$

Proof. Let $G_1 = (1, -2, -3, 4)$ and $G_2 = (1, 2, 3, 4)$. There exists a sorting sequence of length 4 under SCJ_{lin} that makes a fission between -2 and -3, two affix reversals of genes 2 and 3, respectively, and a final fusion. However, one can check that using only affix reversals the sorting distance is 6. □

In some cases, however, additional flexibility does not add power to the model. Consider the SCJ_{uni} model, which differs from the $(\text{SCJ}_{\text{uni}})_{\text{lin}}$ model in that, besides affix reversals, it allows the circularization and linearization of the chromosome. This obviously allows sorting a linear chromosome into a circular one, something that REV does not allow. However, for genomes that are sortable under both models, we can show that circularization cannot help to decrease the distance.

Lemma 5. *For two uni-chromosomal linear genomes G_1 and G_2, we have*

$$d_{\text{SCJ}_{uni}}(G_1, G_2) = d_{(\text{SCJ}_{uni})_{lin}}(G_1, G_2).$$

Proof. We show that for any optimal SCJ_{uni} sorting scenario that creates a circular chromosome in an intermediate step, there exists a sorting scenario of equal length without a circular chromosome. Since the only SCJ operation that can be performed on a uni-chromosomal circular genome is to linearize it, we know that every circularization is immediately followed by a linearization. Thus, w.l.o.g. the situation can be described as an exchange of a prefix A and a suffix B:

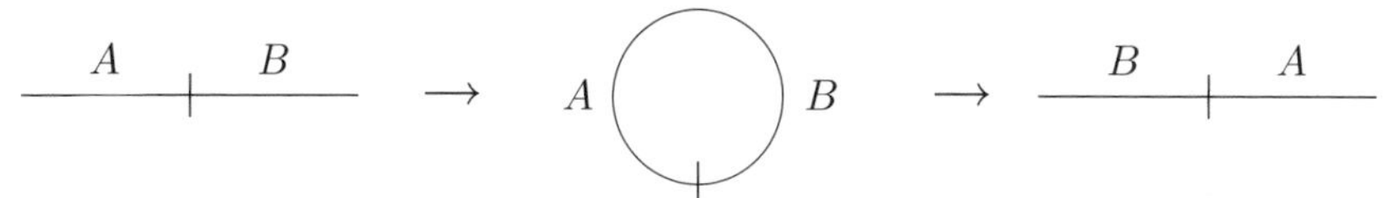

However, the same effect can be achieved by two affix reversals, namely first reversing A and then B:

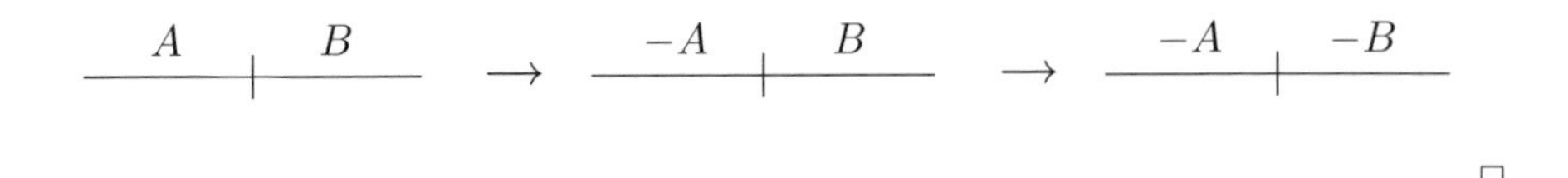

□

A similar situation occurs when we add circularization and linearization to the reversal model, though the proof is more involved:

Lemma 6. *For two uni-chromosomal linear genomes G_1 and G_2, we have*

$$d_{\text{DCJ}_{uni}}(G_1, G_2) = d_{\text{REV}}(G_1, G_2).$$

Proof. Consider an optimal DCJ_{uni} sorting sequence S that has the smallest possible number of circularizations. We show, by contradiction, that this number is zero, proving the lemma. Let L be the genome prior to the first circularization, let C be the one right after it, let C' be the genome right before the first linearization, and let L' be the one right after it. Let d be the length of the sorting sequence between C and C' (these must be reversals).

We will apply Theorem 3.2 from [17], which states that the reversal distance between any two circular chromosomes (C and C' in our case) is the same as the reversal distance between their linearizations, if they share a telomere. If L and L' share a telomere, then there is a sorting sequence with shorter length that replaces $d+2$ events between them with d reversals, which contradicts the optimality of S.

Suppose w.l.o.g that the two telomeres of L are $\{p\}$ and $\{s\}$, that $\{q\}$ is a telomere in L', that $\{q, r\}$ is an adjacency in L, and p, q, r, s are ordered in L. We can perform two reversals on L, the first one replacing $\{p\}$ and $\{q, r\}$ with $\{q\}$ and $\{p, r\}$, and the second replacing $\{p, r\}$ and $\{s\}$ with $\{p, s\}$ and $\{r\}$. The effect on the genome graph can be visualized as follows:

L p q r s ⟹ q p s r **L''**

Note that if you circularize the resulting genome, L'', you get C, and L'' shares a telomere with L'. Therefore, by the theorem, there exist d reversals that sort L'' into L'. We can then get a new sorting sequence that replaces the $d+2$ events between L and L' with the two reversals described above followed by d reversals given by the theorem. This sorting sequence has the same length as S but has one less circularization, a contradiction. □

The results of this section are compactly summarized in Figure 1 by marking the endpoints on the edges of the cube. The Submodel Lemma implies other results which we have not explicitly stated but that can be deduced by looking at the endpoints of the edges. For instance, there are genomes that are sortable under SCJ_{uni} that require less steps under SCJ_{lin}. Additionally some models which are not subsets of each other (like SCJ and HP) are incomparable. That is, there are genomes that are sortable under SCJ but require less steps under HP, and there are other genomes for which the opposite holds. The examples of $(1, 3, 2)$ and $(1, -2, 3)$ are enough to prove incomparability of SCJ with REV and with HP.

There is one edge of the cube with no marked endpoints, which represents the question of whether HP is more powerful to sort uni-chromosomal genomes than REV. Note that this is not trivial, because under HP we can split chromosomes in intermediate steps, which proved to give SCJ_{lin} more power than $(\text{SCJ}_{\text{uni}})_{\text{lin}}$ (Lemma 4). To the best of our knowledge, this question remains open.

6 Single-Cut and Join

We have already motivated the study of the sorting distance under the SCJ model, and in this section we give a linear time algorithm to compute it. Let A and B be two arbitrary genomes with the same underlying set of genes, N. We will use the following potential function in our analysis

$$\Phi(A, B) = |N| - I(A, B)/2 - C_s(A, B) + C_l(A, B)$$

First, we show that the potential function is 0 if and only if the two genomes are the same:

Lemma 7. *$A = B$ if and only if $\Phi(A, B) = 0$.*

Proof. The only if direction follows trivially from the definition of the adjacency graph. The if direction follows from a simple counting argument. Let a be the number of adjacencies in A, and t be the number of telomeres. By definition, $a + t/2 = |N|$. Since each short cycle accounts for one adjacency of A, and each odd path accounts for two telomeres of A, we have that $C_s(A, B) + I(A, B)/2 \leq a + t/2 = |N|$. Therefore, for the equality of the lemma to hold, we must have $C_l(A, B) = 0$, $C_s(A, B) = a$, and $I(A, B) = t$. This further implies, since each path is responsible for at least one telomere, and there are the same number of telomeres as paths, that all the paths must have length 1. Since $AG(A, B)$ contains only paths of length 1 and short cycles, we conclude that the points of A must be the same as the points of B. □

We can show using simple case analysis that Φ can decrease by at most one after any single event (proof omitted).

Lemma 8. *For all SCJ events R, $\Phi(R(A), B) - \Phi(A, B) \geq -1$.*

Combined with the fact that Φ cannot be negative, this gives a lower bound of $\Phi(A, B)$ on the sorting distance. We now consider Algorithm 1, whose cases are also illustrated in Figure 2.

Lemma 9. *Algorithm 1 terminates and outputs an SCJ sorting scenario of length $\Phi(A, B)$.*

Proof. First, we observe that one of the cases always applies since if $A \neq B$ then there must be at least one path of length greater than one or a long cycle. One can also verify that in each case, (A, A') is an SCJ event. Finally, we show that $\Phi(A', B) - \Phi(A, B) = -1$:

Algorithm 1. Algorithm for sorting under SCJ

while $A \neq B$ **do**
  if there exists an A-path P with length > 3 **then**
    Let p, q, r be the first three edges (from an arbitrary A end P).
    Let $A' = A \setminus \{\{p\}, \{q, r\}\} \cup \{\{r\}, \{p, q\}\}$.
  else if there exists an A-path P with length of 2 **then**
    Let p and q be its two edges.
    Let $A' = A \setminus \{\{p\}, \{q\}\} \cup \{\{p, q\}\}$.
  else if there exists a BB-path P **then**
    Let p and q be the first two edges (from an arbitrary end of P).
    Let $A' = A \setminus \{\{p, q\}\} \cup \{\{p\}, \{q\}\}$.
  else if there exists a long cycle **then**
    Let $\{p, q\}$ be a vertex of the cycle in A.
    Let $A' = A \setminus \{\{p, q\}\} \cup \{\{p\}, \{q\}\}$.
  end if
  Print A'.
  Let $A = A'$.
end while

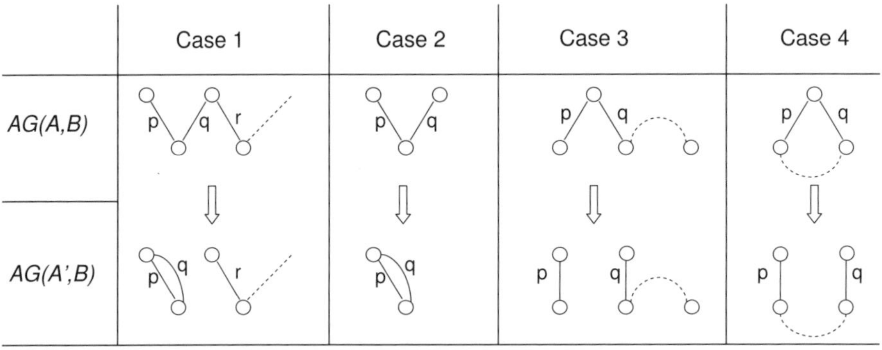

	Case 1	Case 2	Case 3	Case 4
AG(A,B)	p q r	p q	p q	p q
AG(A',B)	p q r	p q	p q	p q

Fig. 2. The four cases of Algorithm 1

Case 1: A short cycle is created and the length of P decreases by two.
Case 2: An even path (P) is removed and a short cycle is created.
Case 3: An even path (P) is replaced by two odd paths.
Case 4: A long cycle is replaced by an even path. □

Thus, we have

Theorem 1. $d_{\text{SCJ}}(A, B) = \Phi(A, B) = |N| - I(A, B)/2 - C_s(A, B) + C_l(A, B)$.

Corollary 2. *$d_{\text{SCJ}}(A, B)$ is computable in $O(|N|)$ time.*

Note the similarity to the formula for the DCJ distance [5]:

$$d_{\text{DCJ}}(A, B) = |N| - I(A, B)/2 - C_s(A, B) - C_l(A, B)$$

Thus the difference between the SCJ and DCJ distances is $2C_l(A, B)$.

7 Experimental Results

We performed six different comparisons, all with respect to the human. In the first, we took the 281 synteny blocks of the mouse-human comparison done in [21]. In the other five, we used the 1359 synteny blocks of the chimp, rhesus monkey, mouse, rat, and dog used in [16]. For each comparison, we computed the SCJ, DCJ, and HP distances. The HP distance was computed using GRIMM [26]. The results are shown in Table 2.

One immediately sees that the SCJ scenarios are far from parsimonious relative to the HP model. However, we stress that the goal of the SCJ model is to explore the importance of double-cut operations in evolution, and not to be a realistic evolutionary model. It can be an indicator of how many double-cut operations are really an advantage and how many are just an alternative that can be avoided. For example, consider that the difference between the HP and SCJ distances is 150% in the chimp-human comparison, and 60% in the mouse[16]-human comparison. This might suggest that somehow single-cut operations play a lesser part in the chimp-human evolution than in the mouse-human evolution. We also notice that the ratio of the SCJ to HP distance in the mouse-human comparison is much lower (1.2 vs. 1.6) using the synteny blocks of [21] than using the synteny blocks of [16]. This suggests the sensitivity of this kind of breakpoint reuse analysis to synteny block partition.

Table 2. Rearrangement distances under different models from different organisms to the human

	Model				Ratio
Organism	#Blocks	DCJ	HP	SCJ	SCJ/ HP
Mouse [21]	281	246	246	300	1.2
Chimp	1359	22	23	58	2.5
Rhesus Monkey	1359	106	110	224	2.0
Mouse [16]	1359	408	409	642	1.6
Rat	1359	707	753	1291	1.7
Dog	1359	291	295	523	1.8

In Section 5, we showed that there exist genomes for which the SCJ distance is smaller than the HP distance (for example $1, 3, 2$). However, in all the experimental results, the HP distance is always much smaller. This suggests that the SCJ operations not allowed by HP, such as excissions, integrations, circularizations, and linearizations, are infrequent relative to fissions, fusions, translocations, and reversals.

8 Conclusion

In this paper, we gave a formal set-theoretic definition of rearrangement models and operations, and used it to compare the power of various submodels of DCJ with unichromosomal and/or linear restrictions. We hope that the formal foundation for the notion of models will eventually lead to further insights into their relationships.

We also initiated the formal study of single-cut operations by giving a linear time algorithm for computing the distance under a new single-cut and join model. Many interesting algorithmic questions remain open, including the complexity of sorting using linear SCJ operations, and sorting using affix reversals.

Acknowledgements. We would like to thank Anne Bergeron, and PM would like to also thank Allan Borodin and Michael Brudno, for useful discussions. Most of the work was done while PM was a member of the International NRW Graduate School in Bioinformatics and Genome Research, and the AG Genominformatik group at Bielefeld, while being additionally funded by a German Academic Exchange Service (DAAD) Research Grant. PM gratefully acknowledges the support of these organizations.

References

1. Adam, Z., Sankoff, D.: The ABCs of MGR with DCJ. Evol. Bioinform. 4, 69–74 (2008)
2. Alekseyev, M.A., Pevzner, P.A.: Are there rearrangement hotspots in the human genome? PLoS Comput. Biol. 3(11), e209 (2007)
3. Alekseyev, M.A., Pevzner, P.A.: Whole genome duplications, multi-break rearrangements, and genome halving problem. In: SODA, pp. 665–679 (2007)
4. Bader, D.A., Moret, B.M.E., Yan, M.: A linear-time algorithm for computing inversion distance between signed permutations with an experimental study. J. Comp. Biol. 8(5), 483–491 (2001)
5. Bergeron, A., Mixtacki, J., Stoye, J.: A unifying view of genome rearrangements. In: Bücher, P., Moret, B.M.E. (eds.) WABI 2006. LNCS (LNBI), vol. 4175, pp. 163–173. Springer, Heidelberg (2006)
6. Bergeron, A., Mixtacki, J., Stoye, J.: HP distance via double cut and join distance. In: Ferragina, P., Landau, G.M. (eds.) CPM 2008. LNCS, vol. 5029, pp. 56–68. Springer, Heidelberg (2008)
7. Bergeron, A., Mixtacki, J., Stoye, J.: On computing the breakpoint reuse rate in rearrangement scenarios. In: Nelson, C.E., Vialette, S. (eds.) RECOMB-CG 2008. LNCS (LNBI), vol. 5267, pp. 226–240. Springer, Heidelberg (2008)
8. Cohen, D.S., Blum, M.: On the problem of sorting burnt pancakes. Discr. Appl. Math. 61(2), 105–120 (1995)
9. Dobzhansky, T., Sturtevant, A.H.: Inversions in the chromosomes of Drosophila Pseudoobscura. Genetics 23, 28–64 (1938)
10. Feijão, P., Meidanis, J.: SCJ: A novel rearrangement operation for which sorting, genome median and genome halving problems are easy. In: Salzberg, S.L., Warnow, T. (eds.) WABI 2009. LNCS (LNBI), vol. 5724, pp. 85–96. Springer, Heidelberg (2009)
11. Gates, W., Papadimitiou, C.: Bounds for sorting by prefix reversals. Discr. Math. 27, 47–57 (1979)
12. Hannenhalli, S., Pevzner, P.A.: Transforming men into mice (polynomial algorithm for genomic distance problem). In: Proceedings of FOCS 1995, pp. 581–592. IEEE Press, Los Alamitos (1995)
13. Hannenhalli, S., Pevzner, P.A.: Transforming cabbage into turnip: Polynomial algorithm for sorting signed permutations by reversals. J. ACM 46(1), 1–27 (1999); First appeared in STOC 1995 Proceedings
14. Jean, G., Nikolski, M.: Genome rearrangements: a correct algorithm for optimal capping. Inf. Process. Lett. 104, 14–20 (2007)
15. Lin, Y., Moret, B.M.E.: Estimating true evolutionary distances under the DCJ model. Bioinformatics 24, i114–i122 (2008); Proceedings of ISMB 2008
16. Ma, J., Zhang, L., Suh, B.B., Raney, B.J., Burhans, R.C., Kent, W.J., Blanchette, M., Haussler, D., Miller, W.: Reconstructing contiguous regions of an ancestral genome. Genome Research 16(12), 1557–1565 (2006)

17. Meidanis, J., Walter, M.E.M.T., Dias, Z.: Reversal distance of signed circular chromosomes. In: Technical Report IC–00-23. Institute of Computing, University of Campinas (2000)
18. Mixtacki, J.: Genome halving under DCJ revisited. In: Hu, X., Wang, J. (eds.) COCOON 2008. LNCS, vol. 5092, pp. 276–286. Springer, Heidelberg (2008)
19. Nadeau, J.H., Taylor, B.A.: Lengths of chromosomal segments conserved since divergence of man and mouse. Proc. Natl. Acad. Sci. USA 81, 814–818 (1984)
20. Ozery-Flato, M., Shamir, R.: Two notes on genome rearrangements. J. Bioinf. Comput. Biol. 1(1), 71–94 (2003)
21. Pevzner, P.A., Tesler, G.: Transforming men into mice: the Nadeau-Taylor chromosomal breakage model revisited. In: Proceedings of RECOMB 2003, pp. 247–256 (2003)
22. Sankoff, D.: Edit distances for genome comparison based on non-local operations. In: Apostolico, A., Galil, Z., Manber, U., Crochemore, M. (eds.) CPM 1992. LNCS, vol. 644, pp. 121–135. Springer, Heidelberg (1992)
23. Sankoff, D., Cedergren, R., Abel, Y.: Genomic divergence through gene rearrangement. In: Doolittle, R.F. (ed.) Molecular Evolution: Computer Analysis of Protein and Nucleic Acid Sequences, Meth. Enzymol., ch. 26, vol. 183, pp. 428–438. Academic Press, San Diego (1990)
24. Sankoff, D., Trinh, P.: Chromosomal breakpoint reuse in genome sequence rearrangement. J. of Comput. Biol. 12(6), 812–821 (2005)
25. Tannier, E., Bergeron, A., Sagot, M.-F.: Advances on sorting by reversals. Discr. Appl. Math. 155(6-7), 881–888 (2007)
26. Tesler, G.: Efficient algorithms for multichromosomal genome rearrangements. J. Comput. Syst. Sci. 65(3), 587–609 (2002)
27. Yancopoulos, S., Attie, O., Friedberg, R.: Efficient sorting of genomic permutations by translocation, inversion and block interchange. Bioinformatics 21(16), 3340–3346 (2005)

Aligning Two Genomic Sequences That Contain Duplications

Minmei Hou[1], Cathy Riemer[2], Piotr Berman[3], Ross C. Hardison[2,4], and Webb Miller[2,3,5]

[1] Dept. of Computer Science, Northern Illinois University, USA
mhou@cs.niu.edu
[2] Center for Comparative Genomics and Bioinformatics, Penn State University, USA
[3] Dept. of Computer Science and Engineering, Penn State University, USA
[4] Dept. of Biochemistry and Molecular Biology, Penn State University, USA
[5] Dept. of Biology, Penn State University, USA

Abstract. It is difficult to properly align genomic sequences that contain intra-species duplications. With this goal in mind, we have developed a tool, called TOAST (two-way orthologous alignment selection tool), for predicting whether two aligned regions from different species are orthologous, i.e., separated by a speciation event, as opposed to a duplication event. The advantage of restricting alignment to orthologous pairs is that they constitute the aligning regions that are most likely to share the same biological function, and most easily analyzed for evidence of selection. We evaluate TOAST on 12 human/mouse gene clusters.

1 Introduction

Richard Owen introduced the term *homology* in 1849 to refer to physical similarities among species [18]. The definition of this term has evolved since then. When applied to sequence analysis, it suggests that the relatedness of sequences stems from their common evolutionary origin. Any two similar genes or genomic sequences having the same evolutionary origin, no matter whether they are from the same genome or different genomes, are *homologous*, and they are a pair of *homologs*. The difference between the two sequences measures their *divergence* since their most recent common origin. The key concept for this paper is the division of all pairs of homologs into two categories, *orthologs* and *paralogs*, as established by Fitch in 1970 [7]. Two genes or genomic segments in different genomes are *orthologous* if they initially diverged as the result of a speciation event, whereas two genes or genomic segments (perhaps in the same genome) are *paralogous* if they initially diverged by an intra-species duplication event. In general the functions of paralogs differ, and hence it is not reliable to infer functionality of a gene from its paralog. In contrast, orthologs are more likely to have similar functionalities [15].

When there is only a one-to-one relationship of orthologs (without confusing paralogs) between two genomes, i.e., one segment of a genome has exactly one ortholog in another genome, it is relatively easy to determine the orthologs. The simple best-hit approach is sufficient to do the job [24]. But from the definition

F.D. Ciccarelli and I. Miklós (Eds.): RECOMB-CG 2009, LNBI 5817, pp. 98–110, 2009.

above, orthology is not necessarily a one-to-one relationship; it can also be one-to-many or many-to-many [8]. A reciprocal best hit would not fully detect such relationships.

Besides one-to-many and many-to-many orthology relationships, many other genetic processes impede orthology assignment. Lineage-specific gene loss is particularly difficult. Suppose species A and B descend from their least common ancestor R, and R had two paralogous copies R1 and R2. The first copy was lost along the speciation process of A, and the second copy was lost along the speciation process of B. In such a case, a reciprocal best hit predicts that A2 and B1 are a pair of orthologs, and there is no way to correct this wrong conclusion based only on the sequences of species A and B. This type of wrong assignment is sometimes acceptable since B2 is still the best counterpart of A1 in the genome of species B, and vice versa. This relationship is called *pseudoorthology* [15]. An incomplete genome sequence can confuse orthology assignment in a similar way, since a missing sequence is equivalent to a gene loss in this scenario. When the related sequences of more species are available, it is possible to identify duplication and gene loss events by the technique of *tree reconciliation* [19], and further to predict more reliable orthology relationships. However, this technique is computationally very expensive, and many uncertainties make it impractical for application to large-scale genomic analysis. Koonin [15] provides a comprehensive review of the issues involving orthologs, paralogs, and homologs, and of the basic techniques to detect orthologs.

Conversion, another kind of genetic process that also complicates the problem of orthology assignment, has been detected frequently in tandem gene clusters [13] . In this process, a segment of genomic sequence is replaced by a homologous sequence from elsewhere in the same genome. It can be regarded as a simultaneous gene loss and duplication. This raises the question of what we should take as the ortholog of the conversed sequence. Should it be the ortholog of the lost sequence, based on gene locations, or the ortholog of the newer sequence, based on gene content? Here, we regard such a conversion as a recent duplication, and assign orthologs based on gene content. This choice is more consistent with Fitch's distinction between orthology and paralogy, and it provides alignments that are appropriate for functional inference and phylogenetic analysis. Furthermore, a segment of genomic sequence that undergoes conversion does not necessarily correspond to an entire gene, which leads to the phenomenon that different parts of a gene may be orthologous to different parts of other genes, and a single gene is no longer the orthologous unit in evolutionary processes.

A traditional orthology assignment is performed on gene (or protein) sequences. Computational tools of this type were pioneered by COG [24], followed by INPARANOID [20], OrthoMCL [16] and OrthoParaMap [5]. Orthology assignment is simpler in certain ways for genes than for genomic sequences, as genes usually have consistent boundaries for sequence comparison. But with genomic sequences, duplications can happen at any level of genome structure, from a partial gene to a whole genome, and cover both coding and non-coding regions, making it unclear which intervals should be considered. Indeed, when dealing with genomic sequences, the problem of orthology assignment of genes from different species is tightly linked to the problem of finding appropriate regions to align among those species.

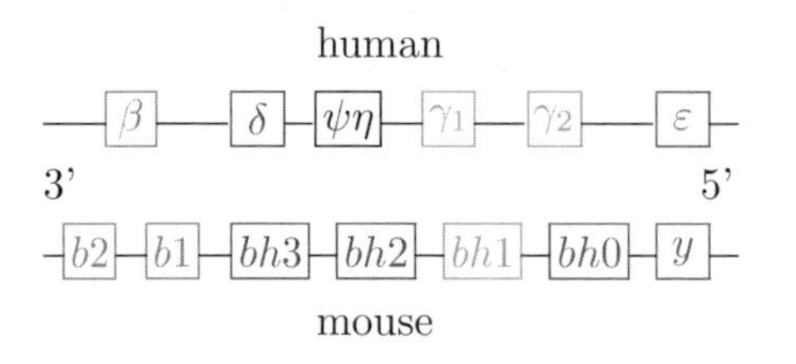

The ENCODE pilot project [17] used three major multi-species genomic sequence aligners: TBA [2], MLAGAN [4] and MAVID [3]; the evolutionary analysis of ENCODE data is based on a combination of the output produced by these aligners. All of these multiple alignment tools start with pairwise alignments. We observe that none of these three tools provides adequate alignment of tandem gene clusters. For example, ENCODE region ENm009 contains the beta globin gene cluster. The genes and their orders in human and mouse are illustrated in the above figure. Genes of the same color are of the same type [9]. Note that the sizes and locations of genes in this figure are not scaled according to their real size and locations in the genome. In the TBA alignment, human δ is split into two parts: one is aligned to part of mouse *bh2*, and the other is aligned to part of mouse *bh3*. Part of human $\psi\eta$ is aligned to mouse y, which is not orthologous. In the MLAGAN alignment, human δ is also split into two parts: one is aligned to part of mouse *b1*, and another part is aligned to part of mouse y, which is not orthologous. The two human γ genes are aligned to mouse y gene. MAVID does not produce any alignment within this region. There exist many other alignment tools, but aligning tandom duplicated gene cluster regions has not been specifically addressed.

Members of a gene family are frequently located in proximity along the same chromosome, although many large gene families have members dispersed on different chromosomes. For some gene families, the duplication events happened very early in their evolutionary history. The HOX gene family is one example; paralogs in this family have low similarity at the DNA level. An ordinary genomic alignment tool is able to do correct orthologous alignment for this case, because of the low similarity between paralogs. In other gene families, such as myosin heavy chain gene cluster, there have been few genomic changes except simple edits since the clusters were formed. Pairwise alignments of such gene clusters fall along a diagonal in a dot-plot, and a chaining program or global alignment tool is probably satisfactory in this case. However, for some gene families, there were new duplications after the speciation split, or large deletions, so the same gene clusters in different species can have different numbers of gene copies. We find that most aligners perform inadequately on clusters of this last type (e.g. the beta globin cluster), and thus they are good test cases for evaluating our new alignment strategy.

2 Methods

Our approach begins with two-way alignments produced by a pairwise local alignmetn tool such as BLASTZ [21]. We align each of the two genomic input sequences with itself, as well as aligning them with each other, to obtain three sets of alignments in which a given position can be aligned several times and/or involve the reverse complement of an input sequence. These "raw" pairwise alignments are then processed by our newly developed program TOAST (Two-way Orthologous Alignment Selection Tool), which retains only putative orthologous alignments. In outline, TOAST collects matches (local alignments) found by BLASTZ into "chains", organizes the chains using a graph, and selects a subset of the graph's edges that constitute the set of putative orthologous matches. We use the pairwise alignment between human and mouse over the beta globin gene cluster region as an example for illustration purpose.

2.1 Chaining

In a genome, some parts of a functional unit can be less conserved than other parts (e.g., introns of a gene), and hence may not be detected by a pairwise alignment tool (BLASTZ in our example). Thus, in a dot-plot graph, the detectable matches between homologous biologically meaningful units may be broken into smaller pieces that lie along roughly the same diagonals, as shown in Fig. 1(a). We want to form longer chains of these small pieces. A k-d tree [1], where $k = 2$ in our case, is used to efficiently identify the chains. For a sketch of how this works, consider the matches that do not involve reverse-complemented intervals. (The chaining procedure is applied independently to both orientations of the second sequence.) Each local alignment has two endpoints, a *start point* at its lower left (in the dot-plot) and an *end point* at its upper right. Each "bucket" (leaf node) of the k-d tree holds a certain number (e.g. 11) of local alignments within a rectangle area in a dot plot. Define a *predecessor* p of an alignment t which is in bucket b, to be an alignment in b or a bucket that is located below b or left of b, such that the end point of p is located more or less to the left of and below the start point of t, but allowing a small overlap (e.g. 50 bases). To reduce the computation time, the distance between p and t must be within a limit d (e.g. 3000 bases). Among all the predecessors of t, the *best predecessor* is the one closest to t. The best predecessor is determined for every alignment, unless there is no best predecessor for an alignment (e.g., because when certain thresholds are enforced, or when actually there is no predecessor). Alignments are connected into chains according to their best predecessors; those without best predecessors start new chains. Fig. 1(b) shows an ideal chaining result.

Sometimes the above chaining procedure fails to form chains long enough to include whole genes. For example, when introns of a gene are very long and diverged, the gap between the diagonal lines that representing adjacent exons may exceed the distance limit, so these two lines are not chained, though they belong to the same pair of orthologous genes. Some genomes, e.g. human, are extensively annotated, in which case it is straightforward to force the chain to contain the fragmented alignment blocks that belong to a complete gene. For example, the search space for the best predecessor of alignment t is restricted to be within the distance d (e.g. 3000 bases). Given the gene annotation, we can extend the search space to include the entire gene that t belongs to.

2.2 A Graph Model

In [11], we defined Duplicate Unit, which is a piece of the longest conserved genomic sequence containing no significant repetitions. In the example of β-like globin gene cluster, each gene including its flanking regions is a duplicate unit. We define the *alignment graph* in which nodes represent duplicate units, and edges represent alignments between them. Let's consider the graph for two species. If we partition the nodes into two parts, such that each part contains nodes (duplicate units) from one species, then the edges within each part represent self-alignments of that species, and the edges crossing between the two parts represent inter-species alignments. Fig. 1(d) is an example of such a graph. We want to select a subset of cross-edges that represent orthologous alignments.

2.3 Determining Graph Nodes from Pairwise Alignments

Fig. 1(a) shows that each of the human genes is aligned to each of the mouse genes. The precise boundaries of each gene are not consistently indicated by all of its alignments. For example, although alignments of human β to each of these mouse genes roughly overlap β, their start positions differ, as do their end positions. We need to determine the boundaries of each duplicate unit. Fig. 1(c) gives a picture of the solution. After projecting the chains onto the human sequence, we get a set of horizontal lines. These lines can be smoothed by an imaginary curve. Each curve peak represents the central point of a potential duplicate unit. The boundaries of the duplicate unit are then determined from the wave hollows. The method is simlar to (but simpler than) the preliminary demarcation of duplicate units in [11]. TOAST uses both self- and cross-alignments to determine the set of such nodes for each species. For example, human-vs.-human self-alignments and human-vs.-mouse cross-alignments are used to determine the set of nodes for human. In cases where adequate gene annotation exists, the annotated boundary information is also incorporated into forming nodes; an annotated gene cannot be broken into several partial genes contained in separate nodes. However it must be emphasized that the duplicate unit is not necessarily a gene. It may be part of a gene or a regulatory unit, or may cover several genes and/or regulatory units. This is in accordance with Fitch's view [8]. Using duplicate units for orthology assignment is different from the approaches taken by most other orthology-assignment tools [25,5,20,16]. These tools target whole genes or gene products, while TOAST is applicable to conserved regions of all kinds.

2.4 Building Edges

The chains of self- and cross-alignments become edges connecting nodes within the same species and between different species respectively. The *score* of each edge is the sum of the scores of all BLASTZ local alignments in the chain associated with the edge. The BLASTZ score penalizes substitutions, insertions and deletions [21]. Each edge is also associated with a parameter of *score per base (spb)*, which is defined to be $2\times score/(length\_A + length\_B)$, where *length_A* and *length_B* are the lengths of the associated chain in species A and B, respectively. (For a self-alignment, A and B are the same species.) To avoid false positive alignments, TOAST ignores chains whose length or *sbp* falls below specified thresholds (e.g. length of 300 bases and *spb* of 5 respectively). Fig. 1(d) depicts the conceptual model of the graph.

2.5 Determining In-paralogous Alignments

At this point, a graph has been formed (Fig. 1(d)). Our goal is to select a set of cross-edges corresponding to orthology relationships. As a preliminary step, we need to select a set of self-edges showing the in-paralogous relationships for each species. (The reason for doing this will be apparent when we explain the next procedure.) We use the definition of *in-paralog* given by Remm et al [20], namely "paralogs that arose after the species split". (Thus, in-paralogs in species A depend on the choice of the second species.) In a normal evolutionary scenario, the similarity level between two in-paralogs in species A is higher than that between either of them and its ortholog in species B, since the orthologous copy in B diverged earlier than the two in-paralogous copies in A. Considering different evolutionary rates, we define a parameter *inflation*. For a given node N, we remove every self-edge whose $spb \times$inflation is below the highest *spb* among N's cross-edges. The remaining self-edges indicate potential in-paralogous relationships. For all results in this paper, inflation=1 is used. Fig. 1(e) shows the graph after selecting these self-edges. It indicates that human $\gamma 1$ and $\gamma 2$, as well as mouse *b1* and *b2*, are recent duplicates (after species split of human and mouse).

2.6 Determining Orthologous Alignments

Our approach for selecting orthologous alignments is based on the following observation: if a gene s in species A has more than one ortholog in species B, say s' and s'', then s' and s'' must be in-paralogs (relative to A). This observation leads to the following procedure: select the set of cross-edges with the maximum total edge score, subject to the condition that any pair of target nodes reached by edges from the same source node (in either species) is connected. For a simple example in Fig. 1(f), we selected edges from human β to mouse *b1* and *b2*, and mouse *b1* and *b2* are connected.

For the above problem, we can prove that it is difficult to get an optimal solution. For a graph $G =< V, E >$, a *cut* is a partition of V to form two sets V_1 and V_2. E_1 denotes the subset of E connecting only nodes in V_1, and E_2 is defined similarly for nodes in V_2. Let G_1 denote $< V_1, E_1 >$ and G_2 denote $< V_2, E_2 >$. An edge $e \in E$ which connects two nodes v_1 and v_2 where $v_1 \in V_1$ and $v_2 \in V_2$ is a *cut edge*. Let C denote the complete set of cut edges for V_1 and V_2. The problem is then to maximize the sum of the scores of a subset of cut edges subject to the constraint that if any two such cut edges share a same node, the nodes at their other ends must be connected directly. Since a *clique* is a graph/subgraph where any two nodes are directly connected by an edge, we describe the above constraint as the property of *clique connection* (the selected

(a) Dot-plot of BLASTZ alignments.

(b) The same alignments, where each oval encloses a chain.

(c) Determining boundaries of duplicate units.

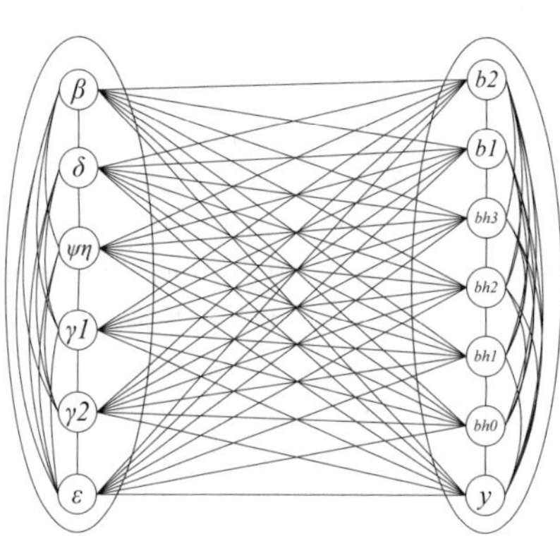

(d) The alignment graph. Each node is named after the gene that the node(duplicate unit) contains.

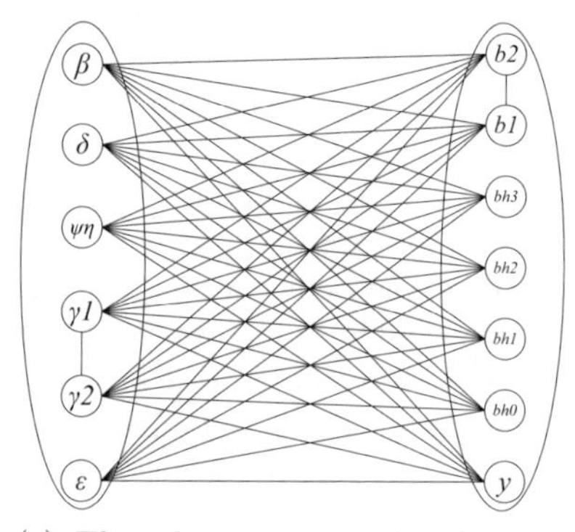

(e) The alignment graph after removing self-edges with scores lower than the score of some cross-edge from the same node.

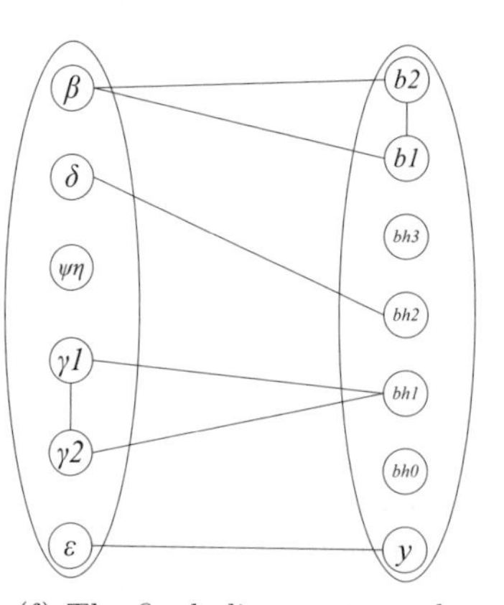

(f) The final alignment graph.

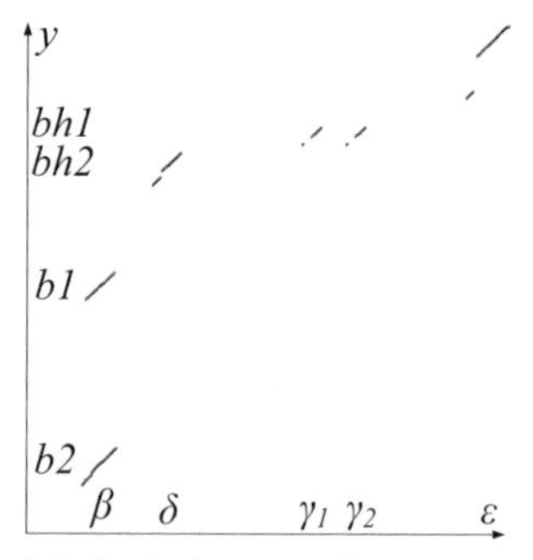

(g) Orthologous alignments obtained by TOAST.

Fig. 1. Major steps in TOAST. In all dot plots, x axis is human sequence, and y axis is mouse sequence. In all bipartition graphs, the left part is for human, and the right part is for mouse.

cut edges connect cliques of G_1 and G_2). The decision version of this problem is actually NP-complete. The proof can be found in [12].

An efficient approximation for the above problem is also known to be difficult [10], and moreover the theoretically optimal solution is not necessarily the biologically correct solution. We found a simple greedy approach, which we proved to be effective by experiments. We sort all cross edges by their scores, and select edges from the highest to the lowest score, subject to the previous condition. Fig. 1(f) shows the orthology assignment generated by this cross-edge selection method. The resulting dot-plot of the corresponding orthologous pairwise alignments is shown in Fig. 1(g).

3 Evaluation and Results

Evaluating the success of our approach proved challenging, due to a shortage of accurate annotation for available vertebrate genome sequences. We present specificity/sensitivity measurements for TOAST obtained on 12 human gene clusters for which the orthologous mouse clusters have a different number of members, and these clusters are the difficult cases to align appropriately, as explained in Introduction.

3.1 Gene Clusters for Testing

We selected the 12 gene clusters as follows. We divided the human genome (repeat masked [22]) into 1-megabase segments, and performed self-alignment on each of them using BLASTZ [21] with default parameters. Many regions with repeated sequences were found. We discarded regions with fewer than five copies, and discarded those without good annotations (from the UCSC Genome Browser [14]) in human. We then manually inspected the sequences in the Genome Browser to determine the orthologous regions of mouse, discarding those without a known orthologous mouse region, those without good annotations in mouse, and those with the same number of gene copies in human and mouse. We then manually determined reasonable cluster boundaries. This process yielded 12 usable gene clusters having 5 to 31 gene/pseudogene copies in human. This selection process is admittedly crude, but our purpose was simply to get some appropriate examples.

For each pair of homologous gene clusters in human and mouse, we applied a traditional maximum-likelihood method with bootstrapping for phylogeny reconstruction [6] to the set of all the encoded protein sequences in human and mouse. In this method, a pair of proteins from different species is considered to be orthologous if they are more closely related (by the tree) to each other than either of them is to another protein in the same species. This is related to the criterion that TOAST uses, but it should be more reliable here because (1) boundaries of the regions under study are clearly defined (they are not aligned regions that overlap in arbitrary ways), (2) in general sequence alignment at the amino-acid level should be more reliable than DNA-based alignment, and

(3) maximum-likelihood is known to be more accurate for this purpose than methods based on alignment scores and/or percent identity. These orthology assignments using phylogenetic analysis formed the benchmarks for evaluating TOAST. Gene cluster locations, phylogenetic gene trees with bootstrapping values, and gene coordinates of each cluster are shown in Chapter 2 of [12].

3.2 The Evaluation Protocol

We use "sensitivity" and "specificity" to quantify the evaluation. Ideally, sensitivity shows the proportion of correctly identified orthologs among all true orthologs, while specificity shows the proportion of correctly identified orthologs among all identified orthologs. Let A = predicted orthologous pairs, and let T = true orthologous pairs. Then specificity = $(|A \cap T|)/|A| \times 100\%$ and sensitivity = $(|A \cap T|)/|T| \times 100\%$. However, in practice, we may not have the complete set of proteins for each cluster, and some of the proteins are not assigned orthologs because of incomplete genome sequence data. Furthermore, we do not know every true orthology, since the bootstrapping values in the phylogenetic analysis vary greatly and many of the pairs do not satisfy the minimum threshold for trustworthiness. Also, the alignment from TOAST includes non-coding regions. Thus, we cannot directly utilize the above definition of specificity and sensitivity. We also need to define the predicted set and true set more carefully.

Let P denote the whole set of protein pairs that are potentially orthologs inferred from the phylogenetic gene tree (proteins without orthologs are denoted as assigned to *null*). Within this set, protein pairs with bootstrap values exceeding a specified threshold (950, 800 or 500) are taken as the true set, denoted as T, a subset of P. A set of pairwise alignments from TOAST is denoted as O, which contains alignments of coding, non-coding, and flanking regions of a gene cluster. A subset A of O is defined to include any alignment block or partial block both of whose rows are within any of the sequences in set P (since we only use gene sequences for evaluation), and at least one of whose rows is within a sequence in T (so at least one sequence is in benchmark). These sets are illustrated in Fig. 2.

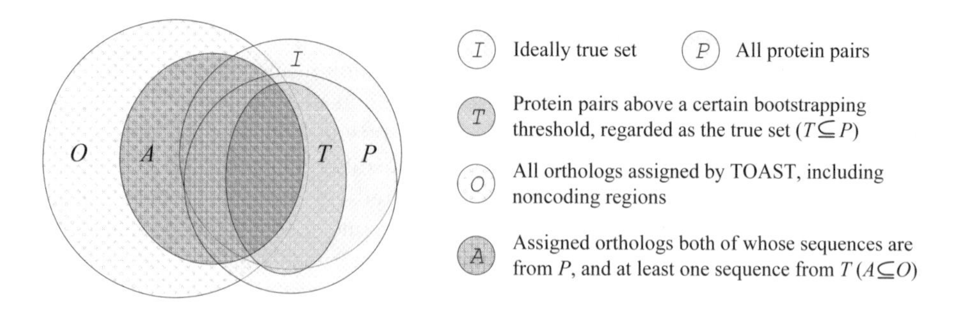

Fig. 2. Abstract structure of the evaluation protocol

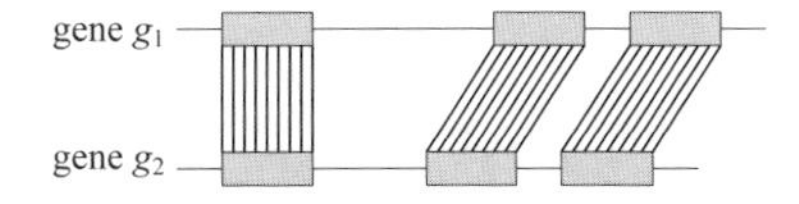

Note that we cannot simply count the pairs of aligned genes in our evaluation. The (local) genomic alignment blocks usually do not correspond to complete genes, since only conserved regions are aligned. To count the number of aligned genes for evaluation, we need some threshold to determine whether two genes are aligned. In the above figure, only partial regions are aligned between genes g_1 and g_2. What fraction should we require to determine that g_1 and g_2 are aligned? Genomic structures of genes vary widely, and there is no reasonable threshold that is good for all gene clusters. So we use the number of aligned positions in our evaluation to count fractions of aligned orthologs. Since there are insertions and deletions in genomic sequences, the numbers of aligned positions may be different between two species.

Now we need to determine the subset $A \cap T$. Suppose bases b_1 and b_2 from different species are aligned together. If one of them does not fall within the boundary of any gene in the set P, or if neither of them falls within the boundary of any gene in the set T, we discard this pair of bases. Suppose the base b_1 falls within gene copy g_1 in the set P, and the base b_2 falls within gene copy g_2 in the set P. If g_1 and g_2 form a pair in set T, we regard b_1 and b_2 as correctly assigned bases. Let c denote the number of correctly assigned positions in one species S. Let a denote the number of aligned positions of species S in set A. Let t denote the number of positions of species S within set T. We then define:

$$\text{Specificity} = c/a \times 100\% \quad \text{Sensitivity} = c/t \times 100\% \tag{1}$$

There are various ways to define whether a DNA pairwise alignment block is within a protein sequence, or to determine the number of DNA bases that are within a protein. One approach is to compare the whole genomic interval of the protein's gene with the DNA sequence. This approach might systematically make the sensitivity appear low, since the whole genomic region of a gene includes noncoding segments, such as the UTRs (Untranslated Regions) and introns, which are usually much less conserved and hence often not aligned. An alternative is to compare only the exon positions of the protein with the DNA sequence. We show results for both measurements.

3.3 Comparison between TOAST and BLASTZ

Fig. 3(a) shows the evaluation result based on exons; Fig. 3(b) shows the evaluation based on whole genes, including exons and introns. When bootstrapping value of 950 or 800 is the criterion, the set of true orthologs is too small. Hence the true orthologs used in Fig. 3 are from branches with bootstrapping values above 500. The evaluation results for each cluster using all bootstrapping thresholds (950, 800 and 500) can be found in [12]. These results show that BLASTZ alignments tend to have high sensitivity and low specificity. TOAST improves the specificity dramatically while maintaining the sensitivity at an acceptable level. Because introns are less conserved than exons, and therefore less detectable by an alignment tool, evaluation of whole gene alignments (Fig. 3(b))

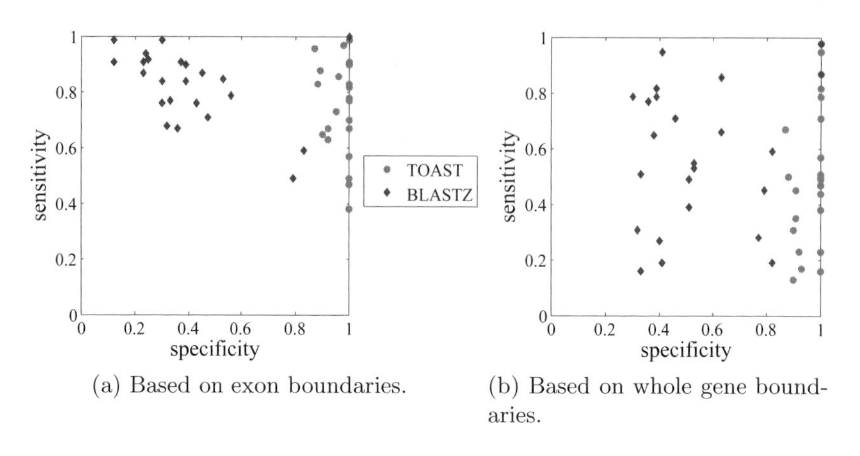

(a) Based on exon boundaries.

(b) Based on whole gene boundaries.

Fig. 3. Comparison of pairwise alignments of 12 gene clusters of human vs. mouse by BLASTZ and TOAST, using 500 as the bootstrapping threshold. Note that there are two plots (for two species) for each gene cluster.

shows generally lower sensitivity than evaluation of exon alignments (Fig. 3(a)), as expected. Introns of some genes are much longer than exons, the sensitivity computed based on the length of a whole gene is therefore significantly less than the one computed by the number of positions of exons for these gene clusters.

4 Conclusion and Discussion

The one-to-many and many-to-many orthology relationships, entangled with the problems of gene loss, conversion, and uncertainties arising from sequencing and assembling, make the orthology assignment problem difficult. The popular solution of using phylogenetic inference is limited by the completeness of related sequences, the expensive computation, and the difficulty in interpreting the phylogenetic tree. Thus it is not practical to apply this technique in large-scale analyses. Methods using gene (or gene product) sequences are also restricted by the availability of annotations for a genome. Here we used a different method and created TOAST to perform orthology assignment by genomic sequence alignment, specifically for regions containing gene clusters. Our initial experiments on complicated gene clusters indicate that it is successful in identifying orthologs at very high specificity with adequate sensitivity (the TOAST alignment is selected from BLASTZ alignment, therefore its sensitivity will not exceed the one of BLASTZ).

Like any alignment program, TOAST has many parameters and internal thresholds. The default values (indicated in Sec. 2) were determined by examining the β-like globin gene cluster. Results used in evaluation were produced by default values. These parameters can be adjusted to exchange sensitivity for specificity or vice versa. For the example of β-like globin gene cluster, some biologists consider both mouse *bh3* and *bh2* genes are orthologous to human δ gene, and both mouse *bh0* and *y* genes are orthologous to human ϵ gene [9].

However, the TOAST result presented in this paper only discovered the orthologous relationship between mouse *bh2* and human δ genes, and mouse y and human ϵ genes. Most probably these in-paralogous pairs of mouse genes evolved at very different rates, and TOAST determined that they were not in-paralogs. By increasing the value of parameter inflation in Sec. 2.5, TOAST is able to find all above in-paralogs. However, it also outputs some non-orthologous alignment such as between mouse y and two human γ genes. For the purpose of our subsequent multi-species orthologous alignment, we prefer to keep high specificity with modest sensitivity of TOAST, since the missing orthologous alignment can still be recovered by the best candidate from BLASTZ alignment.

References

1. Bentley, J.: Multidimensional binary search trees used for associative searching. Comm. of ACM 18, 509–517 (1975)
2. Blanchette, M., Kent, J.W., et al.: Aligning multiple genomic sequences with the threaded blockset aligner. Genome Res. 14, 708–715 (2004)
3. Bray, N., Pachter, L.: MAVID: constrained ancestral alignment of multiple sequences. Genome Res. 14, 693–699 (2004)
4. Brudno, M., Do, C., et al.: LAGAN and Multi-LAGAN: efficient tools for large-scale multiple alignment of genomic DNA. Genome Res. 13, 721–731 (2003)
5. Cannon, S.B., Young, N.D.: OrthoParaMap: distinguishing orthologs from paralogs by integrating comparative genome data and gene phylogenies. BMC Bioinformatics 4, 35–49 (2003)
6. Felsenstein, J.: PHYLIP (Phylogeny Inference Package) version 3.6. Distributed by the author. Dept. of Genome Sciences, Univ. of Washington, Seattle (2005)
7. Fitch, W.M.: Distinguishing homologous from analogous proteins. Syst. Zool 19, 99–113 (1970)
8. Fitch, W.M.: Homology, a personal view on some of the problems. Trends Genet. 16, 227–231 (2000)
9. Hardison, R., Miller, W.: Use of long sequence alignments to study the evolution and regulation of mammalian globin gene clusters. Mol. Biol. Evol. 10, 73–102 (1993)
10. Hastad, J.: Clique is hard to approximate within $n^{1-\epsilon}$. In: 37th Annual IEEE Symposium on Foundations of Computer Science, pp. 627–636 (1999)
11. Hou, M., Berman, P., Hsu, C., Harris, R.: HomologMiner: looking for homologous groups of whole genomes. Bioinformatics 23, 917–925 (2007)
12. Hou, M.: Algorithms for aligning and clustering genomic sequences that contain duplications. Ph.D. thesis, Penn State University, USA (2007)
13. Hsu, C., Zhang, Y., Hardison, R., Miller, W.: Whole-Genome Analysis of Gene Conversion Events. In: RECOMB CG 2009 (accepted 2009)
14. Kent, W.J., Sugnet, C.W., Furey, T.S., Roskin, K.M., Pringle, T.H., Zahler, A.M., Haussler, D.: The human genome browser at UCSC. Genome Res. 12, 996–1006 (2005)
15. Koonin, E.V.: Orthologs, paralogs, and evolutionary genomics. Annu. Rev. Genet. 39, 309–338 (2005)
16. Li, L., Stoeckert, C.J.: Identification of ortholog groups for eukaryotic genomes. Genome Res. 13, 2178–2189 (2003)

17. Margulies, E.H., et al.: Analyses of deep mammalian sequence alignments and constraint predictions for 1% of the human genome. Genome Res. 17(6), 760–774 (2007)
18. Owen, R.: On the Nature of Limbs. van Voorst, London (1849)
19. Page, R.D., Charleston, M.A.: From gene to organismal phylogeny: reconciled trees and the gene tree/species tree problem. Mol. Phylogenet. Evol. 7, 231–240 (1997)
20. Remm, M., Storm, C.E., et al.: Automatic clustering of orthologs and inparalogs from pairwise species comparisons. J. Mol. Biol. 314, 1041–1052 (2001)
21. Schwartz, S., Kent, W.J., et al.: Human-mouse alignments with BLASTZ. Genome Res. 13, 103–107 (2003)
22. Smit, A.F.A., Hubley, R., Green, P.: RepeatMasker Open-3.0 (1996-2004), `http://www.repeatmasker.org`
23. Steinberg, M.H., Forget, B.G., et al. (eds.): Disorders of Hemoglobin: Genetics, Pathophysiology, and Clinical Management. Cambridge University Press, Cambridge (2001)
24. Tatusov, R.L., Koonin, E.V., Lipman, D.J.: A genomic perspective on protein families. Science 278, 631–637 (1997)
25. Yuan, Y.P., Eulenstein, O., Vingron, M., Bork, P.: Towards detection of orthologues in sequence databases. Bioinformatics 14, 285–289 (1998)

Inferring the Recent Duplication History of a Gene Cluster

Giltae Song[1], Louxin Zhang[2], Tomáš Vinař[3], and Webb Miller[1]

[1] Center for Comparative Genomics and Bioinformatics, 506B Wartik Lab, Penn State University, University Park, PA 16802, USA
[2] Department of Mathematics, National University of Singapore, Singapore 117543
[3] Faculty of Mathematics, Physics and Informatics, Comenius University, Mlynska Dolina, 842 48 Bratislava, Slovakia

Abstract. Much important evolutionary activity occurs in gene clusters, where a copy of a gene may be free to evolve new functions. Computational methods to extract evolutionary information from sequence data for such clusters are currently imperfect, in part because accurate sequence data are often lacking in these genomic regions, making the existing methods difficult to apply. We describe a new method for reconstructing the recent evolutionary history of gene clusters. The method's performance is evaluated on simulated data and on actual human gene clusters.

1 Introduction

Gene clusters are formed by duplication, followed by substitution, inversion, deletion, and/or gene conversion events. The resulting copies of genes provide the raw material for rapid evolution, as redundant copies of a gene are free to adopt new functions [1,2]. A copy may take on a novel, beneficial role that is then preserved by natural selection, a process called *neofunctionalization*, or both copies may become partially compromised by mutations that keep their total function equal to that of the original gene, called *subfunctionalization* [3]. Another source of interest in gene clusters is that several human genetic diseases are caused by a tendency for regions between two copies to be deleted [4]. A major finding of the initial sequencing of the human genome is that 5% of the sequence lies in recent duplications [5]. More recently, it has become clear that duplicated regions often vary in copy number between individual humans [6]. A substantial fraction of what distinguishes humans from other primates, as well as the genetic differences among humans, cannot be understood until we have a clearer picture of the contents of gene clusters and of the evolutionary mechanisms that created them.

One impediment to this understanding is that recently duplicated regions, say those that retain over 95% identity (roughly, that duplicated in the last 10 million years) resist assembly by the current whole-genome shotgun approach [7]. Even the so-called "finished" human genome sequence has 300 gaps, most of which are caused by the presence of recent duplications. Moreover, much

F.D. Ciccarelli and I. Miklós (Eds.): RECOMB-CG 2009, LNBI 5817, pp. 111–125, 2009.

available mammalian genomic sequence is only lightly sampled, and hence even further from supporting analysis of gene clusters. Partly because of the lack of accurate sequence data in gene clusters, practical computational tools for their analysis still need to be developed.

We think of the analysis problem for gene clusters as requiring a marriage of two somewhat distinct approaches, one dealing with large-scale evolutionary operations (primarily duplication, inversion, segmental deletion, and gene conversion) and the other with fine-scale evolution (substitutions and very small insertions/deletions). Even the second part, though it is essentially just an extension of the familiar problem of multiple sequence alignment, is currently not handled well by existing tools. Indeed, just defining what is meant by a proper alignment of a gene cluster sequence is a matter of discussion (e.g., see [8,9]). Although a comparison of several multi-genome alignment programs found reasonable accuracy in single-copy portions of the genome [10], we have found their performance on gene clusters to be inadequate [11]. Our observations about the quality of current whole-genome alignments (e.g., those described in [12]) indicate that it may be worthwhile to align gene clusters using methods designed specifically for them, and then splice the results into the whole-genome alignments created by the other methods.

A number of ideas have been explored for reconstructing large-scale evolutionary history (as opposed to the sequence alignment problem, which deals with substitutions and small insertions/deletions). Some of these attempt to reconstruct the history of duplication operations on regions with highly regular boundaries (e.g., [13,14,15]), some allow inversion (e.g. [16,17]), and also deletions (e.g. [18,19,20]). Typically, whenever more than one group has studied a given formulation of the problem, the methods developed have been fundamentally different, often with large differences in the resulting evolutionary reconstructions and the computational efficiency of the methods.

In terms of formulation of the underlying problem, this paper is closest to [18] and particularly [19]. However, the methods described here are quite different. For instance, [19] uses probabilistic techniques whereas our approach is entirely combinatorial.

2 Methods

2.1 Problem Statement and a Basic Algorithm

During genome evolution, a duplication event copies a segment to a new genomic position. Genome sequencing and analysis suggest that a large number of gene clusters in human and other mammalian genomes have been formed by duplication events, often very recent ones. We aim to identify the duplication events that formed a given gene cluster by using a parsimony approach.

Formally, a duplication is an operation that copies a subsequence or its reverse complement into a new position. The original segment is called the *source region* and the inserted copy after the duplication is called the *target region*. Subsequently, the two regions evolve independently by point mutations and small

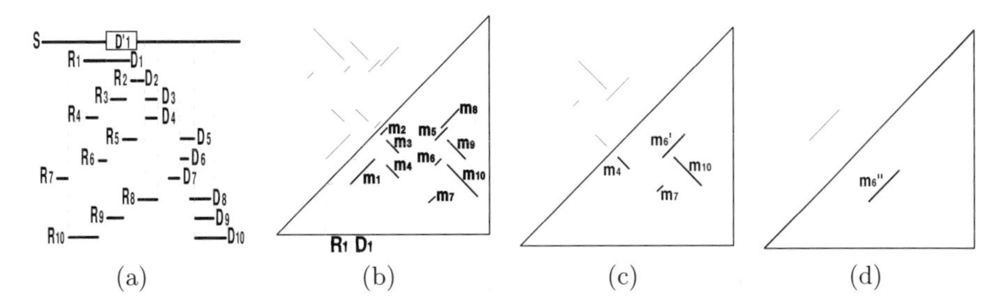

Fig. 1. An example of duplication inference. (a) The original matches in a sequence. (b) The self-alignment of the original sequence. (c) The self-alignment after rolling back a duplication. (d) The self-alignment after rolling back two duplications. See Section 2.3 for details.

insertions and deletions and thus in the present day sequence, the two regions are generally not identical. However, they do form a strong local alignment (called a *match*) in a self-alignment of any genomic region that contains them both.

We detect matches by aligning the genomic sequence with itself using the program BLASTZ [21]. Let S be a genomic sequence containing a gene cluster. First, we run BLASTZ to obtain all local self-alignments of S, which are visualized as a dot plot (e.g. Figure 1(b)). These are then processed in a pipeline that links local alignments separated by small gaps and/or interspersed repeats and adjusts their endpoints to avoid inferring tiny spurious duplications. Finally, we identify a set of matches. The problem of inferring the duplication history of the gene cluster is then to find a duplication-free sequence T and the minimum number of duplication events $O_1, O_2, \cdots, O_k$ such that

(i) The source and target regions of each O_i consist of one or more match regions.

(ii) $S = O_k(O_{k-1}(\ldots(O_1(T))))$, where $O_i(S')$ denotes the resulting sequence after applying O to sequence S'.

Given a sequence of duplications $O_1, O_2, \ldots, O_k$, we call the boundaries of all source and target regions *breakpoints*. If duplication events occur randomly during genome evolution, the two duplications are quite unlikely to share their boundaries. So we assume that no two duplications have a common breakpoint in a duplication history [22], except for tandem duplication. Tandem duplication (with or without reversal) copies a source region into a location adjacent to its boundary. Thus, it is a special exception to the *breakpoint uniqueness* assumption.

The algorithm focuses on identifying the latest duplication event in the duplication history of a gene cluster. Once this is inferred, its target region is eliminated. These steps are repeated until the duplication-free sequence is reconstructed.

Before determining the latest target region, we must identify matches that have been split by a subsequent duplication. Consider a given match (R, R'). If a duplication event inserts a segment A in the region R', then the match (R, R') is split into two small matches (R_1, R'_1) and (R_2, R'_2). If this happens, we can correctly identify the original duplication event forming the match (R, R') only if we first identify the one inserting A. Hence, using a kd-tree data structure, we identify all the pairs of matches (R_1, R'_1) and (R_2, R'_2) such that R_1 and R_2 are adjacent but R'_1 and R'_2 are separated by a region A of some match. To guarantee that (R_1, R'_1) and (R_2, R'_2) are not examined before removing A from the sequence, we place $[R_1, A]$, $[R'_1, A]$, $[R_2, A]$, $[R'_2, A]$ into a *suspend list*. We call the regions R_1, R'_1, R_2, R'_2 the *suspended regions*, and the region A on which they depend the *inserted region*.

Definition 1. *We say a region A is contained in a region B if all bases of A are in B but not vice-versa. We say A and B overlap if they share at least one base but neither contains the other.*

Theorem 1. *Assume a sequence S is transformed from a duplication-free sequence T by a series of duplication events. Then the target region of the latest duplication event does not overlap with the source or target regions of any other duplications, and it is not contained in any match regions of S.*

The proof of Theorem 1 follows from the breakpoint uniqueness assumption. Based on Theorem 1, we determine the latest duplication event in the history of a gene cluster as follows. Suppose there are n matches in genomic sequence S. We define the *constraint graph* $G = (V, E)$ of these matches as follows. G is directed and has $2n$ nodes representing the $2n$ regions of the matches. There are three types of arcs. Let (R, R') be a match. If R overlaps a region B of another match, there is an arc from node R to node R'. Such an arc is called a type-1 arc. If R is contained in another match region C, there is an arc from node R to node C, called a type-2 arc. Finally, if $[R, A]$ is in the suspend list, there is an arc from node R to node A, called a type-3 arc. The constraint graph for Figure 1 is given in Figure 3.

By Theorem 1, there must be at least one node with out-degree 0 in a constraint graph. In each loop of the algorithm, we select a node v with out-degree 0 and remove the region corresponding to v from S. If there are several nodes of out-degree 0, the one with the highest similarity level in the self-alignment is selected as the latest duplicated region. By Theorem 2 below, the following algorithm identifies the true number of duplication events and a plausible sequence of such events in $O(n^2 \log n)$.

procedure INFER-DUPS($\mathcal{S}$)
Input: A set of matches $\mathcal{S}$ in a self-alignment
Output: A set of duplication events
repeat
  for all the pairs of matches (R_1, R'_1) and (R_2, R'_2) **do**
    if R_1 and R_2 are adjacent but R'_1 and R'_2 are separated by a region A of some match **then**
      place $[R_1, A]$, $[R'_1, A]$, $[R_2, A]$, and $[R'_2, A]$ into the suspend list.

end if
end for
$G \leftarrow$ CONSTRUCT-CONSTRAINT-GRAPH ($\mathcal{S}$)
Identify the regions of out-degree 0 in G, and remove the one with the highest similarity value from $\mathcal{S}$.
if the removed region is an inserted region in the suspend list **then**
merge the corresponding suspended regions in $\mathcal{S}$.
end if
$\mathcal{S} \leftarrow \mathcal{S} - M$, where M is the set of matches that disappear with removal of the region
until $\mathcal{S} = \phi$
end procedure

function CONSTRUCT-CONSTRAINT-GRAPH ($\mathcal{S}$)
for all the pairs of matches (A, A') and (B, B') **do**
if A overlaps B **then**
type-1 arc of $A \rightarrow A'$ and $B \rightarrow B'$
else if A is contained in B (or vice versa) **then**
type-2 arc of $A \rightarrow B$ (or $B \rightarrow A$)
else if A (or B) is a suspended region with C in the suspend list **then**
type-3 arc of $A \rightarrow C$ (or $B \rightarrow C$)
end if
end for
end function

Theorem 2. *Suppose a sequence S evolves from a duplication-free sequence T in k duplications. If the breakpoint uniqueness assumption holds, the algorithm identifies a series of k duplications and a duplication-free sequence T' such that T' transforms to S by the identified k duplications.*

2.2 Handling Tandem Duplication

Our model assumption of breakpoint uniqueness may be violated by tandem duplication. Copy-and-paste transposons are an example of frequent reuse of duplication breakpoints. To infer duplication history more accurately, a way of handling this tandem duplication is required. Suppose we have a tandem duplication that copies A into a location adjacent to its boundary. It produces a match (A, A') where A' is adjacent to A. If the copied location of A' is not involved in any other matches, the tandem duplication does not affect the algorithm. But if A' split other matches, the target region is contained in the split matches, which violates Theorem 1. For instance, let m be a match, where $m = (BAD, B''A''D'')$. After the tandem duplication in A occurs, m is split into two matches m_1 and m_2, where $m_1 = (BA, B''A'')$ and $m_2 = (A'D, A''D'')$ (see Figure 2). This causes the algorithm to fail to detect the target region A' because A' is contained in both regions of m_1 and m_2. Fortunately, we observe a property that one region of m_1 has a boundary adjacent to a region of m_2 while

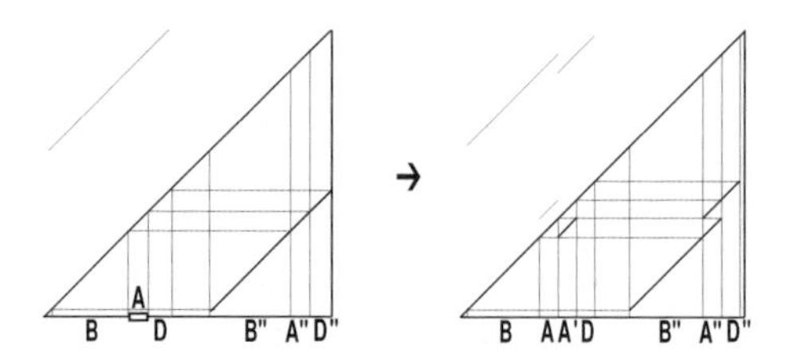

Fig. 2. An example of the self-alignment change caused by a tandem duplication event

the other region of m_1 overlaps the other region of m_2. Also, the boundaries of the overlapped part of m_1 and m_2 correspond to region A''. Thus, the tandem duplication can be detected as follows. If two matches (R_1, R_1') and (R_2, R_2'), where R_1 and R_2 are adjacent but R_1' and R_2' are overlapped (or vice versa), the part where R_1' overlaps R_2' is denoted as a temporary match (A, A'') such that A'' is the overlapped region of R_1' and R_2'. If there exists a match of (A, A') where A and A' are adjacent, then $[R_1, A]$, $[R_1', A]$, $[R_2, A]$, $[R_2', A]$, $[R_1, A']$, $[R_1', A']$, $[R_2, A']$, and $[R_2', A']$ are all placed in the suspend list. In addition, while constructing the constraint graph, if A is contained in a suspended region whose inserted region is A, the drawing of a type-2 arc from A is skipped if A forms a match with the adjacent region.

One potential problem is whether the case we detect as tandem duplication can be generated by other scenarios. Suppose we have three matches m_1, m_2, and m_3 where $m_1 = (BA, B''A'')$, $m_2 = (A'D, A''D'')$, $m_3 = (A, A')$, and that A and A' are adjacent. To simplify, we assume that the out-degree of m_1, m_2, and m_3 in the constraint graph constructed with the other matches is 0. If we consider only parsimonious solutions for this case, the only other scenario is separate duplications of both m_1 and m_2. But m_1 and m_2 cause the violation of the breakpoint uniqueness assumption. If tandem duplication is regarded as a special case of breakpoint reuse, inferring a tandem duplication of m_3 and a duplication of a merged match of m_1 and m_2 makes more sense.

This modification can be extended for tandem duplications which copy a segment more than once into its adjacent location. The tandem duplications of more than one copy are detected as follows. Assume the same source region is involved in all of the copies. If there are $n(\geq 2)$ matches such that $m_1 = (A_1, A_{n+1})$, $m_2 = (A_1A_2, A_nA_{n+1})$, ... , $m_n = (A_1...A_n, A_2...A_{n+1})$, then $m_i(2 \leq i \leq n)$ is converted into m_i' where $m_i' = (A_i, A_{n+1})$. Then, the latest region for this iteration can be identified by running the rest of the algorithm normally.

There is another event that may be confused with tandem duplication. This is a duplication that copies a source region into a location within itself. To handle these correctly, we replace the two matches formed by this type of duplication with one match. This step is motivated by the following. Let m_1 and m_2 be two matches, and suppose we observe $AA'B'B$ or $A\overline{B'A'}B$, where $m_1 = (A, A')$, $m_2 = (B', B)$ and $\overline{B'A'}$ is the reverse complement of sequence $A'B'$. $AA'B'B$ might arise from two duplication events: an event duplicating A and another duplicating B. It could also arise from a single duplication that copies AB within itself. The two-event explanation violates the breakpoint uniqueness hypothesis, and is also less parsimonious than a single event. Therefore, our algorithm infers that $AA'B'B$ arose from a single event that copied and inserted AB within itself. In the same manner, $A\overline{B'A'}B$ also arose from a single event that copied and inserted AB within itself in the reverse orientation. In order to infer one event for two matches in $AA'B'B$, the two matches m_1 and m_2 are replaced with a new match $m' = (AB, A'B')$. The two matches in $A\overline{B'A'}B$ are also replaced in the same way. We call this type of duplication *intraposed duplication.*

2.3 Illustration of the Method

To demonstrate how the method works, we consider a genomic sequence S containing 10 matches m_i, $1 \leq i \leq 10$. The dot plot of the self-alignment of S is shown in Figure 1(b).

First, we observe that the regions of m_1 and m_2 form a segment $AA'B'B$, where $m_1 = (A, A')$ and $m_2 = (B', B)$, so we infer that they were formed by an intraposed duplication event that inserted a copy of segment AB within itself. We replace m_1 and m_2 with a new match m'_1, whose regions are $R'_1 = AB$ and $D'_1 = A'B'$ respectively. Furthermore, the following two facts are true. Let $m_i = (R_i, D_i)$ for $3 \leq i \leq 10$.

- D_6 and D_8 are adjacent, while R_6 and R_8 are separated by D'_1. Hence, $[R_6, D'_1], [D_6, D'_1], [R_8, D'_1], [D_8, D'_1]$ are added to the suspend list.
- D_6 and D_7 are adjacent, while R_6 and R_7 are separated by R_{10}. Hence, $[R_6, R_{10}], [D_6, R_{10}], [R_7, R_{10}], [D_7, R_{10}]$ are added to the suspend list.

The constraint graph G for the 9 matches is shown in Figure 3(a). Note that there are no arcs leaving node D'_1, so m'_1 is selected as the latest duplication event. After D'_1 is removed from the sequence S, m_6 and m_8 are merged into a match m'_6 in the resulting sequence S'. In addition, since R_3, R_5 and R_9 are contained in D'_1, the matches m_3, m_5 and m_9 do not exist in S'. Overall, in the self-alignment of S' shown in Figure 1(c), only four matches remain, which are m_4, m'_6, m_7, m_{10}. The constraint graph for these four matches is shown in Figure 3(b). Since there are no arcs leaving node R_{10}, we select m_{10} as the latest duplication event. After removal of R_{10}, m'_6 and m_7 are merged into a match m''_6 and m_4 disappears in the resulting sequence S''. As a result, only m''_6 remains in the self-alignment of S''. In summary, we identify 3 duplication events that give rise to the matches in the given genomic sequence S.

2.4 Influence of Deletion and Inversion Events

Deletion events can affect the inference of duplications, so it is important to consider them simultaneously. In order to infer deletions, we use the following procedure. Assume an input sequence S has two segments ABC and $A'C'$ for some non-empty segments A, B, C, A', and C'. We may infer two duplication events that copy A and C respectively, or one duplication that copies ABC and one hypothetical deletion event that deletes B'. Since our goal is to find

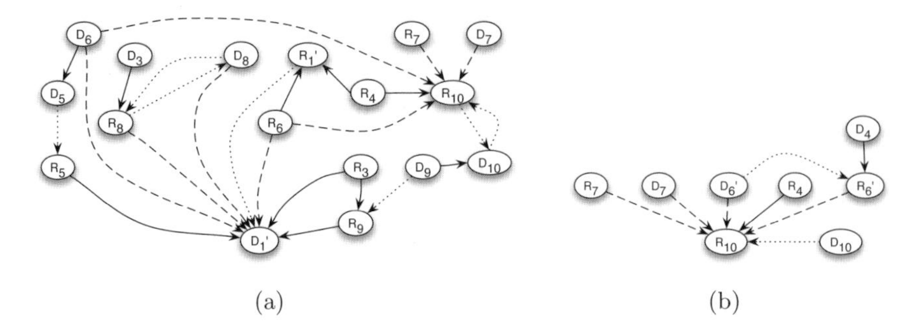

Fig. 3. The constraint graphs of matches in Fig. 1(b) and 1(c). Arcs of type 1, 2, and 3 are represented by dotted, solid, and dashed lines, respectively. In (a), matches m_1 and m_2 have been replaced with m_1' according to the procedure for intraposed duplications discussed in Section 2.2, node D_4 is omitted because it is identical to D_3. D_4 reappears in (b) after m_3 has been eliminated.

a parsimonious duplication history for the cluster, we infer a duplication event and a deletion event from segments ABC and $A'C'$ when B is relatively short compared to the length of A and C. In our implementation, we detect all possible deletion events by using k-d tree data structure before entering each loop of inferring duplication events.

In the case of inversions, if the inversion does not split any matches generated by duplication events (i.e., it contains the whole region of one or more other matches, or occurs in a region that does not have any matches), then it does not affect the inference of duplication events. Moreover, if the inversion occurs within a match, it can be detected. If it splits other matches by involving source or target regions of duplication events, two duplications are inferred rather than one, but the duplicated regions can be removed correctly.

3 Results

3.1 Human Gene Clusters

We first applied our method to 25 gene clusters in the human genome. For a genomic region containing each gene cluster, we constructed its self-alignment and identified matches using five different thresholds of similarity level: 98%, 93%, 89%, 85%, and 80%, in the same way as [19]. These five similarity levels correspond roughly to the sequence divergence between humans and great apes (GA), old-world monkeys (OWM), new-world monkeys (NWM), lemurs and galagos (LG), and dogs (DOG), respectively. With the matches at different similarity levels, we inferred duplication and deletion events that occurred in different periods in the human lineage. The inferred numbers of duplication and deletion events in these periods are summarized in Table 1.

The human leukocyte antigen (HLA) gene cluster is known to be involved in narcolepsy [23] and celiac disease [24]. In addition, HLA has been observed in

Table 1. Inferred numbers of large-scale duplications and deletions in human gene clusters following divergence from various mammalian clades. GA = great apes (at least 98%); OWM = old world monkeys (93%); NWM = new world monkeys (89%); LG = lemurs and galagos (85%); DOG = dogs (80%). Cluster locations are indicated as coordinates in the March 2006 human genome sequence assembly.

Name	Location	GA	OWM	NWM	LG	DOG
CYP4	chr1:47048227-47411959	1, 0	4, 1	4, 1	5, 1	10, 1
LCE	chr1:150776235-151067237	0, 0	0, 0	5, 1	6, 1	9, 2
CR, DAP3	chr1:153784948-154023311	0, 0	3, 2	12, 2	19, 5	21, 7
FC	chr1:159742726-159915333	1, 2	1, 2	3, 2	4, 2	4, 2
CR1	chr1:205701588-205958677	1, 0	7, 0	8, 3	8, 3	10, 3
CCDC, CFC1	chr2:130461934-131153411	3, 0	5, 0	7, 5	7, 5	10, 5
CXCL, IL8	chr4:74781081-75209572	0, 0	0, 0	0, 0	2, 1	17, 8
PCDH	chr5:140145736-140851366	0, 0	0, 0	0, 0	1, 0	36, 0
HLA	chr6:29786467-30568761	0, 0	2, 0	27, 3	38, 5	50, 6
HLA-D	chr6:33082752-33265289	0, 0	0, 0	0, 0	4, 3	6, 8
OR2	chr7:143005241-143760083	6, 1	8, 1	8, 1	10, 1	16, 1
AKR1C	chr10:4907977-5322660	0, 0	3, 2	4, 3	9, 4	28, 4
GAD2	chr10:26458036-27007198	0, 0	4, 0	6, 1	15, 2	17, 2
PNLIP	chr10:118205218-118387999	0, 0	0, 0	1, 0	2, 1	5, 2
OR5, HB, TRIM	chr11:4124149-6177952	6, 0	7, 0	9, 0	9, 0	24, 2
LST3, SLCO1B	chr12:20846959-21313050	0, 0	0, 0	0, 0	9, 2	21, 3
C14orf	chr14:23177922-23591420	1, 0	4, 0	5, 2	6, 2	8, 3
CYP1A1	chr15:71687352-74071019	2, 0	12, 2	21, 3	23, 4	25, 4
ACSM	chr16:20234773-20711192	2, 0	4, 2	4, 2	4, 2	5, 2
LGALS9, NOS2A	chr17:22979762-23370074	0, 0	2, 0	2, 2	2, 2	3, 2
OR, ZNF	chr19:8569586-9765797	4, 0	6, 0	14, 0	23, 0	33, 0
NPHS1, ZNF	chr19:40976726-43450858	0, 0	6, 1	12, 1	16, 2	23, 4
CYP2A	chr19:46016475-46404199	0, 0	5, 0	11, 3	14, 3	16, 3
DGCR6L, ZNF74	chr22:18594272-19312230	3, 0	4, 0	5, 1	5, 1	5, 1
SLC5A, YWHAH	chr22:30379202-31096691	0, 0	3, 1	4, 1	6, 1	7, 1

the association with prostate cancer [25] and breast cancer [26]. For the HLA gene cluster, the MCMC method of [19] estimated 15 duplications in the lineage between NWM and LG, but our method inferred only 11 duplications, 4 fewer events than the MCMC method (The numbers of the inferred events in the lineage are highlighted in bold in Figure 4(a) and Figure 4(b)).

The aldo-keto reductase (AKR) 1C gene cluster is involved in steroid hormone and nuclear receptors and associated with prostate disease, endometrial cancer, and mammary carcinoma [27]. For this gene cluster, Figure 4(c) and Figure 4(d) show the inference results; our method identified 3 duplications between GA and OWM while the MCMC inferred 5 duplications.

Another interesting observation is that several gene clusters were probably formed by recent gene duplications. For instance, we examined three Cytochrome P450 (CYP) gene clusters, which are associated with lung cancer [28] and esophageal cancer [29]. About 65% of the duplication events inferred for these

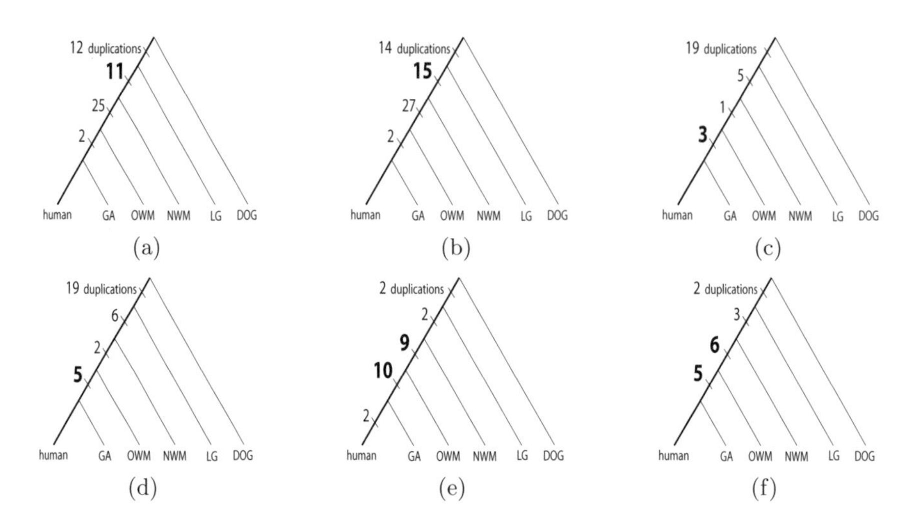

Fig. 4. Duplication events inferred (a) for the HLA gene cluster by the deterministic method, (b) for the HLA gene cluster by the MCMC method, (c) for the AKR1C gene cluster by the deterministic method, (d) for the AKR1C gene cluster by the MCMC method, and (e) for the CYP1A1 gene cluster and (f) the CYP2A gene cluster by the deterministic method

clusters occurred in the evolutionary period between the divergence of humans from new-world monkeys and from great apes. Duplication events inferred for CYP1A1 and CYP2A are mapped onto the phylogeny in Figure 4(e) and Figure 4(f), respectively.

3.2 Validation Test on Simulated Data

Starting from a 500-kb duplication-free sequence, we generated gene clusters by applying a series of duplication events based on the length and distance distributions for duplication and deletion events that we observed in the human genome. We generated 50 gene clusters formed from n duplications for each $n = 10, 20, \ldots, 100$.

On these clusters, our method outperformed the MCMC method reported in [19] in terms of both the total number of inferred duplication events and the number of true duplications detected correctly as indicated in Figure 5(a) and Figure 5(b). On average, our method estimated only 3% events more than true events. A duplication event is expressed as a 3-tuple consisting of a source interval, a target location, and an orientation. If these values for an inferred event exactly match one of the true events, the event is defined to be *correctly detected*, i.e. a true event. We count how many of the inferred events were correctly detected. The fraction of true events detected correctly by our method (91% on average) is much higher than the MCMC method (80% on average).

It is worth noting that duplication breakpoints can be reused in the simulation dataset, since it is generated according to the observed distributions (Figure 5(c))

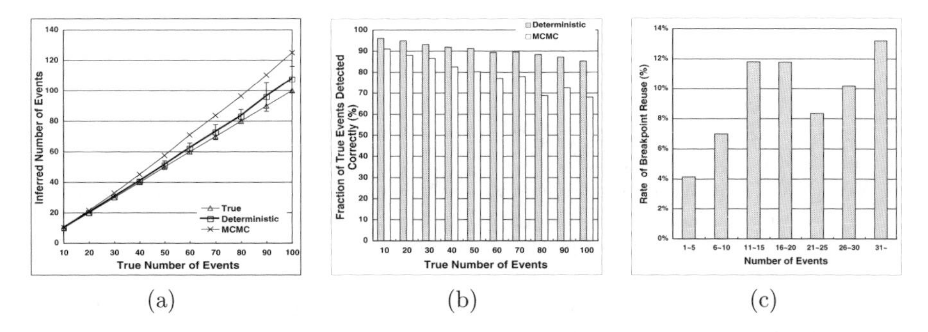

Fig. 5. (a) and (b) show the simulation results to evaluate detection of duplication; (a) numbers of reconstructed events and (b) fraction of true events detected correctly. (c) is the rate of breakpoint reuse by the inferred duplications in the human gene clusters.

without constraining the breakpoints to avoid reuse. However, the inferred events are still very close to true events.

4 Discussion and Conclusion

We have developed a combinatorial algorithm that reconstructs recent duplication and deletion operations in a gene cluster from a single present-day sequence. We have compared our combinatorial method with a probabilistic method for the same problem [19], and shown that the relative performance of the combinatorial algorithm is quite good. In addition, a simulation study has validated that our method is very effective for identifying the duplication history.

We are exploring several extensions of this method. The results should be cross-checked against other primate genomes; another, more ambitious goal is to identify the orthology relationship among genes in each gene cluster in the species. Gene conversion should be also considered for more accurate orthology detection rather than using only the overall similarity in the alignment data.

Our goal is to find methods for analyzing large-scale evolutionary operations that integrate well with the specific needs of our current approach for producing whole-genome alignments [12]. We are still in an exploratory stage where the aim is to investigate as many promising avenues as possible. This paper describes a new method whose accuracy, computational efficiency, and focus on an individual species make it a particularly strong contender.

Acknowledgement. GS and WM are supported by Grant HG02238 from NHGRI and LX Zhang is supported by NUS ARF grant R-146-000-109-112. The authors thank Yu Zhang, Cathy Riemer, and Bob Harris for helpful discussions.

References

1. Ohno, S.: Evolution by Gene Duplication. Springer, Berlin (1970)
2. Lynch, M., Conery, J.S.: The evolutionary fate and consequences of duplicate genes. Science 290, 1151–1155 (2000)

3. Force, A., et al.: Preservation of duplicate genes by complementary, degenerative mutations. Genetics 151, 1531–1545 (1999)
4. Lupski, J.R.: Genomic rearrangements and sporadic disease. Nat. Genet. 39(suppl. 7), 43–47 (2007)
5. Lander, E.S., et al.: Initial sequencing and analysis of the human genome. Nature 409(6822), 860–921 (2001)
6. Wong, K.K., et al.: A comprehensive analysis of common copy-number variations in the human genome. Am. J. Hum. Genet. 80(1), 91–104 (2007)
7. Green, E.D.: Strategies for the systematic sequencing of complex genomes. Nat. Rev. Genet. 2(8), 573 (2001)
8. Blanchette, M., et al.: Aligning multiple genomic sequences with the threaded blockset aligner. Genome Res. 14(4), 708–715 (2004)
9. Raphael, B., et al.: A novel method for multiple alignment of sequences with repeated and shuffled elements. Genome Res. 14(11), 2336 (2004)
10. Margulies, E., et al.: Analyses of deep mammalian sequence alignments and constraint predictions for 1% of the human genome. Genome Res. 17(6), 760–764 (2007)
11. Hou, M.: unpublished data (2007)
12. Miller, W., et al.: 28-way vertebrate alignment and conservation track in the UCSC genome browser. Genome Res. 17, 1797–1808 (2007)
13. Elemento, et al.: Reconstructing the duplication history of tandemly repeated genes. Mol. Biol. Evol. 19(3), 278 (2002)
14. Zhang, L., et al.: Greedy method for inferring tandem duplication history. Bioinformatics 19, 1497–1504 (2003)
15. Sammeth, M., Stoye, J.: Comparing tandem repeats with duplications and excisions of variable degree. TCBB 3, 395–407 (2006)
16. Bertrand, D., Lajoie, M., El-Mabrouk, N., Gascuel, O.: Evolution of tandemly repeated sequences through duplication and Inversion. In: Bourque, G., El-Mabrouk, N. (eds.) RECOMB-CG 2006. LNCS (LNBI), vol. 4205, pp. 129–140. Springer, Heidelberg (2006)
17. Ma, J., et al.: The infinite sites model of genome evolution. PNAS 105(38), 14254–14261 (2008)
18. Jiang, Z., et al.: Ancestral reconstruction of segmental duplications reveals punctuated cores of human genome evolution. Nat. Genet. 39(11), 1361–1368 (2007)
19. Zhang, Y., et al.: Reconstructing the evolutionary history of complex human gene clusters. In: Vingron, M., Wong, L. (eds.) RECOMB 2008. LNCS (LNBI), vol. 4955, pp. 29–49. Springer, Heidelberg (2008)
20. Zhang, Y., et al.: Simultaneous history reconstruction for complex gene clusters in multiple species. In: Pacific Symposium on Biocomputing 2009, pp. 162–173 (2009)
21. Schwartz, S., et al.: Human-mouse alignments with BLASTZ. Genome Res. 13(1), 103–107 (2003)
22. Nadeau, J.H., Taylor, B.A.: Lengths of chromosomal segments conserved since divergence of man and mouse. Proc. Natl. Acad. Sci. USA 81(3), 814–818 (1984)
23. Nakayama, J., et al.: Linkage of human narcolepsy with HLA association to chromosome 4p13-q21. Genomics 65, 84–86 (2000)
24. Sollid, L., et al.: Evidence for a primary association of celiac disease to a particular HLA-DQ α/β heterodimer. J. Exp. Med. 169, 345–350 (2000)
25. Haque, A., et al.: HLA class II protein expression in prostate cancer cells. Journal of Immunology 178, 48.22 (2007)
26. Chaudhuri, S., et al.: Genetic susceptibility to breast cancer: HLA DQB*03032 and HLA DRB1*11 may represent protective alleles. PNAS 97, 11451–11454 (2000)

27. Penning, T., et al.: Aldo-keto reductase (AKR) 1C3: Role in prostate disease and the development of specific inhibitors. Mol. Cell. Endocrinol. 248, 182–191 (2006)
28. Crofts, F., et al.: Functional significance of different human CYP1A1 genotypes. Carcinogenesis 15, 2961–2963 (1994)
29. Sato, M., et al.: Genetic polymorphism of drug-metabolizing enzymes and susceptibility to oral cancer. Carcinogenesis 20, 1927–1931 (1999)

Appendix

A Types of Duplication

Let

$$S = s_1 s_2 \dots s_n$$

be a genomic sequence of length n, where $s_i \in \{A, C, G, T\}$. For any a, b and c such that $1 \leq a \leq b \leq n$ and $1 \leq c \leq n$, a forward duplication that copies the segment $s_a s_{a+1} \dots s_b$ and inserts it between s_{c-1} and s_c is written $[a, b] + c$. It transforms S into the following sequence

$$S' = s_1 s_2 \dots s_{c-1} \; \underline{s_a s_{a+1} \dots s_b} \; s_c s_{c+1} \dots s_n .$$

If $c = b + 1$, the forward duplication $[a, b] + c$ forms a tandem duplication in the resulting sequence

$$s_1 s_2 \dots s_{a-1} \; \underline{s_a s_{a+1} \dots s_b} \; \underline{s_a s_{a+1} \dots s_b} \; s_{b+1} \dots s_n .$$

Tandem duplications are observed in many important gene clusters in eukaryotic genomes. If $a < c < b$, then $[a, b] + c$ copies within itself and produces a segment $AABB$ where A and B are non-empty segments in the resulting sequence

$$s_1 s_2 \dots s_{a-1} \; \underline{s_a \dots s_{c-1}} \; \underline{s_a \dots s_{c-1}} \; \underline{s_c \dots s_b} \; \underline{s_c \dots s_b} \; s_{b+1} \dots s_n .$$

A backward duplication inserting the reverse-complement sequence $-s_b - s_{b-1} \dots - s_a$ between s_{c-1} and s_c is written $[a, b] - c$. If $c = b + 1$, the backward duplication $[a, b] - c$ produces a palindrome. Let $\overline{A}$ denote the reverse complement of sequence A. If $a < c < b$, $[a, b] - c$ produces a segment $A\overline{B}\overline{A}B$.

B Overlap Relationship

To explain the algorithm, we consider the overlap relationship between two duplication events. Let A be a region of a duplication event and let B be a region of another duplication event in S. A base of location i in S is denoted s_i.

1. If there exist i, j, k, l $(1 \leq i < k \leq j < l \leq n)$ such that

$$A = s_i s_{i+1} \dots s_j , \; B = s_k s_{k+1} s_{k+2} \dots s_l$$

   or vice versa, we say that A *overlaps* B.

2. If there exist i, j, k, l $(i < k < j < l)$ such that

$$B = B_1 B_2 = \underline{s_i s_{i+1} \dots s_k} \ \underline{s_{j+1} s_{j+2} \dots s_l} \ ,$$
$$A = s_{k+1} s_{k+2} \dots s_j,$$

   A is said to be *inserted* into B.

3. If there exist i, j, k, l $(i \leq k < j \leq l)$ such that

$$A = s_i s_{i+1} \dots s_k \ \underline{s_{k+1} s_{k+2} \dots s_j} \ s_{j+1} s_{j+2} \dots s_l$$
$$= s_i s_{i+1} \dots s_k B_{j+1} s_{j+2} \dots s_l,$$

   we say that A *contains* B.

4. A region is said to be *disjoint* from other regions if it does not overlap with any other regions, is not inserted into and does not contain any other regions.

C Proof of Theorem 2

We prove it by induction on the number n duplications. The results are trivial when $n = 1$ and 2. Assume it is true for $n \leq k - 1$ and S evolves from a duplication-free sequence T by k duplications $O_1, O_2, ..., O_k$. Let D be the region first selected by the algorithm such that D forms match $M = (R, D)$ or $M = (D, R)$. Then, D does not overlap with other matches and is not contained in any other matches since the out-degree of D is 0 in the constraint graph. Let M be generated by duplication O_i. We consider the following cases.

Case 1. If $i = k$, it means that M is generated by the latest duplication. Assume the resulting sequence is S'' when D is removed from S. If D is a target region of O_k, T transforms into S'' by $O_1, O_2, ..., O_{k-1}$. By induction, the algorithm will reduce S'' to a duplication-free sequence T'' such that T'' transforms into S'' by $k - 1$ duplications. Therefore, T'' transforms into S by k duplications.

If D is a source region of O_k, R is a target region. Since D does not overlap with any other matches and D is not contained in other matches, D cannot be involved in any other duplications. Thus, T includes D, i.e. D is in the duplication-free sequence. Let S' be the resulting sequence after removing D from S and let T' be the resulting sequence after removing D from T and inserting R in the corresponding position of S. By assumption, T' transforms into S' by $O_1, O_2, ..., O_{k-1}$. By induction on S', the algorithm outputs a duplication-free sequence T'' such that it transforms into S' by $k - 1$ duplications. Since S' transforms into S by duplication that creates D, the removal of D guarantees to find a solution of the same number of true events.

Case 2. If $i < k$, we consider the following sub-cases.

Sub-case 2.1. Suppose both of D and R in M have out-degree 0. If D is removed by the algorithm, all the regions involved in duplications $o_j, i < j \leq k$ do not overlap D. Thus, $o_1, o_2, ..., o_{i-1}, o_{i+1}, ..., o_k, o_i$ also transform T into S. This reduces to Case 1.

Sub-case 2.2. If D is a target region of O_i and its out-degree is 0 and R does not have out-degree 0, D is removed by the algorithm. Then, since D is a target region, the resulting sequence S' after the removal of D in S can be generated from T by $k-1$ duplications $O_1, O_2, ..., O_{i-1}, O_{i+1}, ..., O_k$. Thus, by induction, algorithm reduces S' to a duplication-free sequence T''. Obviously, T'' evolves into S in k duplications.

Sub-case 2.3. If D is a source region of O_i and its out-degree is 0 and R does not have out-degree 0, D is removed by the algorithm. Since D is not contained in other match regions, D is a subsequence of the original duplication-free sequence T. In this case, since the breakpoint uniqueness assumption holds, D is not inserted in any match region, and hence it must be in T. Let T' be the resulting sequence after removal of D and insertion of R in T. Then, T' evolves into S by $k-1$ duplications. By induction, the algorithm identifies a duplication-free sequence T'' that evolves into S by $k-1$ operations. By modifying T'' by inserting R and removing D, we derive a duplication-free sequence that evolves into S in k duplications.

D Figures

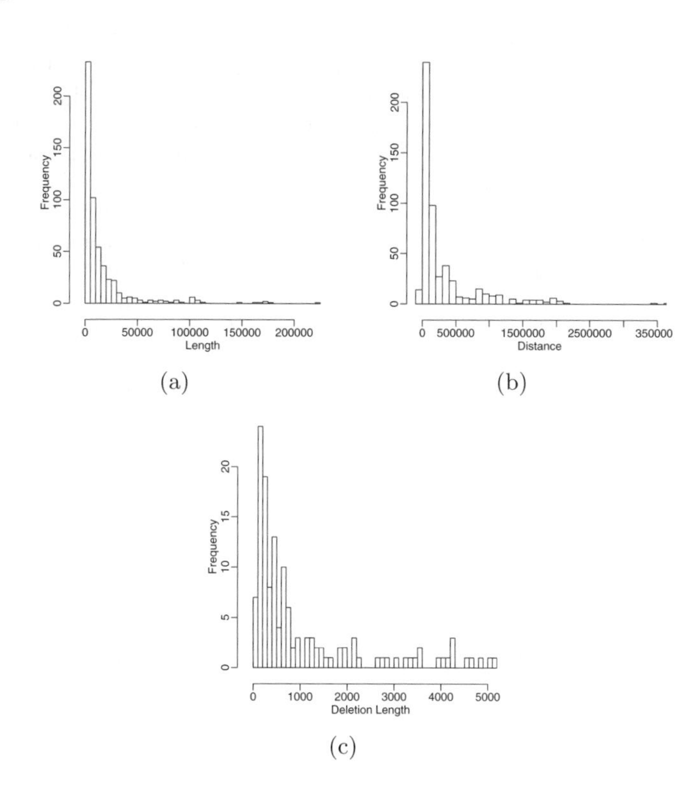

Fig. A-1. Distribution of (a) duplication length, (b) distance between the source and target regions for duplications, and (c) deletion length

Pseudo Boolean Programming for Partially Ordered Genomes

Sébastien Angibaud[1], Guillaume Fertin[1], Annelyse Thévenin[2], and Stéphane Vialette[3]

[1] Laboratoire d'Informatique de Nantes-Atlantique (LINA), UMR CNRS 6241
Université de Nantes, 2 rue de la Houssinière, 44322 Nantes Cedex 3 - France
[2] Laboratoire de Recherche en Informatique (LRI), UMR CNRS 8623
Université Paris-Sud, 91405 Orsay - France
[3] IGM-LabInfo, UMR CNRS 8049, Université Paris-Est,
5 Bd Descartes 77454 Marne-la-Vallée, France
{Sebastien.Angibaud,Guillaume.Fertin}@univ-nantes.fr, thevenin@lri.fr, vialette@univ-mlv.fr

Abstract. Comparing genomes of different species is a crucial problem in comparative genomics. Different measures have been proposed to compare two genomes: number of common intervals, number of adjacencies, number of reversals, etc. These measures are classically used between two totally ordered genomes. However, genetic mapping techniques often give rise to different maps with some unordered genes. Starting from a partial order between genes of a genome, one method to find a total order consists in optimizing a given measure between a linear extension of this partial order and a given total order of a close and well-known genome. However, for most common measures, the problem turns out to be **NP**-hard. In this paper, we propose a $(0, 1)$-linear programming approach to compute a linear extension of one genome that maximizes the number of common intervals (resp. the number of adjacencies) between this linear extension and a given total order. Next, we propose an algorithm to find linear extensions of two partial orders that maximize the number of adjacencies.

1 Introduction

Genetic mapping techniques often give rise to different maps with some unordered genes. In that case, maps are combined in the form of a partially ordered sequence of genes, and thus a genome is modeled by a *poset* (i.e., a partially ordered set), as was done in [10]. In this model, any linear extension of a poset represents a possible total order of the genome. In order to find a total order from a partial order, Sankoff et al. [10] suggested the following method: confront the partial order with a given total order of a close and well-known genome. More precisely, this method asks for a linear extension of the partially ordered genome that optimizes a (dis)similarity measure between this total order and a given totally ordered genome. Several measures can be used: number of common

F.D. Ciccarelli and I. Miklós (Eds.): RECOMB-CG 2009, LNBI 5817, pp. 126–137, 2009.

intervals [8], of adjacencies [1], of breakpoints [9], of conserved intervals [3], etc. Computing these measures between two total orders, whenever no gene is duplicated, is a polynomial-time solvable problem. Unfortunately, concerning the number of adjacencies, the number of common intervals and the reversal distance, finding a linear extension that optimizes one of these measures between this extension and a given total order is **NP**-hard [4,6]. In this paper, we first present an approach to compute a linear extension of a partially ordered genome with respect to a given totally ordered genome without duplicated genes. Our method lies on a transformation of the initial problem into a $(0, 1)$-linear program (i.e., a linear program with boolean variables) [2,1]. We focus here on two similarity measures: the *number of common intervals* and the *number of adjacencies*. We also extend our approach and consider the problem of comparing *two* partially ordered genomes with respect to the number of adjacencies. After presenting our three algorithms, we evaluate our method by using the same simulated data as in [4].

This paper is organized as follows. In Section 2, we present some preliminaries and definitions. We focus in Section 3 on the problem of finding a total order of a partially ordered genome that maximizes either the number of common intervals or the number of adjacencies with a given totally ordered genome. Next, we extend the case of adjacencies to find two totally ordered genomes from two partially ordered genomes. For each problem, we give its formulation in terms of a $(0, 1)$-linear program, together with some reduction rules that ail at speeding-up the process. Section 4 is devoted to experimental results on simulated data.

2 Preliminaries

In the literature, a (totally ordered) duplication-free genome having n oriented genes is usually represented by a signed permutation, one where each element in the set $\{1, 2, \ldots, n-1, n\}$ has either a sign, + or −. Let T_1 and T_2 be two duplication-free genomes of size n. Wlog, we may assume that T_2 is the identity positive permutation, i.e., $T_2 = +1\ +2\ \ldots\ +n$. An interval of T_1 is a set of m consecutive genes $\{\mathtt{g}_1, \mathtt{g}_2, \ldots, \mathtt{g}_m\}$ in T_1. A common interval between T_1 and T_2 is an interval which occurs in T_1 and in T_2. Observe that the notion of common interval does not take into account the sign of the genes. We say that there is a *breakpoint* after gene $T_1[i]$, $1 \leq i \leq n-1$, in T_1 if neither $T_1[i]$ and $T_1[i+1]$ nor $-T_1[i+1]$ and $-T_1[i]$ are consecutive genes in T_2, otherwise we say that there is an *adjacency* after gene $T_1[i]$. Notice, that in order to take into account common intervals and adjacencies that may occur at the extremities of a genome, we artificially add to a genome with n genes the gene $+0$ (resp. $+(n+1)$) to its left (resp. right). We now introduce the required material for partially ordered genomes. A *partial order* on a set P is a binary relation $\preceq$ such that, for x, y and z in P, (i) $x \preceq x$, (ii) $x \preceq y$ and $y \preceq x$ imply $x = y$, and (iii) $x \preceq y$ and $y \preceq z$ imply $x \preceq z$. These three relations are referred to as *reflexivity*, *antisymmetry*, and *transitivity*, respectively. A set P equipped with a partial order relation is said to be a *partially ordered set* (also called a *poset*).

When it is necessary to specify the partial order relation, we write $(P, \preceq)$. A partial order relation $\preceq$ on P gives rise to a relation $\prec$ of strict inequality: $x \prec y$ in P iff $x \preceq y$ and $x \neq y$. Strict partial orders differ from partial orders only in whether each element is required to be unrelated, or required to be related, to itself. A totally ordered set (or linearly ordered set) is a poset P which has the property that every two elements of P are comparable (i.e., for all $x, y \in P$ either $x \preceq y$ or $y \preceq x$). A poset Q is called an *extension* of a poset P if the elements of P and Q are the same, and the set of relations of P is a subset of the set of relations of Q (i.e., for all $x, y \in P$, if $x \prec y$ in P, then $x \prec y$ in Q, but not necessarily conversely). Q is called a *linear extension* of P if Q is an extension of P and also a linear order. In our context, a totally (resp. partially) ordered genome is thus a set P that contains exactly one occurrence of i or $-i$, $0 \leq i \leq n+1$, equipped with some total (resp. partial) order. For the sake of brevity, the totally ordered genome $+0\ +1\ +2\ \ldots\ +n+(n+1)$ is abbreviated as Id. Let P_1 and P_2 be two partially ordered genomes, and let $x \in \{1, 2\}$. The sign of a gene g in P_x is written $s_x(\mathrm{g})$. We write $\mathrm{g}_1 \prec_x \mathrm{g}_2$ if gene g_1 precedes gene g_2 in P_x. The number of genes g_1 such that $\mathrm{g}_1 \prec_x \mathrm{g}_2$ is denoted by $\mathrm{prec}_x(\mathrm{g}_2)$. We write $\mathrm{g}_2 \succ_x \mathrm{g}_1$ if gene g_2 follows g_1 in P_x. The number of genes g_2 such that $\mathrm{g}_1 \prec_x \mathrm{g}_2$ is denoted by $\mathrm{succ}_x(\mathrm{g}_1)$. In P_x, two genes g_1 and g_2 are said to be *incomparable* (written $\mathrm{g}_1 \|_x \mathrm{g}_2$) if neither $\mathrm{g}_1 \prec_x \mathrm{g}_2$ nor $\mathrm{g}_2 \prec_x \mathrm{g}_1$. The *width* of a partially ordered genome is the size of a largest subset of incomparable genes. Two genes g_1 and g_2 are said to be *adjoining* in P_x if there does not exist g_3 such that $\mathrm{g}_1 \prec_x \mathrm{g}_3 \prec_x \mathrm{g}_2$ or $\mathrm{g}_2 \prec_x \mathrm{g}_3 \prec_x \mathrm{g}_1$ or $\mathrm{g}_1 \|_x \mathrm{g}_3$ or $\mathrm{g}_2 \|_x \mathrm{g}_3$. Let T_x be a linear extension of P_x. The set of allowed positions of a gene g in T_x is written by $POS_x(\mathrm{g})$; one can easily check that $POS_x(\mathrm{g}) = \{\mathrm{prec}_x(\mathrm{g})+1, \ldots, \mathrm{succ}_x(\mathrm{g})-1\}$. The position of g in T_x is written $T_x(\mathrm{g})$ (by definition, $T_x(\mathrm{g}) \in POS_x(\mathrm{g})$). We say that a gene g_1 is *i-nail$_x$* if $POS_x(\mathrm{g}_1) = \{i\}$ (i.e., there are $i-1$ genes g_2 such that $\mathrm{g}_2 \prec_x \mathrm{g}_1$ and no gene is incomparable with g_1 in P_x). In this paper, we are interested in three combinatorial problems. The first two problems are concerned with confronting a partially ordered genome with a reference totally ordered genome: given a partially ordered genome P_1 and a reference totally ordered genome T_2, problem `MCIL-1PO` (resp. `MAL-1PO`) asks for a linear extension T_1 of P_1 that yields a maximum number of common intervals (resp. maximum number of adjacencies) between any linear extension of P_1 and T_2. Observe that there is no loss of generality in assuming here that $T_2 = \mathrm{Id}$. The third problem is concerned with confronting two partially ordered genomes: given two partially ordered genomes P_1 and P_2, problem `MAL-2PO` asks for a linear extension T_1 of P_1 and a linear extension T_2 of P_2 s.t. the number of adjacencies between T_1 and T_2 is maximized among all linear extensions of P_1 and P_2. Note that problems `MCIL-1PO`, `MAL-1PO` and `MAL-2PO` have been proved to be **NP**-hard in [4].

3 An Exact (0, 1)-Linear Programming Approach

We present in this section an exact generic approach. The main idea is to transform our problems into (0, 1)-linear programs [7] and use a powerful solver to obtain optimal solutions. All computations conducted here use the `minisat+` solver [5]. In order to solve problems `MCIL-1PO`, `MAL-1PO` and `MAL-2PO`, we present the $(0,1)$-linear programs called `CI-1PO`, `Adj-1PO` and `Adj-2PO`, respectively. Programs `CI-1PO` and `Adj-1PO` take as input a partial order P_1 while `Adj-2PO` takes two partial orders P_1 and P_2. We first present the common part of these three programs and next give a complete description of each program. Then, we present some data reduction rules for reducing the size of the programs.

The common part of the three programs. We present here the common part of programs `CI-1PO`, `Adj-1PO` and `Adj-2PO`. Let P_1 and P_2 be two partially ordered genomes over $[0 : n+1]$. Fix x to be 1 if we consider either `CI-1PO` or `Adj-1PO`, or 1 or 2 if we consider `Adj-2PO`. For each problem, we seek for a linear extension T_x of P_x that maximizes the number of common intervals or the number of adjacencies. Our programs are divided into two parts: (i) definition of a linear extension T_x, and (ii) maximization of the measure. The first part is common to the three programs whereas the second part is problem dependent and needs specific variables and constraints. To define a linear extension of a partially ordered genome, we use the same set $\mathcal{A}^x = \{a^x_{\mathrm{g},i} : \mathrm{g} \in [0 : n+1] \text{ and } i \in [0 : n+1]\}$ of boolean variables. For each $a^x_{\mathrm{g},i} \in \mathcal{A}^x$, $a^x_{\mathrm{g},i} = 1$ iff $T_x(\mathrm{g}) = i$, i.e., g is at position i in the resulting linear extension. To this aim, we define the three constraints presented in Figure 1:

- **(C.a)** ensures that each gene is assigned to exactly one position in T_x,
- **(C.b)** ensures that no two genes are assigned to the same position in T_x,
- **(C.c)** checks genes order in T_x. Let g_1 and g_2 be two genes of P_x such that $\mathrm{g}_1 \prec_x \mathrm{g}_2$. We must certainly check that g_1 precedes g_2 in T_x. Thus, for all $0 \leq j < i \leq n+1$ we impose that $a^x_{\mathrm{g}_1,i} = 0$ or $a^x_{\mathrm{g}_2,j} = 0$.

Now, we present specific variables and constraints for each problem.

Common constraints of programs `CI-1PO`, `Adj-1PO` and `Adj-2PO`.

Constraints:

`C.a` $\forall x \in \{1,2\}, \forall\, 0 \leq \mathrm{g} \leq n+1, \sum_{0 \leq i \leq n+1} a^x_{\mathrm{g},i} = 1$

`C.b` $\forall x \in \{1,2\}, \forall\, 0 \leq i \leq n+1, \sum_{0 \leq \mathrm{g} \leq n+1} a^x_{\mathrm{g},i} = 1$

`C.c` $\forall x \in \{1,2\}, \forall\, 0 \leq \mathrm{g}_1 \leq n+1,\ 0 \leq \mathrm{g}_2 \leq n+1,\ \mathrm{g}_1 \prec_x \mathrm{g}_2,\ 0 < j \leq i \leq n+1,$
$x a^x_{\mathrm{g}_1,i} + a^x_{\mathrm{g}_2,j} \leq 1$

Fig. 1. Common constraints for `CI-1PO`, `Adj-1PO` and `Adj-2PO`. A gene g is at position i in T_x iff $a^x_{\mathrm{g},i}$. Constraints `C.a` to `C.c` ensure that T_x is a valid linear extension of P_x, $x \in \{1,2\}$.

Confronting a partially ordered genome with a totally ordered genome: maximizing the number of common intervals. We give here a (0, 1)-linear program called CI-1PO that computes a linear extension T_1 of a partially ordered genome P_1 that maximizes the number of common intervals between T_1 and Id. Recall that common intervals do not take into account the sign of the genes. Program CI-1PO is given in Figure 2.

We describe now the variables used in CI-1PO. Let $\mathrm{g} \in [0 : n+1]$ be a gene, $i \in [0 : n+1]$ be a position and $t \in [0 : n+1]$. After having defined the total order T_1 with the set $\mathcal{A}^1$, we define two sets of boolean variables: $\mathcal{B} = \{b_{\mathrm{g},i,t} : \mathrm{g} \in [0 : n+1],\ i \in [0 : n+1] \text{ and } t \in [0 : n+1]\}$ and $\mathcal{C} = \{c_{\mathrm{g},i,t} : \mathrm{g} \in [0 : n+1],\ i \in [0 : n+1] \text{ and } t \in [0 : n+1]\}$. For each $b_{\mathrm{g},i,t} \in \mathcal{B}$, we have $b_{\mathrm{g},i,t} = 1$ iff $T_1(\mathrm{g}) \in [i : i+t]$, i.e., g is located between positions i and $i+t$ in T_1. The set $\mathcal{C}$ corresponds to the set of all possible common intervals. For each $c_{\mathrm{g},i,t} \in \mathcal{C}$, we have $c_{\mathrm{g},i,t} = 1$ iff, for all $k \in [0 : t]$, we have $T_1(\mathrm{g}+k) \in [i : i+t]$, i.e., the set of genes $\mathrm{g}+k$ ($k \in [0 : t]$) is an interval of T_1. In this case, we certainly have a common interval between T_1 and the identity. Therefore, the objective function is the sum of all variables $c_{\mathrm{g},i,t} \in \mathcal{C}$.

Constraints **(C.01)** to **(C.05)** are needed to ensure correctness of CI-1PO. First, we obviously need constraints **(C.01)** to **(C.03)**, that correspond to the common constraints **(C.a)** to **(C.c)**, with $x = 1$. Then, we add constraint **(C.04)** to check variables $b_{\mathrm{g},i,t}$: $b_{\mathrm{g},i,t} = 1$ iff there exists j in $[0 : t]$ such that $a_{\mathrm{g},i+j} = 1$. Finally, we add constraint **(C.05)** to check variables $c_{\mathrm{g},i,t}$: $c_{\mathrm{g},i,t} = 1$ iff, for all k in $[0 : t]$, we have $b_{\mathrm{g}+k,i,t} = 1$.

Program CI-1PO

Objective: Maximize $\sum_{0 \leq \mathrm{g} \leq n+1} \sum_{0 \leq i \leq n+1} \sum_{0 \leq t \leq n+1-i} c_{\mathrm{g},i,t}$

Constraints:

C.01 $\forall\, 0 \leq \mathrm{g} \leq n+1, \quad \sum_{0 \leq i \leq n+1} a^1_{\mathrm{g},i} = 1$

C.02 $\forall\, 0 \leq i \leq n+1, \quad \sum_{0 \leq \mathrm{g} \leq n+1} a^1_{\mathrm{g},i} = 1$

C.03 $\forall\, 0 \leq \mathrm{g}_1 \leq n+1, 0 \leq \mathrm{g}_2 \leq n+1, \mathrm{g}_1 \prec_1 \mathrm{g}_2, 0 < j \leq i \leq n+1, a^1_{\mathrm{g}_1,i} + a^1_{\mathrm{g}_2,j} < 1$

C.04 $\forall\, 0 \leq \mathrm{g} \leq n+1,\ 0 \leq i \leq n+1,\ 0 \leq t \leq n+1-i,\ \sum_{0 \leq j \leq t} a^1_{\mathrm{g},i+j} - b_{\mathrm{g},i,t} = 0$

C.05 $\forall\, 0 \leq \mathrm{g} \leq n+1,\ 0 \leq i \leq n+1,\ 0 \leq t \leq n+1-i,$

$$\sum_{0 \leq k \leq t} b_{\mathrm{g}+k,i,t} - c_{\mathrm{g},i,t} < t+1 \text{ and } \sum_{0 \leq k \leq t} b_{\mathrm{g}+k,i,t} - t.c_{\mathrm{g},i,t} \geq 0$$

Fig. 2. Program CI-1PO computes a linear extension L of a partial order maximizing the maximum number of common intervals between L and the identity

Confronting a partially ordered genome with a totally ordered genome: maximizing the number of adjacencies. We now give a (0, 1)-linear program (`Adj-1PO`) that computes a linear extension T_1 of a partial order P_1 that maximizes the number of adjacencies between T_1 and Id. Oppositely to common intervals, the sign of genes is now relevant. Program `Adj-1PO` is presented in Figure 3.

Let $\mathtt{g} \in [0 : n+1]$ be a gene, and $i \in [0 : n+1]$ be a position. After having defined the total order T_1 with the set of variables $\mathcal{A}^1$, we define the set $\mathcal{D} = \{d_{\mathtt{g},i} \ : \ \mathtt{g} \in [0 : n+1[\text{ and } i \in [0 : n+1]\}$ of boolean variables that corresponds to the set of possible adjacencies. For each $d_{\mathtt{g},i} \in \mathcal{D}$ ($\mathtt{g} \neq n+1$), we have $d_{\mathtt{g},i} = 1$ iff (i) $T_1(\mathtt{g}) = i$ and (ii) $\mathtt{g}$ and $(\mathtt{g}+1)$ create an adjacency. Therefore, we define the objective function as the sum of all variables $d_{\mathtt{g},i} \in \mathcal{D}$.

We now define the constraints **(C'.01)** to **(C'.06)** to ensure the validity of program `Adj-1PO`. We first add constraints **(C'.01)** to **(C'.03)**, that correspond

Program `Adj-1PO`

Objective: `Maximize` $\sum_{0 \le \mathtt{g} \le n} \sum_{0 \le i \le n+1} d_{\mathtt{g},i}$

Constraints:

`C'.01` $\forall\, 0 \le \mathtt{g} \le n+1, \ \sum_{0 \le i \le n+1} a^1_{\mathtt{g},i} = 1$

`C'.02` $\forall\, 0 \le i \le n+1, \ \sum_{0 \le \mathtt{g} \le n+1} a^1_{\mathtt{g},i} = 1$

`C'.03` $\forall\, 0 \le \mathtt{g}_1 \le n+1,\ 0 \le \mathtt{g}_2 \le n+1,\ \mathtt{g}_1 \prec_0 \mathtt{g}_2,\ 0 < j \le i \le n+1,$

$a^1_{\mathtt{g}_1,i} + a^1_{\mathtt{g}_2,j} \le 1$

`C'.04` $\forall\, 0 \le \mathtt{g} \le n, 1 \le i \le n,$

if $s_1(\mathtt{g}) =$ "+" and $s_1(\mathtt{g}+1) =$ "+" then $a^1_{\mathtt{g},i} + a^1_{\mathtt{g}+1,i+1} - d_{\mathtt{g},i} \le 1$ and $3d_{\mathtt{g},i} - a^1_{\mathtt{g},i} - a^1_{\mathtt{g}+1,i+1} \le 1$

if $s_1(\mathtt{g}) =$ "−" and $s_1(\mathtt{g}+1) =$ "−" then $a^1_{\mathtt{g},i} + a^1_{\mathtt{g}+1,i-1} - d_{\mathtt{g},i} \le 1$ and $3d_{\mathtt{g},i} - a^1_{\mathtt{g},i} - a^1_{\mathtt{g}+1,i+1} \le 1$

else $d_{\mathtt{g},i} = 0$

`C'.05` $\forall\, 0 \le \mathtt{g} \le n,$

if $s_1(\mathtt{g}) =$ "+" and $s_1(\mathtt{g}+1) =$ "+" then $a^1_{\mathtt{g},0} + a^1_{\mathtt{g}+1,1} - d_{\mathtt{g},0} \le 1$ and $3d_{\mathtt{g},0} - a^1_{\mathtt{g},0} - a^1_{\mathtt{g}+1,1} \le 1$

else $d_{\mathtt{g},0} = 0$

`C'.06` $\forall\, 0 \le \mathtt{g} \le n,$

if $s_1(\mathtt{g}) =$ "−" and $s_1(\mathtt{g}+1) =$ "−" then $a^1_{\mathtt{g},n+1} + a^1_{\mathtt{g}+1,n} - d_{\mathtt{g},n+1} \le 1$ and $3d_{\mathtt{g},n+1} - a^1_{\mathtt{g},n+1} - a^1_{\mathtt{g}+1,n} \le 1$

else $d_{\mathtt{g},n+1} = 0$

Fig. 3. Program `Adj-1PO` computes a linear extension of a partially ordered genome that maximizes the number of adjacencies between this linear extension and the identity

to the common constraints **(C.a)** to **(C.c)** with $x = 1$, to obtain a valid linear extension T_1. Next, we add constraints **(C'.04)** to connect the assignment of variables in $\mathcal{D}$ to the assignment of variables in $\mathcal{A}^1$. To this aim, for each $d_{\mathtt{g},i} \in \mathcal{D}$, $\mathtt{g} \neq n+1$, $i \notin \{0, n+1\}$, we have $d_{\mathtt{g},i} = 1$ iff the two following conditions hold: (i) $a^1_{\mathtt{g},i} = 1$ (i.e., $\mathtt{g}$ is at the position i in T_1) and (ii) $a^1_{\mathtt{g}+1,i+1} = 1$ (case 1) or $a^1_{\mathtt{g}+1,i-1} = 1$ (case 2) (depending on the sign of $\mathtt{g}$ and $\mathtt{g}+1$). Finally, constraints **(C'.05)** and **(C'.06)** check variables $d_{\mathtt{g},0}$ and $d_{\mathtt{g},n+1}$ that correspond to the two possible adjacencies located at the extremities. These two constraints are defined as constraints **(C'.04)**, **(C'.05)** (case 1 only) and **(C'.06)** (case 2 only).

Confronting two partially ordered genomes: maximizing the number adjacencies. The (0, 1)-linear program `Adj-2PO` proposed here computes two linear extensions T_1 and T_2 of two partially ordered genomes P_1 and P_2 such that the number of adjacencies between T_1 and T_2 is maximized. Recall that the sign of the genes is relevant here. Oppositely to `MAL-1PO` where one genome is totally ordered, we do not know the relation between two genes that will eventually create an adjacency. Program `Adj-2PO` is given in Figure 4.

Let $\mathtt{g}_1 \in [0 : n+1]$ and $\mathtt{g}_2 \in [0 : n+1]$ be two genes, and let $i \in [0 : n+1]$ and $j \in [0 : n+1]$ be two positions. After having defined the total orders T_1 and T_2 with the sets of variables $\mathcal{A}^1$ and $\mathcal{A}^2$, we introduce the set of variables $\mathcal{E} = \{e_{\mathtt{g}_1,i,j,\mathtt{g}_2} : \mathtt{g}_1 \in [0 : n+1], i \in [0 : n+1], j \in [0 : n+1] \text{ and } \mathtt{g}_2 \in [0 : n+1]\}$ that correspond to the possible adjacencies. A boolean variable $e_{\mathtt{g}_1,i,j,\mathtt{g}_2} \in \mathcal{E}$ has of four indices: genes $\mathtt{g}_1$ and $\mathtt{g}_2$, and positions i and j. For each $e_{\mathtt{g}_1,i,j,\mathtt{g}_2} \in \mathcal{E}$, we have $e_{\mathtt{g}_1,i,j,\mathtt{g}_2} = 1$ iff (i) genes $\mathtt{g}_1$ and $\mathtt{g}_2$ create an adjacency and (ii) $T_1(\mathtt{g}_1) = i$, $T_2(\mathtt{g}_1) = j$, and $T_1(\mathtt{g}_2) = i + 1$. The objective function is defined as the sum of variables $e_{\mathtt{g}_1,i,j,\mathtt{g}_2} \in \mathcal{E}$.

We define constraints **(C".01)** to **(C".09)** to check correctness of program `Adj-2PO`. First, we use again the common constraints **(C.a)** to **(C.c)**. Here, we must use these constraints for x in $\{1, 2\}$ in order to define two linear extensions. We refer to these constraints as constraints **(C".01)** to **(C".06)**. Then, we check the assignment of variables $e_{\mathtt{g}_1,i,j,\mathtt{g}_2}$ for $i \neq n+1$ and $j \notin \{0, n+1\}$ with constraint **(C".07)**. To this aim, for each $e_{\mathtt{g}_1,i,j,\mathtt{g}_2}$, we have $e_{\mathtt{g}_1,i,j,\mathtt{g}_2} = 1$ iff the two following conditions hold: (i) $a^1_{\mathtt{g}_1,i} = 1$, $a^2_{\mathtt{g}_1,j} = 1$, and $a^1_{\mathtt{g}_2,i+1} = 1$, and (ii) $a^2_{\mathtt{g}_2,j+1} = 1$ (case 1) or $a^2_{\mathtt{g}_2,j-1} = 1$ (case 2) (these two cases depend on the sign of $\mathtt{g}_1$ and $\mathtt{g}_2$). Finally, we add constraints **(C".08)** and **(C".09)** to ensure the validity of variables $e_{\mathtt{g}_1,i,0,\mathtt{g}_2}$ and $e_{\mathtt{g}_1,i,n+1,\mathtt{g}_2}$ that correspond to possible adjacencies located at the extremities. These two constraints are similar to **(C".07)** but constraints **(C".08)** is for case 1 only and **(C".09)** is for case 2 only.

Speeding-up the program. Program `CI-1PO` has $O(n^3)$ variables and $O(n^3)$ constraints. Program `Adj-1PO` has $O(n^2)$ variables and $O(n^2)$ constraints, while program `Adj-2PO` has $O(n^4)$ variables and $O(n^4)$ constraints. In order to speed-up the running time of the programs, we present here some simple rules for reducing the number of variables and constraints involved in the programs or for computing in a fast preprocessing step some of the variables.

Program `Adj-2PO`

Objective: `Maximize` $\sum_{0\leq g\leq n+1} \sum_{0\leq i\leq n+1} \sum_{0\leq j\leq n+1} \sum_{0\leq y\leq n+1} e_{g,i,j,y}$

Constraints:

C''.01 $\forall\, 0 \leq g \leq n+1, \sum_{0\leq i\leq n+1} a^1_{g,i} = 1$

C''.02 $\forall\, 0 \leq i \leq n+1, \sum_{0\leq g\leq n+1} a^1_{g,i} = 1$

C''.03 $\forall\, 0 \leq g_1 \leq n+1, 0 \leq g_2 \leq n+1, g_1 \prec_0 g_2, 0 < j \leq i \leq n+1, a^1_{g_1,i} + a^1_{g_2,j} \leq 1$

C''.04 $\forall\, 0 \leq g \leq n+1, \sum_{0\leq i\leq n+1} a^2_{g,i} = 1$

C''.05 $\forall\, 0 \leq i \leq n+1, \sum_{0\leq g\leq n+1} a^2_{g,i} = 1$

C''.06 $\forall\, 0 \leq g_1 \leq n+1, 0 \leq g_2 \leq n+1, g_1 \prec_1 g_2, 0 < j \leq i \leq n+1, a^2_{g_1,i} + a^2_{g_2,j} \leq 1$

C''.07 $\forall\, 0 \leq g_1 \leq n+1, 0 \leq i \leq n+1, 1 \leq j \leq n, 0 \leq g_2 \leq n+1,$
if $s_1(g_1) = s_2(g_1)$ and $s_1(g_2) = s_2(g_2)$ then
$a^1_{g_1,i} + a^2_{g_1,j} + a^1_{g_2,i+1} + a^2_{g_2,j+1} - 4e_{g_1,i,j,g_2} \geq 0$
and $a^1_{g_1,i} + a^2_{g_1,j} + a^1_{g_2,i+1} + a^2_{g_2,j+1} - 4e_{g_1,i,j,g_2} \leq 3$
if $s_1(g_1) = -s_2(g_1)$ and $s_1(g_2) = -s_2(g_2)$ then
$a^1_{g_1,i} + a^2_{g_1,j} + a^1_{g_2,i+1} + a^2_{g_2,j-1} - 4e_{g_1,i,j,g_2} \geq 0$
and $a^1_{g_1,i} + a^2_{g_1,j} + a^1_{g_2,i+1} + a^2_{g_2,j-1} - 4e_{g_1,i,j,g_2} \leq 3$
else $e_{g_1,i,j,g_2} = 0$

C''.08 $\forall\, 0 \leq g_1 \leq n+1, 0 \leq i \leq n+1, 0 \leq g_2 \leq n+1,$
if $s_1(g_1) = s_2(g_1)$ and $s_1(g_2) = s_2(g_2)$ then
$a^1_{g_1,i} + a^1_{g_1,0} + a^2_{g_2,i+1} + a^2_{g_2,1} - 4e_{g_1,i,0,g_2} \geq 0$
and $a^1_{g_1,i} + a^1_{g_1,0} + a^2_{g_2,i+1} + a^2_{g_2,1} - 4e_{g_1,i,0,g_2} \leq 3$
else $e_{g_1,i,0,g_2} = 0$

C''.09 $\forall\, 0 \leq g_1 \leq n+1, 0 \leq i \leq n+1, 0 \leq g_2 \leq n+1,$
if $s_1(g_1) = -s_2(g_1)$ and $s_1(g_2) = -s_2(g_2)$ then
$a^1_{g_1,i} + a^1_{g_1,n+1} + a^2_{g_2,i+1} + a^2_{g_2,n} - 4e_{g_1,i,n+1,g_2} \geq 0$
and $a^1_{g_1,i} + a^1_{g_1,n+1} + a^2_{g_2,i+1} + a^2_{g_2,n} - 4e_{g_1,i,n+1,g_2} \leq 3$
else $e_{g_1,i,n+1,g_2} = 0$

Fig. 4. Program `Adj-2PO` computes two linear extensions of two partial orders maximizing the number of adjacencies

Restricted position. In T_x, for each gene g, $T_x(g)$ must be in $POS_x(g)$. According to this remark, some variables can be pre-computed:

- $\forall x \in \{1,2\}$, $\forall a^x_{g,i} \in \mathcal{A}^x$, $a^x_{g,i} = 0$ if $i \notin POS_x(g)$;
- $\forall b_{g,i,t} \in \mathcal{B}$, $b_{g,i,t} = 0$ if $T_1(g)$ cannot be in $[i : i+t]$, and $b_{g,i,t} = 1$ if $T_1(g)$ is necessarily in $[i : i+t]$;
- $\forall d_{g,i} \in \mathcal{D}$, $d_{g,i} = 0$ if $i \notin POS_1(g)$;

- $\forall e_{g_1,i,j,g_2} \in \mathcal{E}$, $e_{g_1,i,j,g_2} = 0$ if one of the four following conditions holds: (i) $i \notin POS_1(g_1)$, (ii) $i+1 \notin POS_1(g_2)$, (ii) $i+1 \notin POS_2(g_2)$ and $s_1(g_1) = s_1(g_2)$, (iv) $i-1 \notin POS_2(g_2)$ and $s_1(g_1) = -s_1(g_2)$.

For each gene g that is i-$nail_x$ (i.e., g is necessarily at a known position i in the linear extension), some variables can also be pre-computed:

- $\forall x \in \{1,2\}$, $\forall a^x_{g,i} \in \mathcal{A}^x$, $a^x_{g,i} = 1$ if g is i-$nail_x$;
- $\forall d_{g,i} \in \mathcal{D}$, $d_{g,i} = 1$ if gene g is i-$nail_1$ and if one of the following conditions is satisfied: (i) gene (g+1) is $(i+1)$-$nail_2$ with $s_1(g) =$ "+" and $s_1(g+1) =$ "+", or (ii) gene $(g+1)$ is $(i-1)$-$nail_2$ with $s_1(g) =$ " − " and $s_1(g+1) =$ " − ".
- $\forall e_{g_1,i,j,g_2} \in \mathcal{E}$, $e_{g_1,i,j,g_2} = 1$ if gene g_1 is i-$nail_1$ and j-$nail_2$, gene g_2 is a $(i+1)$-$nail_1$, and one of the two following conditions holds: (i) g_2 is $(j+1)$-$nail_2$ with $s_1(g_1) = s_1(g_2)$, (ii) g_2 is $(j-1)$-$nail_2$ with $s_1(g_1) = -s_1(g_2)$.

Specific rules for `CI-1PO`. By definition, each interval of size 1 and $n+2$ is a common interval. Therefore, $c_{g,i,0} = 1$ and $c_{g,0,n+2} = 1$ for $0 \leq g \leq n+1$, and we add $n+2$ to the objective function. Also, we can delete all variables $c_{g,i,0}$ and $c_{g,i,n+1}$ in the program as well as their related constraints. The second reduction consists in removing some variables $c_{g,i,t}$. We do not generate a variable $c_{g,i,t}$ and its related constraints if $c_{g,i,t} = 1$ is not possible. Indeed, if there exists an integer $k \in [0:t]$ such that $T_1(g+k)$ cannot be in $[i : i+t]$ (i.e., gene $g+k$ cannot be between i and $i+t$ in the linear extension), then the interval $\{g, g+1, \ldots, g+t\}$ is certainly not a common interval.

Specific rules for `Adj-1PO`. Constraint `C'.06` defines the adjacency created by genes g and $(g+1)$ at positions $n+1$ and n, respectively, in T_1. Since we have artificially added gene $n+1$ at the end of partial order, then the gene at position $n+1$ is $n+1$. Therefore, $d_{g,n+1} = 0$ for g in $[0:n]$ and the constraints `C'.06` are deleted. Moreover, if no adjacency is possible between two genes that are *nails*, then the total order between these genes has no impact on the number of adjacencies. Specifically, let g_1 and g_2 be two genes such that g_1 is i_1-$nail_1$ and g_2 is i_2-$nail_1$ ($i_1 < i_2$). Notice that no adjacency can be create between the positions i_1 and i_2 between a linear extension of P_1 and the identity if, for each gene g_3 such that $POS_1(g_3) \subseteq [i_1 : i_2]$, one of the three following conditions hold: (i) $s_1(g_3) \neq s_1(g_3+1)$, (ii) $s_1(g_3) =$ " + " and $s_1(g_3+1) =$ " + " and (g_3, g_3+1) is not adjoining in P_1, or (iii) $s_1(g_3) =$ " − " and $s_1(g_3+1) =$ " − " and (g_3+1, g_3) is not adjoining in P_1. Hence, if each such gene g_3 satisfies at least one condition, we can define arbitrarily a total order between i_1 and i_2 without adding an adjacency.

4 Experimental Results

Following the example of [4], we have tested our algorithms on simulated data to assess the performance of our programs for different parameters inherent to partial orders. The rationale for this choice is two-fold. For one, the presented

Table 1. Results of `CI-1PO` obtained in less than thirty minutes

Size	Number of results	Number of instances	Running time
30	95	114	$1h$ $48m$
40	85	114	$1h$ $19m$
50	79	114	$1h$ $44m$
Total	259	342	$4h$ $41m$

Table 2. Impact of width for `CI-1PO`

Width	1	2	3	4	5	6	7	8	9	10	11
Number of instances	5	63	63	50	44	31	26	18	13	7	9
Number of solved instances	5	63	63	44	36	26	15	5	1	1	0
% of solved instances	100%	100%	100%	88%	82%	84%	58%	28%	8%	15%	0%
Average running time (sec)	2	13	28	60	88	154	201	311	73	490	

algorithms are quite complicated and using parameterized data allows us to tune the programs to best solve the problems. For another, using the dataset of [4] makes comparisons easier with previous works. Our linear program solver engine is powered by the `minisat+` solver [5]. All computations were carried out on a *Duo Intel* CPU GHz with 2 Gb of memory running under Linux. The reference data set is from [4]. Blin *et al.* have generated partial orders P_x according to three parameters: the *size* n, the *order rate* p that determines the number of adjoining in the expression, and the *gene distribution rule* q that corresponds to the probability of possible adjacencies with respect to the identity. We use 19 different instances for each triplet of parameters (n, p, q) with $n \in \{30, 40, 50, 60, 70, 80, 90\}$, $p \in \{0.7, 0.9\}$ and $q \in \{0.4, 0.6, 0.8\}$, so that we have 798 genomes for each program. The results of the three programs on this dataset are presented below. We evaluated two criteria: the running time and the given measure (number of common intervals or number of adjacencies) induced by the returned linear extension. The width of partial orders will be also considered. Then, we study the advantages of program `Adj-1PO` to `Adj-2PO` for the comparison of a partial order and the identity.

Results for CI-1PO. For program `CI-1PO`, the `minisat+` engine resolves 264 instances out of 342 inputs (77.2%) with $n \in \{30, 40, 50\}$, $p \in \{0.7, 0.9\}$, and $q \in \{0.4, 0.6, 0.8\}$ (Table 1) in less than 5 hours. The 83 remaining cases have been stopped after 30 minutes. We note that `CI-1PO` reaches its limits even for small instances. As shown in Table 2, the width of an instance is an important parameter that certainly contributes to the complexity of the instances. Indeed, we remark that the number of obtained results decreases according to the width of the instances. In the same way, the average running time increases with the width. We also notice that the width is correlated with the complexity of the problem, which is not surprising since combinatorial difficulty is clearly contained in the greatest sets of non-comparable genes.

Results for Adj-1PO. We applied `Adj-1PO` on 19 genomes for each triplet (n, p, q) with $n \in \{30, 40, 50, 60, 70, 80, 90\}$, $p \in \{0.7, 0.9\}$ and $q \in \{0.4, 0.6, 0.8\}$. We obtained 778 results out of 798 (97,5%) after 2 months and 13 days. Due to huge memory requirements, a Quadri Intel(R) Xeon(TM) CPU 3,00 GHz with 16Gb of memory running Linux was required for 14 of the runs. We note that parameter p has the largest impact on the running time. It is indeed significantly higher when $p = 0.7$ rather than when $p = 0.9$, i.e., when the partial orders have less adjoining. The width also affects the running time. Parameter q seems, however, to have no impact on the running time. For the 70 cases where the running time exceeds 1 hour, we note that the corresponding genomes contain 50 genes or more, with an order rate p equals to 0.7 (58 times out of 70), but without specific gene distribution (21 times 0.4, 27 times 0.6 and 22 times 0.8). Their width ranges from 5 to 22 and is on average equal to 11.6. For 19 genomes, the linear program could not been solved by `minisat+` because the number of constraints and variables is too large for this solver. For these genomes, we observe that their size is greater than or equal to 70, that p is equal to 0.7 and that q is variable (9 times 0.4, 3 times 0.6 and 8 times 0.8). Their width is also large: 17.5 on average (from 11 to 29), compared to 6 for all genomes.

Results for Adj-2PO. For each triplet, we compare 19 genomes by pairs. At the present time, we have obtained the results for the following triplets: $\{30, 0.7, 0.8\}$, $\{30, 0.9, 0.4\}$, $\{30, 0.9, 0.6\}$, $\{30, 0.9, 0.8\}$, $\{40, 0.9, 0.4\}$, $\{40, 0.9, 0.6\}$, $\{40, 0.9, 0.8\}$, $\{50, 0.9, 0.6\}$, $\{50, 0.9, 0.8\}$, $\{60, 0.9, 0.6\}$, $\{60, 0.9, 0.8\}$, and $\{70, 0.9, 0.8\}$. Before discussing the 11 triplets for which $p = 0.9$, let us look at the results obtained with the triplet $\{30, 0.7, 0.8\}$. The running time is clearly more important when $p = 0.7$ rather than $p = 0.9$ (2 days and 21 hours for the triplet $\{30, 0.7, 0.8\}$ and 40 minutes for the triplet $\{30, 0.9, 0.8\}$). Moreover, fewer results were obtained (90 results for the triplet $\{30, 0.7, 0.8\}$ and 171 for the triplet $\{30, 0.9, 0.8\}$). For the 11 triplets for which $p = 0.9$, we made 1881 comparisons (19 genomes compared with 18 genomes of the same triplet). We manage to obtain in 1668 results (i.e., 88.6%) in a little over 38 days. For the 213 unfinished runs, two types of problems have emerged: either the linear program has too many variables and constraints for the solver `minisat+`, or the memory requested is too large for the Duo Intel 2Gb of memory. The width clearly influences the running time. Indeed, when the sum of width of both partials orders P_1 and P_2 is less than 5, the average running time is 14 seconds, while it is of 56 minutes when this sum is more than 5. Parameter q has an impact on the number of adjacencies only.

Comparison between `Adj-1PO` and `Adj-2PO`. To evaluate `Adj-1PO`, we compare its running time with `Adj-2PO` when P_2 is the identity. For the 19 instances of each triplet (n, p, q) such that n is in $\{30, 40\}$, p is in $\{0.7, 0.9\}$ and q is in $\{0.4, 0.6, 0.8\}$, the sum of running times is 2 hours and 30 minutes (minimum = 0.1 second, maximum = 15 minutes 25s) for `Adj-1PO` and 67 hours (minimum = 0.1 second, maximum = 14 hours 42m) for `Adj-2PO`. Clearly, `Adj-1PO` is much faster than `Adj-2PO` from the viewpoint of running time (a diminution of 96.3%). If we compare the running time for each case, we note that both programs have a running time more important for the same genome. We can infer that both programs are facing the same difficulties. This is not surprising in view of the similarities of variables and constraints of linear programming of `Adj-1PO` and `Adj-2PO`.

5 Conclusion and Future Works

Our results are quite preliminary and there is still a great amount of work to be done. Among other things, one can cite: (i) for each case, determine other strong and relevant rules for speeding-up the process by avoiding the generation of too many variables and constraints, (ii) generalize the program `CI-1PO` to compute the number of common intervals between two partial orders, and (iii) define and evaluate heuristics for these problems. Indeed, we do think that our approach is useful for providing exact reference results to which new developed heuristics can be compared.

References

1. Angibaud, S., Fertin, G., Rusu, I., Thévenin, A., Vialette, S.: Efficient tools for computing the number of breakpoints and the number of adjacencies between two genomes with duplicate genes. J. Computational Biology 15(8), 1093–1115 (2008)
2. Angibaud, S., Fertin, G., Rusu, I., Vialette, S.: A pseudo-boolean general framework for computing rearrangement distances between genomes with duplicates. J. Computational Biology 14(4), 379–393 (2007)
3. Bergeron, A., Stoye, J.: On the similarity of sets of permutations and its applications to genome comparison. In: Warnow, T.J., Zhu, B. (eds.) COCOON 2003. LNCS, vol. 2697, pp. 68–79. Springer, Heidelberg (2003)
4. Blin, G., Blais, E., Hermelin, D., Guillon, P., Blanchette, M., El-Mabrouk, N.: Gene maps linearization using genomic rearrangement distances. J. Computational Biology 14(4), 394–407 (2007)
5. Eén, N., Sörensson, N.: Translating pseudo-boolean constraints into SAT. Journal on Satisfiability, Boolean Modeling and Computation 2, 1–26 (2006)
6. Fu, Z., Jiang, T.: Computing the breakpoint distance between partially ordered genomes. J. Bioinformatics and Computational Biology 5(5), 1087–1101 (2007)
7. Schrijver, A.: Theory of Linear and Integer Programming. J. Wiley & Sons, Chichester (1998)
8. Uno, T., Yagiura, M.: Fast algorithms to enumerate all common intervals of two permutations. Algorithmica 26(2), 290–309 (2000)
9. Watterson, G.A., Ewens, W.J., Hall, T.E., Morgan, A.: The chromosome inversion problem. Journal of Theoretical Biology 99(1), 1–7 (1982)
10. Zheng, C., Lenert, A., Sankoff, D.: Reversal distance for partially ordered genomes. Bioinformatics 21(1), 502–508 (2005)

Computing the Summed Adjacency Disruption Number between Two Genomes with Duplicate Genes Using Pseudo-Boolean Optimization

João Delgado, Inês Lynce, and Vasco Manquinho

IST/INESC-ID, Technical University of Lisbon, Portugal
{jbd,ines,vmm}@sat.inesc-id.pt

Abstract. The increasing number of fully sequenced genomes has led to the study of genome rearrangements. Several approaches have been proposed to solve this problem, all of them being either too complex to be solved efficiently or too simple to be applied to genomes of complex organisms. The latest challenge has been to overcome the problem of having genomes with duplicate genes. This led to the definition of matching models and similarity measures. The idea is to find a matching between genes in two genomes, in order to disambiguate the data of duplicate genes and calculate a similarity measure. The problem becomes that of finding a matching that best preserves the order of genes in two genomes, where gene order is evaluated by a chosen similarity measure. This paper presents a new pseudo-Boolean encoding for computing the exact summed adjacency disruption number for two genomes with duplicate genes. Experimental results on a γ-Proteobacteria data set illustrate the approach.

1 Introduction

Traditional methods for comparing genomes in terms of genetic evolution are based in the comparison of homologous versions of genes in two genomes [8,3,18]. These methods capture *point mutations* in the sequence of genes, but are limited in disregarding differences in the order by which genes occur in the considered genomes [18].

In *genome rearrangements*, the idea is to compare two genomes in terms of gene order evolution. Several approaches have been proposed to solve this problem, each modeling genomes in a different way and considering different types of *rearrangement events* [12,4,11]. Their aim is to find a minimum number of the considered *rearrangement events* necessary to transform the order of genes in one genome into the order of genes in the other genome. However, they all share the main bottleneck of not allowing the presence of duplicate genes in genomes (i.e., the presence of several homologous versions of genes in genomes), which has been found to be common [16,14].

One of the proposed approaches to overcome this limitation is the use of *matching models* and *similarity measures*, an idea first proposed by David Sankoff [15].

F.D. Ciccarelli and I. Miklós (Eds.): RECOMB-CG 2009, LNBI 5817, pp. 138–149, 2009.

A *matching* is a one-to-one correspondence between homologous genes in two genomes, which considers the order of genes in one genome to Be a permutation of the order of genes in the other genome. The idea is to find a *matching* for which this induced order is best preserved, where order preservation is evaluated by a chosen *similarity measure.*

In this paper, we propose a novel generic pseudo-Boolean optimization encoding for the problem of finding a *matching* between two genomes, of the exemplar [15] or the maximum [7] model, that minimizes the *summed adjacency disruption number* [17]. Additionally, we test this encoding on the same data set of γ-Proteobacteria used in [2,1].

2 Matching Models and Similarity Measures

This section provides several definitions and formalisms used in the remaining of the paper. First, genomes are introduced as sequences over an alphabet of gene families, where each gene family represents a group of homologous genes. Later, matchings are defined, as well as how to obtain the summed adjacency disruption number for a given matching between two genomes.

Definition 1 (genome). *Consider a finite alphabet $\mathcal{A}$ of gene families. A gene is an element of $\mathcal{A}$ having a + or − sign prefixed to it, standing for transcriptional reading direction. If g is a gene then: (i) $-g$ denotes the gene obtained from g by switching its sign, (ii) $s(g)$ denotes the sign of g and (iii) $|g|$ denotes the gene family of g. A genome G over $\mathcal{A}$ is a sequence $G[1]\cdots G[n_G]$ of genes, where n_G denotes the length of G. For all $a \in \mathcal{A}$ and for all $i, j \in \mathbb{N}$ such that $1 \leq i \leq j \leq n_G$, $occ_G(a, i, j)$ denotes the number of occurrences of genes g in a genome G, between positions i and j (inclusively), such that $|g| = a$. $occ_G(a, 1, n_G)$ is abbreviated to $occ_G(a)$. A genome G over $\mathcal{A}$ is said to be* duplicate-free *if for all $a \in \mathcal{A}$, $occ_G(a) \leq 1$. G is said to have* duplicate genes *otherwise.*

Definition 2 (matching). *A matching $\mathcal{M}$ between two genomes G_0 and G_1 over the same finite alphabet of gene families $\mathcal{A}$, is a set of pairs $\mathcal{M} \subseteq \{1, \ldots, n_{G_0}\} \times\{1, \ldots, n_{G_1}\}$ that are pairwise disjoint and establish a correspondence between genes of the same gene family in both genomes, i.e: (i) every two distinct elements $(i, j), (i', j') \in \mathcal{M}$ verify that $i \neq i'$ and $j \neq j'$ and (ii) every element $(i, j) \in \mathcal{M}$ verifies that $|G_0[i]| = |G_1[j]|$. The number of elements in a matching $\mathcal{M}$ is denoted as $|\mathcal{M}|$. Genes $G_0[i]$ and $G_1[j]$ that are matched by a pair (i, j) in a matching $\mathcal{M}$ are said to be $\mathcal{M}$-saturated.*

A matching is said to be an *exemplar matching* if, for each gene family $a \in \mathcal{A}$, such that $occ_{G_0}(a) \geq 1$ and $occ_{G_1}(a) \geq 1$, it saturates exactly one gene, in each genome, that belongs to that family. It is said to be a *maximum matching* if it saturates as many genes of each gene family as possible, i.e., for each gene family $a \in \mathcal{A}$ the number of matched genes is $\min(occ_{G_0}(a), occ_{G_1}(a))$.

A fundamental concept for calculating genomic distances is that of $\mathcal{M}$-reduced genomes that can be induced by a given matching. The definition follows:

Definition 3 ($\mathcal{M}$-reduced genomes). *Let G_0 and G_1 be two genomes over a finite alphabet of gene families $\mathcal{A}$. Furthermore, consider a matching $\mathcal{M}$ between the two genomes of size $|\mathcal{M}|$. The $\mathcal{M}$-reduced genomes $\mathcal{M}(G_0)$ and $\mathcal{M}(G_1)$ of G_0 and G_1 are obtained by first deleting unmatched genes and then renaming the matched genes according to the matching $\mathcal{M}$, maintaining their original signs. The resulting $\mathcal{M}$-reduced genomes are duplicate free and can be viewed as signed permutations.*

The signs of genes are disregarded for computing SAD, therefore and for the sake of simplicity, we will only consider unsigned genomes, i.e., genomes where genes have no signs.

For example, let us consider the following genomes: $G_0 = a \quad a \quad b \quad c \quad e \quad d$ and $G_1 = a \quad f \quad c \quad d \quad b$. Suppose a matching $\mathcal{M}$ between these two genomes where genes e in G_0 and f in G_1 are unmatched, as well as the first first gene of G_0. All other genes are matched. In this case, the size of the matching is 4. One can generate $\mathcal{M}$-reduced genomes $\mathcal{M}(G_0)$ and $\mathcal{M}(G_1)$ by removing the unmatched genes, renaming the matched genes in G_0 to the identity permutation and then renaming the genes in G_1 according to the matching $\mathcal{M}$. As a result, the $\mathcal{M}$-reduced genomes would be: $\mathcal{M}(G_0) = 1 \quad 2 \quad 3 \quad 4$ and $\mathcal{M}(G_1) = 1 \quad 3 \quad 4 \quad 2$.

Definition 4 (summed adjacency disruption number). *Let G_0 and G_1 be two genomes and $\mathcal{M}$ a matching between them of size $|\mathcal{M}|$ Let G_0^0 and G_1^0 be the resulting $\mathcal{M}$-reduced genomes where $\mathcal{M}(G_0)$ is renamed to the identity permutation. Furthermore, let G_0^1 and G_1^1 be the resulting $\mathcal{M}$-reduced where $\mathcal{M}(G_1)$ is renamed to the identity permutation. Then, the summed adjacency disruption number between G_0 and G_1 is given by the sum of $S_{0,1}$ and $S_{1,0}$ defined as follows:*

$$S_{0,1} = \sum_{i=1}^{|\mathcal{M}|-1} \left| G_1^0[i] - G_1^0[i+1] \right| \tag{1}$$

$$S_{1,0} = \sum_{i=1}^{|\mathcal{M}|-1} \left| G_0^1[i] - G_0^1[i+1] \right| \tag{2}$$

Consider again the genomes G_0 and G_1 and the same matching. The $\mathcal{M}$-reduced genomes induced by renaming the matched genes in G_0 to the identity permutation are $G_0^0 = 1 \quad 2 \quad 3 \quad 4$ and $G_1^0 = 1 \quad 3 \quad 4 \quad 2$. Therefore, the value of $S_{0,1} = 5$ is obtained from G_1^0. If the genes in G_1 are renamed to the identity permutation we have $G_0^1 = 1 \quad 4 \quad 2 \quad 3$ and $G_1^1 = 1 \quad 2 \quad 3 \quad 4$. As a result, $S_{1,0} = 6$ is obtained from G_0^1, resulting in a summed adjacency disruption number of 11 between G_0 and G_1, considering the matching as defined.

3 Algorithm for Computing the Summed Adjacency Disruption Number

In this section, an algorithm for computing the Summed Adjacency Disruption (SAD) number is presented. Our proposal is based on encoding the SAD problem into a Pseudo-Boolean Optimization (PBO) problem instance and subsequently using a PBO solver. Furthermore, genome preprocessing steps are applied, in order to obtain a smaller problem instance to be dealt by the PBO solver.

3.1 Pseudo-Boolean Optimization

Pseudo-Boolean Optimization (PBO) is a Boolean formalism where constraints can be any linear inequality with integer coefficients (also known as pseudo-Boolean constraints) defined over the set of problem variables.

The objective in PBO is to find an assignment to problem variables such that all problem constraints are satisfied and the value of a linear objective function is optimized. One can define PBO formulations as:

$$\begin{array}{ll} \text{minimize} & \sum_{j \in N} c_j \cdot x_j \\ \text{subject to} & \sum_{j \in N} a_{ij} \cdot x_j \rhd b_i, \\ & x_j \in \{0, 1\}, \end{array} \tag{3}$$

where $a_{ij}, b_i, c_j \in \mathbb{Z}$, $\rhd$ can be any of the $=, >, \geq, <, \leq, \neq$ operators and $a_{ij} \cdot x_j$ denotes the usual integer multiplication of a_{ij} by x_j. Other PBO formulations are possible, but all being easily translated into this one [5]. Observe that PBO formulations can also be viewed as a special case of integer linear programming.

3.2 PBO Encoding for the Summed Adjacency Disruption Number

Pseudo-Boolean optimization encodings have already been used for the problems of finding an exemplar or maximum matching that maximizes the number of common intervals [2] and finding an exemplar or maximum matching that minimizes the number of breakpoints [1]. In this section, PBO is used to minimize the Summed Adjacency Disruption (SAD) number between two unsigned genomes G_0 and G_1.

Let G_0 and G_1 be two unsigned genomes over an alphabet of gene families $\mathcal{A}$ such that for all $a \in \mathcal{A}$, $occ_{G_0}(a) \geq 1$ and $occ_{G_1}(a) \geq 1$[1]. For simplicity of notation, let $n_0 = n_{G_0}$ and $n_1 = n_{G_1}$ be the lengths of G_0 and G_1 and $\overline{x}$ denote $1-x$, for $x \in \{0, 1\}$. Consider the following sets of Boolean variables in our PBO encoding for SAD:

- $A = \{a(i, k) : 1 \leq i \leq n_0 \wedge 1 \leq k \leq n_1 \wedge G_0[i] = G_1[k]\}$
- $B = \{b_x(i) : x \in \{0, 1\} \wedge 1 \leq i \leq n_x\}$
- $C = \{c_x(i, j) : x \in \{0, 1\} \wedge 1 \leq i < j \leq n_x\}$
- $D = \{d_x(i, j, k, l) : x \in \{0, 1\} \wedge 1 \leq i < j \leq n_x \wedge 1 \leq k < l \leq n_{\overline{x}}\}$
- $E = \{e_x(i, j, p) : x \in \{0, 1\} \wedge 1 \leq i < j \leq n_x \wedge i < p < j\}$

The idea behind each set of variables is as follows:

[1] It is assumed that genes that only occur in one genome can be readily eliminated, since they are not relevant to compute the SAD number.

- $a(i, k)$: these variables define the matching $\mathcal{M}$ used to compute the SAD number. We define a variable $a(i, k)$ for each pair (i, k) such that $G_0[i] = G_1[k]$. If $a(i, k) = 1$ it means that the gene at position i in G_0 is matched with the gene at position k in G_1. Otherwise, if $a(i, k) = 0$, then these genes are not matched.
- $b_x(i)$: these variables denote if gene i from genome G_x is matched with another gene in the other genome.
- $c_x(i, j)$: these variables verify if genes at positions i and j can be considered to be consecutive in $\mathcal{M}(G_x)$. According to a given matching $\mathcal{M}$, we say that genes i and j can be consecutive in $\mathcal{M}(G_x)$ if all genes strictly between positions i and j are unmatched in genome G_x. In this case, variable $c_x(i, j)$ is assigned value 1. Otherwise, it is assigned value 0.
- $d_x(i, j, k, l)$: these variables are assigned value 1 if and only if genes at positions i and j in genome G_x are matched with genes at positions k and l (by any order) in genome $G_{\bar{x}}$. Furthermore, genes k and l must be consecutive in $\mathcal{M}(G_{\bar{x}})$. Otherwise, variable $d_x(i, j, k, l)$ must be assigned value 0.
- $e_x(i, j, p)$: these variables are used to count the number of unmatched genes strictly between positions i and j in genome G_x. Variable $e_x(i, j, p)$ is assigned value 1 if there is at least one variable $d_x(i, j, k, l)$ assigned value 1. Moreover, gene at position p in G_x cannot be matched. Otherwise, $e_x(i, j, p)$ must be assigned value 0.

Table 3 presents the pseudo-Boolean constraints for a correct encoding of the different variable sets. The first sets of constraints $(C.1)$ and $(C.2)$ define the value of variables $b_x(i)$, $x \in \{0, 1\}$, in terms of the set of variables A. Moreover, since the value of $b_x(i)$ is at most 1, $(C.1)$ and $(C.2)$ also establish that a given gene in a genome can only be matched at most to a single gene in the other genome from the same gene family.

Constraint set $(C.3)$ defines that in order to have a valid matching in the maximum model, for each gene family $a \in \mathcal{A}$ there must be $\min\{occ_{G_0}(a), occ_{G_1}(a)\}$ genes matched in each genome. For the exemplar model, the right-hand side of the $(C.3)$ constraints is just 1, since we can have only 1 gene matched for each gene family.

Sets $(C.4)$ and $(C.5)$ constrain the value of C variables in terms of the B variables. Constraints $(C.4)$ force variables $c_x(i, j)$ to have value 1 if none of the genes strictly between positions i and j are matched in genome G_x. Otherwise, if there is at least one variable $b_x(p)$, with $i < p < j$, that is assigned value 1, then genes at positions i and j are not consecutive in $\mathcal{M}(G_x)$ and by constraints $(C.5)$, variable $c_x(i, j)$ must be assigned value 0.

The values of D variables are defined in constraints $(C.6)$ to $(C.10)$. Constraints $(C.6)$ and $(C.7)$ set the value of $d_0(i, j, k, l)$ to 1 when the gene at position i in G_0 matches gene at position k (or l) in G_1 and gene at position j in G_0 matches the gene at position l (or k) in G_1. Moreover, for $d_0(i, j, k, l)$ to have value 1, genes at positions k and l must be consecutive in $\mathcal{M}(G_1)$. If any of these conditions do not hold, then $d_0(i, j, k, l)$ must have value 0.

Constraints $(C.8)$ and $(C.9)$ set the value of $d_1(i, j, k, l)$ by applying the same reasoning. Next, constraints $(C.10)$ force variables $d_x(i, j, k, l)$, with $x \in \{0, 1\}$, to have value 0 when it is not possible to have a matching considering genes at positions i and j in G_x and positions k and l in $G_{\overline{x}}$.

Constraints $(C.11)$ define the value for the E variables. The first constraint of set $(C.11)$ defines that if there is one variable $d_x(i, j, k, l)$ set to 1 and gene at position p is not matched, then variable $e_x(i, j, p)$ must be set to 1. The other constraints in $(C.11)$ force variables $e_x(i, j, p)$ to be assigned value 0 if these conditions do not hold.

Finally, it is necessary to ensure that the objective function correctly computes the SAD number given the matching $\mathcal{M}$ induced by the A variables in our PBO formulation. Recall that the SAD number is defined as the sum of two numbers $S_{0,1}$ and $S_{1,0}$ (see Definition 4). Moreover, these numbers are computed using the $\mathcal{M}$-reduced genomes of G_0 and G_1 defined by the A variables.

We start by considering $S_{0,1}$. Suppose that there are two genes $G_0[i], G_0[j]$ (with $i < j$) in G_0 such that they match (by any order) with genes $G_1[k], G_1[l]$ (with $k < l$) in G_1 and also that $G_1[k]$ and $G_1[l]$ are consecutive in $\mathcal{M}(G_1)$. This adds to $S_{0,1}$ the absolute difference of the values given to $G_1[k]$ and $G_1[l]$ in $\mathcal{M}(G_1)$ when the matched genes in G_0 are renamed to the identity permutation. It can be shown that this equals the value of $(j - i)$ minus the number of unmatched genes in G_0, strictly between positions i and j.

Considering our pseudo-Boolean model, for $S_{0,1}$, the idea is to consider all possible pairs of positions i and j, where $1 \leq i < j \leq n_0$, and determine which ones verify the following condition:

(a) there exists a variable $d_0(i, j, k, l) \in D$ such that $d_0(i, j, k, l) = 1$.

If a given pair of positions i and j verify condition (a), one knows that $G_0[i]$ and $G_0[j]$ are matched to genes $G_1[k]$ and $G_1[l]$ that are consecutive in $\mathcal{M}(G_1)$. Every such pair will add $(j - i)$ minus the number of unmatched genes in genome G_0, strictly between positions i and j. However, a pair of positions i and j that does not verify condition (a) should simply add 0 to $S_{0,1}$. Both cases are attained by adding to $S_{0,1}$ the following value:

$$S_{0,1}(i, j) = \left(\sum_{k=1}^{n_1 - 1} \sum_{l=k+1}^{n_1} (j - i) \cdot d_0(i, j, k, l) \right) - \sum_{p=i+1}^{j-1} e_0(i, j, p) \tag{4}$$

1. If a pair of positions i and j, where $1 \leq i < j \leq n_0$, verifies condition (a), then it holds that $\sum_{k=1}^{n_1-1} \sum_{l=k+1}^{n_1} d_0(i, j, k, l) = 1$ and, consequently, variables $e_0(i, j, p)$, for $i < p < j$, will verify that $e_0(i, j, p) = 1$ if, and only if, $G_0[p]$ is not matched. So, when i and j verify condition (a), $S_{0,1}(i, j)$ equals $(j - i)$ minus the number of unmatched genes in genome G_0, strictly between positions i and j.
2. If a pair of positions i and j, where $1 \leq i < j \leq n_0$, does not verify condition (a), then, it holds that $\sum_{k=1}^{n_1-1} \sum_{l=k+1}^{n_1} d_0(i, j, k, l) = 0$ and consequently, variables $e_0(i, j, p)$, with $i < p < j$, verify that $e_0(i, j, p) = 0$ and so, $S_{0,1}(i, j)$ equals 0.

Therefore, and since there is a one to one correspondence between variables $d_0(i,j,k,l)$ for which $d_0(i,j,k,l) = 1$ and consecutive genes in $\mathcal{M}(G_1)$, we have that:

$$\begin{aligned} S_{0,1} &= \sum_{i=1}^{n_0-1} \sum_{j=i+1}^{n_0} S_{0,1}(i,j) = \\ &= \sum_{i=1}^{n_0-1} \sum_{j=i+1}^{n_0} \left(\left(\sum_{k=1}^{n_1-1} \sum_{l=k+1}^{n_1} (j-i) \cdot d_0(i,j,k,l) \right) - \sum_{p=i+1}^{j-1} e_0(i,j,p) \right) \end{aligned} \tag{5}$$

By similar reasoning over all possible pairs of positions i and j, where $1 \leq i < j \leq n_1$, we have that:

$$S_{1,0} = \sum_{i=1}^{n_1-1} \sum_{j=i+1}^{n_1} \left(\left(\sum_{k=1}^{n_0-1} \sum_{l=k+1}^{n_0} (j-i) \cdot d_1(i,j,k,l) \right) - \sum_{p=i+1}^{j-1} e_1(i,j,p) \right) \tag{6}$$

Therefore, the objective function of our PBO problem is the following:

$$\sum_{x=0}^{1} \left(\sum_{i=1}^{n_x-1} \sum_{j=i+1}^{n_x} \left(\left(\sum_{k=1}^{n_{\overline{x}}-1} \sum_{l=k+1}^{n_{\overline{x}}} (j-i) \cdot d_x(i,j,k,l) \right) - \sum_{p=i+1}^{j-1} e_x(i,j,p) \right) \right) \tag{7}$$

An optimal solution to the proposed PBO formulation defines a valid matching (exemplar or maximum) between G_0 and G_1 for which the summed adjacency disruption number between the $\mathcal{M}$-reduced genomes of G_0 and G_1 is minimum. The proposed encoding has $O(n_0^2 n_1^2)$ variables and $O(n_0^2 n_1^2)$ constraints.

3.3 Genome Preprocessing

Prior to the generation of each pseudo-Boolean optimization instance, some preprocessing steps are presented for reducing the size of genomes. The preprocessing of the input genomes G_0 and G_1 will result in two new genomes G'_0 and G'_1, for which the SAD number in an exemplar or maximum matching is not changed.

Step one. A first step is applied for both exemplar and maximum models. It consists in deleting from the input genomes G_0 and G_1 the genes that belong to a gene family with occurrences in only one of the genomes, while maintaining the order of the remaining genes. Clearly, no such gene will be matched in any valid matching between G_0 and G_1, and therefore these genes can simply be deleted from the genomes.

Step two. A second step is applied if the exemplar model is considered. In a genome with consecutive occurrences of genes of the same gene family, these can be reduced to just one, since in an exemplar matching $\mathcal{M}$ one must match exactly one gene of each gene family in each of the genomes.

3.4 Variable Reduction Rules

Here we introduce variable reduction rules for variables $d_x(i,j,k,l) \in D$ and $e_x(i,j,p) \in E$. The rules regarding variables $d_x(i,j,k,l) \in D$ are similar to those already used in other pseudo-Boolean encodings for common intervals [2], and breakpoints distance [1].

If the exemplar model is considered, then the following rules can be applied:

- Delete from D all variables $d_x(i,j,k,l)$ for which $G_x[i] = G_x[j]$ or $G_{\overline{x}}[k] = G_{\overline{x}}[l]$.
- If all occurrences of a given gene family $a \in \mathcal{A}$ in G_x occur between i and j ($occ_{G_x}(a, i+1, j-1) = occ_{G_x}(a)$), then variable $d_{\overline{x}}(i,j,k,l)$ can be removed.
- Delete from E all variables $e_x(i,j,p)$ for which $occ_{G_x}(G_x[p]) = 1$.

If the maximum model is considered, then the following rules can be safely applied:

- Delete from D all variables $d_x(i,j,k,l)$ for which there exists $p \in \mathbb{N}$ such that $k < p < l$ and any of the following conditions hold:
  1. $occ_{G_{\overline{x}}}(G_{\overline{x}}[p]) \leq occ_{G_x}(G_{\overline{x}}[p])$.
  2. $occ_{G_{\overline{x}}}(G_{\overline{x}}[p], k+1, l-1) > |occ_{G_0}(G_{\overline{x}}[p]) - occ_{G_1}(G_{\overline{x}}[p])|$.
- Delete from E all variables $e_x(i,j,p)$ for which $occ_{G_x}(G_x[p]) \leq occ_{G_{\overline{x}}}(G_x[p])$.

3.5 Additional Reduction Rules

The variable value setting rules used in the PBO encoding, which have been proposed for the number of breakpoints [1] for variables $a(i,k) \in A$ and $b_x(i) \in B$, can also be applied to our encoding for SAD. However, for other variables, new rules are proposed that are to be applied after the variable reduction rules.

Let $d_x(i,j,k,k+1) \in D$ be such that one of the following conditions hold:

- $G_x[i] = G_{\overline{x}}[k]$ and $G_x[j] = G_{\overline{x}}[k+1]$.
- $G_x[i] = G_{\overline{x}}[k+1]$ and $G_x[j] = G_{\overline{x}}[k]$.

In this case, if $occ_{G_x}(G_{\overline{x}}[k]) = occ_{G_{\overline{x}}}(G_{\overline{x}}[k]) = occ_{G_x}(G_{\overline{x}}[k+1]) = occ_{G_{\overline{x}}}(G_{\overline{x}}[k+1]) = 1$ (i.e., $G_x[i]$, $G_x[j]$, $G_{\overline{x}}[k]$ and $G_{\overline{x}}[k+1]$ are genes that belong to gene families without duplicates in G_x and $G_{\overline{x}}$), then $d_x(i,j,k,k+1)$ can be assigned value 1.

Finally, let $e_x(i,j,p) \in E$ be such that there is a variable $d_x(i,j,k,k+1)$ that must be assigned value 1. In this case, the following constraint can be added: $e_x(i,j,p) + b_x(p) = 1$.

4 Experimental Results

The experimental evaluation of the PBO encodings for the SAD distance was performed using the genomes of γ-Proteobacteria, which correspond to the same data set used in [2,1]. The data set was originally studied by Lerat *et al.* [13]

Table 1. Results for the summed adjacency disruption number using the exemplar model. (The greatest gap in sub-optimal solutions was of 3.46%.)

	Ecoli	Haein	Paeru	Pmult	Salty	Wglos	Xaxon	Xcamp	Xfast	Y-CO92	Y-KIM
Baphi	32058	66510	(61600)	64894	31982	37082	55614	53018	56036	(42544)	(43742)
Wglos	(63348)	72908	(75942)	74676	(63000)		70272	68628	68304	(62344)	(59348)

Table 2. Results for the summed adjacency disruption number using the maximum model. (The greatest gap in sub-optimal solutions was of 5.48%.)

	Ecoli	Haein	Paeru	Pmult	Salty	Wglos	Xaxon	Xcamp	Xfast	Y-CO92	Y-KIM
Baphi	34126	69462	(64573)	68336	34054	37906	58348	55546	59172	(44974)	(45737)
Wglos	(66787)	76669	(80461)	78286	(66450)		(74014)	72045	72140	(65764)	(62859)

and is composed of thirteen genomes. Only twelve of these genomes were considered, given that the thirteenth genome (*Vibrio cholerae*) is separated in two chromosomes, and therefore cannot be represented using the proposed encoding. The partition of the whole gene set of these genomes into gene families and the representation of each genome as a signed sequence over the obtained alphabet of gene families was done by Blin *et al.* [6].

The problem of computing genomic distances between genomes, regardless the matching model being either the exemplar or the maximum model, was proved to be NP-hard when any of the genomes have duplicate genes and, in the case of SAD, also to be APX-hard [7]. In case no gene family has duplicates in any of the genomes, then there exists only one possible solution for the problem. Hence, some of the variable reduction and value setting rules assume the existence of gene families that occur only once in both genomes. On the other hand, increasing the presence of duplicate genes in the genomes can greatly increase the number of valid matchings between them. Due to these facts, the number of genes belonging to gene families that occur in both genomes and occur more than once in one of them is believed to be a feature of some interest for determining the difficulty of solving the corresponding instance. For a more detailed analysis we refer to [9].

All possible 66 pairwise genome comparisons instances were generated using the proposed PBO encoding for computing the SAD number. From the resulting set of problem instances, only those not exceeding 1GB of size have been considered, which resulted in 21 instances. In practice, these are the instances involving the genomes *Baphi* and *Wglos*. Although this represents a clear drawback to the extent of the evaluation, it is simply a consequence of the complexity of the distance measure being used. We should note, however, that this has been the case even after applying the reduction rules described in section 3.5.

The solutions for the exemplar and maximum models for the genomes Baphi and Wglos are presented in tables 1 and 2. Experiments were run on an Intel Xeon 5160 server (3.0GHZ, 1333Mhz, 4GB) running Red Hat Enterprise Linux

Table 3. PBO encoding for the summed adjacency disruption number

minimize: $\sum_{x=0}^{1}\left(\sum_{i=1}^{n_x-1}\sum_{j=i+1}^{n_x}\left(\sum_{k=1}^{n_{\overline{x}}-1}\sum_{l=k+1}^{n_{\overline{x}}}(j-i)\cdot d_x(i,j,k,l)-\sum_{p=i+1}^{j-1}e_x(i,j,p)\right)\right)$		
subject to:		
$(C.1)$	$\sum_{\substack{1\leq k\leq n_1\\ G_0[i]=G_1[k]}} a(i,k)=b_0(i)$	$1\leq i\leq n_0$
$(C.2)$	$\sum_{\substack{1\leq i\leq n_0\\ G_0[i]=G_1[k]}} a(i,k)=b_1(k)$	$1\leq k\leq n_1$
$(C.3)$	$\sum_{\substack{1\leq i\leq n_x\\ G_x[i]=a}} b_x(i)=\min\{occ_{G_0}(a),occ_{G_1}(a)\}$	$a\in\mathcal{A}$ $0\leq x\leq 1$ $1\leq i\leq n_x$
$(C.4)$	$c_x(i,j)+\sum_{i<p<j} b_x(p)\geq 1$	$0\leq x\leq 1$ $1\leq i<j\leq n_x$
$(C.5)$	$c_x(i,j)+b_x(p)\leq 1$	$0\leq x\leq 1$ $1\leq i<j\leq n_x$ $i<p<j$
$(C.6)$	if $G_0[i]=G_1[k]\wedge G_0[j]=G_1[l]$ then $a(i,k)+a(j,l)+c_1(k,l)-d_0(i,j,k,l)\leq 2$ $a(i,k)-d_0(i,j,k,l)\geq 0$ $a(j,l)-d_0(i,j,k,l)\geq 0$ $c_1(k,l)-d_0(i,j,k,l)\geq 0$	$1\leq i<j\leq n_0$ $1\leq k<l\leq n_1$
$(C.7)$	if $G_0[i]=G_1[l]\wedge G_0[j]=G_1[k]$ then $a(i,l)+a(j,k)+c_1(k,l)-d_0(i,j,k,l)\leq 2$ $a(i,l)-d_0(i,j,k,l)\geq 0$ $a(j,k)-d_0(i,j,k,l)\geq 0$ $c_1(k,l)-d_0(i,j,k,l)\geq 0$	$1\leq i<j\leq n_0$ $1\leq k<l\leq n_1$
$(C.8)$	if $G_0[i]=G_1[k]\wedge G_0[j]=G_1[l]$ then $a(i,k)+a(j,l)+c_0(i,j)-d_1(k,l,i,j)\leq 2$ $a(i,k)-d_1(k,l,i,j)\geq 0$ $a(j,l)-d_1(k,l,i,j)\geq 0$ $c_0(i,j)-d_1(k,l,i,j)\geq 0$	$1\leq i<j\leq n_0$ $1\leq k<l\leq n_1$
$(C.9)$	if $G_0[i]=G_1[l]\wedge G_0[j]=G_1[k]$ then $a(i,l)+a(j,k)+c_0(k,l)-d_1(k,l,i,j)\leq 2$ $a(i,l)-d_1(k,l,i,j)\geq 0$ $a(j,k)-d_1(k,l,i,j)\geq 0$ $c_0(i,j)-d_1(k,l,i,j)\geq 0$	$1\leq i<j\leq n_0$ $1\leq k<l\leq n_1$
$(C.10)$	if $(G_x[i]\neq G_{\overline{x}}[k]\vee G_x[j]\neq G_{\overline{x}}[l])$ and $(G_x[i]\neq G_{\overline{x}}[l]\vee G_x[j]\neq G_{\overline{x}}[k])$ then $d_x(i,j,k,l)=0$	$0\leq x\leq 1$ $1\leq i<j\leq n_x$ $1\leq k<l\leq n_{\overline{x}}$
$(C.11)$	$\sum_{1\leq k<n_{\overline{x}}}\sum_{k<l\leq n_{\overline{x}}} d_x(i,j,k,l)-b_x(p)-e_x(i,j,p)\leq 0$ $e_x(i,j,p)-\sum_{1\leq k<n_{\overline{x}}}\sum_{k<l\leq n_{\overline{x}}} d_x(i,j,k,l)\leq 0$ $e_x(i,j,p)+b_x(p))\leq 1$	$0\leq x\leq 1$ $1\leq i<j\leq n_x$ $i<p<j$

WS 4. For each instance, the cutoff CPU time was set to 10 hours. For the instances which were not solved within the time limit, the respective value appears enclosed in parentheses and corresponds to the best solution found. The PBO solver used was *cplex* version 12.1, a well-known commercial solver for general Linear Programming. PBO solvers were also tried, namely *minisat+* [10], but *cplex* performed much better than *minisat+* in our instance set. Clearly, the techniques used in *cplex*, namely linear programming relaxation and cutting planes, results in large cuts on the search space.

The PBO instances for the SAD number have shown to be harder to solve than the PBO instances for common intervals or breakpoints [2,1]. This comes as no surprise since the variable reduction rules are much less strict than the ones for common intervals or breakpoints. Therefore, the encoding for SAD results in much larger instances due to a larger number of variables involved. Nevertheless, we were able to obtain the optimum value for 13 instances using the example model and 12 instances using the maximum model. Let us note, that these results are the first exact results obtained for this APX-hard problem and can be used to evaluate the accuracy of non-optimal algorithms which are not guaranteed to provide optimal solutions. Moreover, the sub-optimal solutions found are valid and can be used as upper bounds on the optimal solutions. Additionally, *cplex* also provides a lower bound, for every sub-optimal solution.

5 Conclusions

This paper presents a new complete algorithm to compute the Summed Adjacency Disruption (SAD) number between two genomes with common duplicate genes. To the best of our knowledge, this is the first complete algorithm for computing the SAD number. The new approach is based on encoding the SAD problem into the Pseudo-Boolean Optimization (PBO) formalism and subsequently using a pseudo-Boolean solver. Moreover, techniques to reduce the PBO encoding are proposed.

Experimental results show that finding the SAD number is harder than using other similarity measures, such as common intervals or breakpoints. Nevertheless, optimal solutions were obtained for some instances and these can be used to validate the accuracy of approximation algorithms. Additionally, the sub-optimal solutions found are valid, and also provide tight upper bounds on the optimal solutions.

Acknowledgments. We thank Guillaume Fertin and Stéphane Vialette for introducing us to the use of PBO in comparative genomics. This work is partially supported by FCT under research projects BSOLO (PTDC/EIA/76572/2006) and SHIPs (PTDC/EIA/64164/2006).

References

1. Angibaud, S., Fertin, G., Rusu, I., Thévenin, A., Vialette, S.: Efficient tools for computing the number of breakpoints and the number of adjacencies between two genomes with duplicate genes. Journal of Computational Biology 15(8), 1093–1115 (2008)

2. Angibaud, S., Fertin, G., Rusu, I., Vialette, S.: A general framework for computing rearrangement distances between genomes with duplicates. Journal of Computational Biology 14(4), 379–393 (2007)
3. Bafna, V., Pevzner, P.A.: Sorting by reversals: Genome rearrangements in plant organneles and evolutionary history of x chromosome. Molecular Biology and Evolution 12(2), 239–246 (1995)
4. Bafna, V., Pevzner, P.A.: Genome rearrangements and sorting by reversals. SIAM Journal on Computing 25(2), 272–289 (1996)
5. Barth, P.: A Davis-Putnam enumeration algorithm for linear pseudo-Boolean optimization. Technical Report MPI-I-95-2-003, Max Plank Institute for CS (1995)
6. Blin, G., Chauve, C., Fertin, G.: Genes order and phylogenetic reconstruction: Application to γ-protobacteria. In: McLysaght, A., Huson, D.H. (eds.) RECOMB 2005. LNCS (LNBI), vol. 3678, pp. 11–20. Springer, Heidelberg (2005)
7. Blin, G., Chauve, C., Fertin, G., Rizzi, R., Vialette, S.: Comparing genomes with duplications: a computational complexity point of view. IEEE/ACM Transactions on Computational Biology and Bioinformatics 4(4), 523–534 (2007)
8. Caprara, A., Lancia, G., Ng, S.K.: A column-generation based branch-and-bound algorithm for sorting by reversals. In: Mathematical Support for Molecular Biology. DIMACS Series in Discrete Mathematics and Theoretical Computer Science, vol. 47, pp. 213–226 (1999)
9. Delgado, J.: Pseudo-boolean approaches to comparative genomics. Master's thesis, Instituto Superior Técnico, Technical University of Lisbon, Portugal (June 2009)
10. Eén, N., Sörensson, N.: Translating pseudo-boolean constraints into sat. Journal on Satisfiability, Boolean Modeling and Computation 2, 1–26 (2006)
11. Kececioglu, J., Ravi, R.: Of mice and men: Evolutionary distances between genomes under translocations. In: 6th Annual ACM-SIAM Symposium on Discrete Algorithms, pp. 604–613 (1995)
12. Kececioglu, J., Sankoff, D.: Exact and approximation algorithms for sorting by reversals, with application to genome rearrangements. Algorithmica 13, 180–210 (1995)
13. Lerat, E., Daubin, V., Moran, N.A.: From gene trees to organismal phylogeny in prokaryotes: the case of the γ-proteobacteria. PLoS Biology 1(1), 101–109 (2003)
14. Li, W.-H., Gu, Z., Wang, H., Nekrutenko, A.: Evolutionary analyses of the human genome. Nature 409, 847–849 (2001)
15. Sankoff, D.: Genome rearrangement with gene families. Bioinformatics 15(11), 909–917 (1999)
16. Sankoff, D.: Gene and genome duplication. Current Opinion in Genetics & Development 11(6), 681–684 (2001)
17. Sankoff, D., Haque, L.: Power boost for cluster tests. In: McLysaght, A., Huson, D.H. (eds.) RECOMB 2005. LNCS (LNBI), vol. 3678, pp. 121–130. Springer, Heidelberg (2005)
18. Sankoff, D., Leduc, G., Antoine, N., Paquin, B., Lang, B.F., Cedergren, R.: Gene order comparisons for phylogenetic inference: evolution of the mitochondrial genome. Proceedings of the National Academy of Sciences of the United States of America 89(14), 6575–6579 (1992)

Reconstructing Histories of Complex Gene Clusters on a Phylogeny

Tomáš Vinař[1], Broňa Brejová[1], Giltae Song[2], and Adam Siepel[3]

[1] Faculty of Mathematics, Physics and Informatics, Comenius University, Mlynská Dolina, 842 48 Bratislava, Slovakia
[2] Center for Comparative Genomics and Bioinformatics, 506B Wartik Lab, Penn State University, University Park, PA 16802, USA
[3] Dept. of Biological Statistics and Comp. Biology, Cornell University, Ithaca, NY 14853, USA

Abstract. Clusters of genes that have evolved by repeated segmental duplication present difficult challenges throughout genomic analysis, from sequence assembly to functional analysis. These clusters are one of the major sources of evolutionary innovation, and they are linked to multiple diseases, including HIV and a variety of cancers. Understanding their evolutionary histories is a key to the application of comparative genomics methods in these regions of the genome. We propose a probabilistic model of gene cluster evolution on a phylogeny, and an MCMC algorithm for reconstruction of duplication histories from genomic sequences in multiple species. Several projects are underway to obtain high quality BAC-based assemblies of duplicated clusters in multiple species, and we anticipate use of our methods in their analysis. Supplementary materials are located at http://compbio.fmph.uniba.sk/suppl/09recombcg/

1 Introduction

Segmental duplications cover about 5% of the human genome (Lander et al., 2001). When multiple segmental duplications occur at a particular genomic locus they give rise to complex gene clusters. Many functionally important families residing in such clusters are linked to various diseases, e.g. UGT1A (colorectal cancer; Tang et al. (2005)), UGT2 (prostate cancer; Hajdinjak and Zagradisnik (2004)), APOBEC3 (HIV; An et al. (2004)), CCL3 (HIV; Degenhardt et al. (2009)), HLA (multiple sclerosis; Bitti et al. (2001)), CST (Alzheimer's disease; Finckh et al. (2000)). Gene duplication is often followed by functional diversification (Ohno, 1970), and, indeed, genes overlapping segmental duplications have been shown to be enriched for positive selection (Gibbs et al., 2007).

In this paper, we describe a probabilistic model of evolution of gene clusters on a phylogeny, and devise an algorithm for inference of highly probable duplication histories and ancestral sequences. We apply our algorithm to simulated sequences on the human-chimp-macaque phylogeny, as well as to real clusters assembled from available BAC sequencing data.

F.D. Ciccarelli and I. Miklós (Eds.): RECOMB-CG 2009, LNBI 5817, pp. 150–163, 2009.

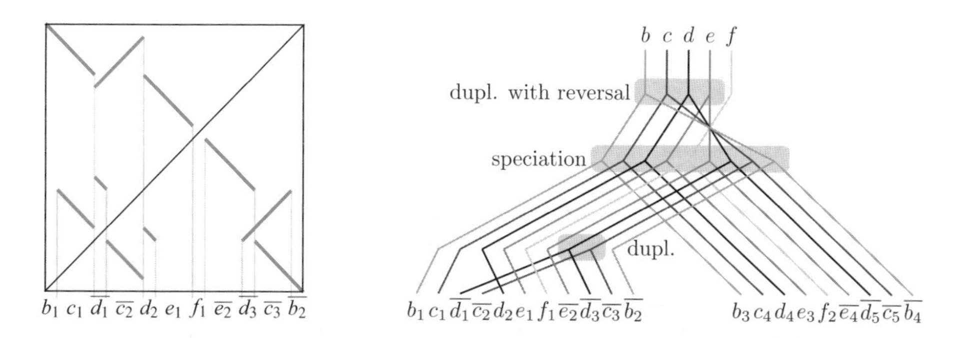

Fig. 1. Sequence atomization and segment trees. (left) Self-alignment of a sequence resulting from two duplication events. Lines represent local alignments. There are five types of atomic segments (b, c, d, e, f). For example, type d has three copies: one on the forwards strand (d_2) and two on the reverse strand $(\overline{d_1}, \overline{d_3})$. (right) Segment trees of individual atom types organized in a tube tree (inspired by similar graphics of Brian Raney). Tube trees visualize a duplication history of several atomic segments in the context of the species tree, and their locations in the extant and ancestral sequences. The figure shows a tube tree with two duplications and one speciation.

Previously, Elemento et al. (2002) and Lajoie et al. (2007) studied the reconstruction of gene family histories in a simplified model, where gene clusters contain several tandemly repeated copies of a single gene. Elemento et al. (2002) consider tandem duplications only, while Lajoie et al. (2007) also consider inversions of variable lengths. However, more complex models are necessary to address evolution of gene clusters in the human genome. In recent work, genes have been replaced by generic *atomic segments* (Zhang et al., 2008; Ma et al., 2008; Bertrand et al., 2008). Briefly, a self-alignment is constructed by a local alignment program (e.g., blastz (Schwartz et al., 2003)), and only alignments above certain threshold (e.g., 93% for human-macaque split) are kept. Alignment boundaries mark *breakpoints*, and the sequences between neighboring breakpoints are considered atomic segments (Fig.1). Under reasonable evolutionary models, the sequence similarities between atomic segments are *transitive*, and the set of atomic segments can be decomposed into equivalence classes, or *atom types*, such that all segments of the same type have similar sequence. In this way, the nucleotide sequence is transformed into a simpler sequence of atoms.

The task of *duplication history reconstruction* is to find a sequence of events (duplications, deletions, and speciations) that starts with an ancestral sequence of atoms, where each atom occurs once, and ends with atomic sequences of extant species. Such a history directly implies *segment trees* of individual atomic types, implicitly reconciled with the species tree (Fig.1). Each segment tree represents duplication and speciation events concerning one atom type, similarly as gene tree represents history of a single gene. A common way of looking at these histories is from the most recent events back in time. In this context, we can start from the extant sequences, and *unwind* events one-by-one, until the ancestral sequence is reached.

Zhang et al. (2008) sought parsimonious solutions of this problems given the sequence from a single species. In particular, they proved a necessary condition to identify candidates for the latest duplication operation, assuming no reuse of breakpoints. After unwinding the latest duplication, the same step is repeated to identify the second latest duplication, etc. Zhang *et al.* showed that any sequence of candidate duplications leads to a history with the same number of duplication events under no-breakpoint-reuse assumption. As a result, there may be an exponential number of most parsimonious solutions to the problem, and it may be impossible to reconstruct a unique history. Recently, Zhang et al. (2009) extended the same approach to simultaneous inference in multiple species. A similar parsimony problem has also been recently explored by Ma et al. (2008) in the context of much larger sequences (whole genomes) and a broader set of operations (including inversions, translocations, etc.). In their algorithm, Ma *et al.* reconstruct phylogenetic trees for every atomic segment, and reconcile these segment trees with the species tree to infer deletions and rooting. The authors give a polynomial-time algorithm for the history reconstruction, assuming no-breakpoint-reuse and correct atomic segment trees. Both methods make use of extensive heuristics to overcome violations of their assumptions in real data.

The no-breakpoint-reuse assumption is often justified by the argument that in long sequences, it is unlikely that the same breakpoint is used twice (Nadeau and Taylor, 1984). However, there is evidence that breakpoints do not occur uniformly throughout the sequence, and that breakpoint reuse is in fact frequent (Peng et al., 2006; Becker and Lenhard, 2007). Moreover, breakpoints located close to each other lead to short atoms that cannot be reliably identified by sequence similarity algorithms and categorized into atom types. For example, in our simulated data (Section 4), approximately 2% of atoms are shorter than 20bp. These short atoms may appear as additional breakpoint reuses. Thus, no-breakpoint-reuse can be a useful guide, but cannot be relied on in application to real data sets. We have also examined the assumption of correctness of segment trees inferred from sequences of individual segments. For segments shorter than 500bp (39% of all segments in our simulations) 69% of the trees were incorrectly reconstructed, and even for segments 500-1,000bp long, a substantial fraction (46%) is incorrect (Fig.2).

Here, we propose a probabilistic model for sequence evolution by duplication, and we design a sampling algorithm that explicitly accounts for uncertainty in the estimation of segment trees and allows for breakpoint reuse. The results of Zhang et al. (2008) suggest that, in spite of an improved model, there may still be many solutions of similar likelihood. The stochastic sampling approach allows us to examine multiple solutions in the same framework and extract expectations for quantities of interest (e.g., the expected number of events on individual branches of the phylogeny, or local properties of the ancestral sequences).

Our problem is also closely related to the problem of reconstruction of gene trees and their reconciliation with species trees. Even though the recent algorithms for gene tree reconstruction (e.g., Wapinski et al. (2007)) consider genomic context of individual genes as an additional piece of information, our

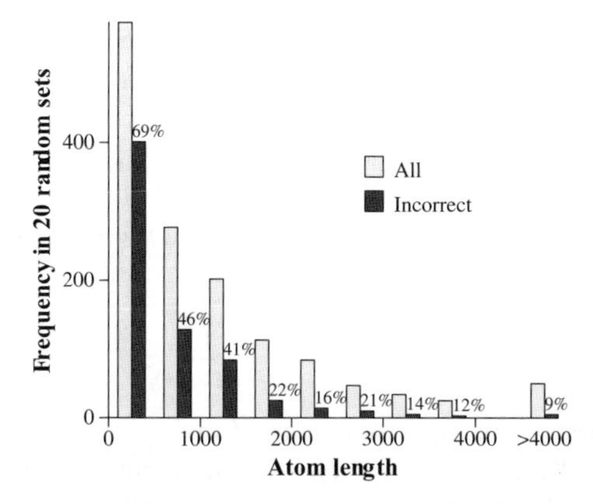

Fig. 2. Distribution of atomic segment lengths and accuracy of segment tree inference in 20 simulated fast-evolving clusters (see Section 4). The gray bars show the numbers of segment types. The black bars show the percentages of segment types for which the highest posterior probability unrooted segment tree inferred by MrBayes (Ronquist and Huelsenbeck, 2003) does not match the correct segment tree.

new methods aim to fully explain genomic context of individual genes through reconstructed duplication histories.

2 Probabilistic Model

In this section, we give a probabilistic model of evolution of gene clusters through segmental duplication on a given species tree T. Later, we use this model for inference of duplication histories and to generate simulated gene clusters.

We start with an ancestral sequence of length N. The sequence evolves by duplications, deletions and substitutions. A *duplication* copies a source region and inserts the new copy at a target position in the sequence, either on the forward strand (with probability $1 - P_i$) or on the reverse strand (with probability P_i). A duplication can be characterized by four coordinates: a *centroid* (the midpoint of the region between the leftmost and rightmost end of the duplication), the *length* of the source region, the *distance* between the source and the target, and a *direction* (from left to right or from right to left). The centroid is chosen uniformly, and the length and distance are chosen from given distributions (see below). Note that some centroid, distance, and length combinations are invalid; those combinations are rejected. Similarly, a *deletion* removes a portion of the sequence, and can be characterized by its *centroid* and *length*. Each event is a deletion with probability P_x, and a duplication with probability $(1 - P_x)$. This process straightforwardly defines the probability $P(E \mid \text{len})$ of any duplication or deletion event E. Here, len is the length of the sequence just before the event E. The number of events on each branch is governed by a Poisson process with rate λ, and thus the probability of observing k events on a branch of length ℓ is $P_n(k, \ell) = (\lambda\ell)^k e^{-\lambda\ell}/k!$.

A duplication history H generated in this way implies a set $\sigma(H)$ of *atomic segments* of several types and a segment tree T_x for each atom type x. The substitutions in the nucleotide sequences of atom x are governed by the HKY substitution model along the segment tree T_x.

We can compute the joint probability $p(H, X)$ of a given set of extant sequences X and a history H (up to a normalization constant) as follows. Let T be a species tree with branches $b_1, b_2, \ldots$ Then:

$$p(H, X) \propto \prod_{b_i \in T} P(H, b_i) \times \prod_{x \in \sigma(H)} P(X_x \mid T_x), \quad (1)$$

where $P(H, b_i)$ is the probability of events of history H that occur on branch b_i of the species tree, X_x represents nucleotide sequences of atoms of type x, and $P(X_x \mid T_x)$ is the probability of these sequences given tree T_{x_i}. For a sequence of events $E_1, \ldots, E_k$ on branch b_i, the probability $P(H, b_i)$ is simply:

$$P(H, b_i) = P_n(k, \ell) \prod_{j=1}^{k} P(E_j \mid \text{len}(j-1)) \quad (2)$$

where $\text{len}(j-1)$ is the length of the sequence before event E_j.

To reduce the number of model parameters, we use geometric distributions to model lengths and distances of duplication events. To estimate these distributions, we have used the lengths and distances estimated by Zhang et al. (2008) from human genome gene clusters (mean length 14,307, mean distance 306,718). The geometric distributions seem to approximate the observed length distributions reasonably well. Similarly, we estimated the probability of duplication with inversion as $P_i = 0.39$ from the same data, we set the probability of deletion as $P_x = 0.05$, and the length distribution of deletions matches the distribution of duplication lengths. In our model, we do not allow inversions that are not accompanied by a duplication.

Note that for our application, the normalization constant for $p(H, X)$ does not need to be computed. We assume a uniform prior on length distribution of ancestral lengths. This has only a small effect for fixed extant sequences, since the ancestral sequence should contain a single occurrence of each segment type, and therefore the ancestral length is determined mostly by the length of individual atomic segment types. Some combinations of centroids, distances, and lengths will be rejected, but we assume that in long enough sequences, the effect of this rejection step will be negligible and we ignore it altogether.

In the inference algorithm below, we compute likelihood $P(X_x \mid T_x)$ and branch lengths for each segment tree separately. This independence assumption simplifies computation and allows variation of rates and branch lengths between atoms. This is desirable, since sequences of different functions may evolve at different substitution rates, and selection pressures may change the proportions of individual branch lengths. Nonetheless, branch lengths tend to be correlated among segment trees when individual atoms are duplicated together, and this information is lost by separating the likelihood computations. We are working on a more systematic solution to the problem of rate and branch length variation.

3 Metropolis–Hastings Sampling

For inference of duplication histories, we use the Metropolis–Hastings Markov chain Monte Carlo algorithm (Hastings, 1970) to sample from the posterior probability distribution $p(H \mid X)$ defined in the previous section, conditional on the extant sequences X and their atomization. The result of the algorithm is a series of samples that can be used to estimate expectations of quantities of interest (e.g., the number of events on individual branches, posteriors of individual segment trees, and particular ancestral sequences), or to examine high likelihood histories.

Briefly, the Metropolis–Hastings algorithm defines a Markov chain whose stationary distribution is the target distribution, but the moves of the Markov chain are defined through a different *proposal distribution* due to the difficulties of sampling from the target distribution directly. We start by initializing sample history H_0. In each iteration, we use a randomized *proposal algorithm* described below to propose a candidate history H_i' according to a distribution conditional on sample H_{i-1}. Sample H_i' is either *accepted* ($H_i := H_i'$) with probability $\alpha(H_{i-1}, H_i')$, or *rejected* ($H_i := H_{i-1}$) otherwise. The acceptance probability $\alpha(H, H')$ is used to ensure that the stationary distribution of the Markov chain is indeed the target distribution (Hastings, 1970):

$$\alpha(H, H') = \min\left(1, \frac{p(H'|X)q(H \mid H')}{p(H|X)q(H' \mid H)}\right), \tag{3}$$

where $q(H' \mid H)$ is the probability of proposing history H' if the previous history was H.

The proposal algorithm starts by sampling an unrooted *guide tree* T_x for every atom type x. The segment trees implied by the proposed history will be later rooted and refined from these guide trees. Guide tree T_x is sampled from the posterior distribution of the trees conditional on a fixed multiple alignment of all instances of atom type x. Sampling guide trees accounts for uncertainty in the estimation of segment trees. We collapse branches with fewer than 5 expected substitutions over the length of the atom sequence, since such short branches usually cannot be estimated reliably. Thus, the guide trees for shorter atoms, where uncertainty is high, will be close to uninformative star trees, while the trees for longer atoms will remain more resolved.

The proposal algorithm samples a history consistent with the given set of guide trees, starting at the leaves of the trees, and progressively sampling groups of atom pairs to merge, until only a single copy of each atom remains. Merging of two groups of atoms corresponds to unwinding one duplication. To obtain a valid history consistent with the guide trees, each of the two groups has to be a contiguous subsequence in the current atomic sequence. Also, the corresponding atoms of the two groups must be of the same type. Finally, the corresponding atoms must be cherries in their guide trees (see also Fig.3; the leaves x_i and x_j are cherries in T_x if they have the same parent.)

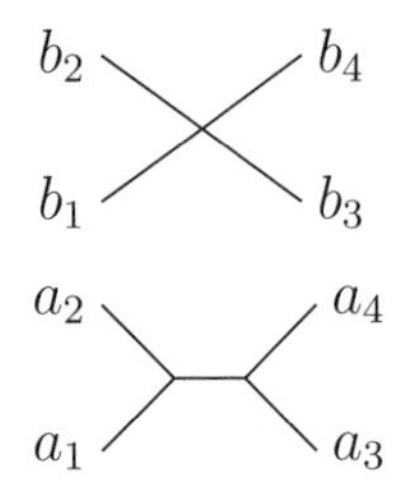

Fig. 3. Consistency of histories with the guide trees. Example of guide trees for an atomic sequence $a_1b_1a_2b_2a_3b_3a_4b_4$. The segment tree T_b was collapsed into an uniformative star tree, perhaps since atom b is too short. Duplication (a_1b_1, a_2b_2) is consistent with the guide trees since (a_1, a_2) and (b_1, b_2) are cherries in the corresponding guide trees, but duplication (a_1b_1, a_3b_3) is inconsistent, as is $(a_1b_1a_2b_2, a_3b_3a_4b_4)$.

For example, if the most recent duplication copied atoms a_1b_1 to atoms a_2b_2 then a_1 and a_2 must be cherries in the tree T_a, and b_1 and b_2 must be cherries in the tree T_b. Unwinding of this duplication will correspond to removal of a_2 and b_2 from the trees T_a and T_b and from the atomic sequence. Now, the same conditions can be applied to the second latest duplication. In this way, a particular set of guide trees can significantly restrict the set of possible histories.

The sampling distribution over candidate groups of atoms is determined by a series of heuristic penalties described in Appendix A, favoring longer duplications, those not inducing breakpoint reuse, and those that were used in the previous sample H_{i-1}. Even though this algorithm employs a number of heuristics to improve the overall performance of the sampler, they only affect the mixing rate and convergence properties of the sampling, not the asymptotic correctness of the MCMC algorithm.

The multiple alignment for each segment type is created by MUSCLE (Edgar, 2004). Even though it is possible to sample multiple alignments to prevent potential alignment errors from propagating throughout the whole analysis (Holmes and Bruno, 2001), such sampling is by itself computationally intensive. Given that in this paper we consider sequences of greater than 90% similarity, we do not expect multiple alignments to be a major source of error in our reconstructions. Trees, branch lengths, and HKY nucleotide substitution model parameters are sampled by MrBayes (Ronquist and Huelsenbeck, 2003) with uniform prior over tree topologies, and default priors for the other parameters. For each segment type x, all the tree samples are precomputed in a run of 10,000 iterations with a burn-in of 2,500 samples, keeping every 10th sampled tree. In every iteration of the history proposal algorithm, we keep the previous guide tree with 95% probability, otherwise we choose a new tree randomly from the pre-computed samples.

Deletion operations cannot be easily dated, and some of them cannot be even observed in the extant sequence. To address this problem, we attach each deletion to the most recent overlapping duplication or speciation and in the proposal algorithm, described in Appendix A, we always propose duplications and corresponding deletions together. To keep the running time feasible, we assume that there is at most one deletion following each duplication and that it does not extend beyond the boundaries of the corresponding duplication segment.

4 Experiments

We have implemented the MCMC sampler described above and verified its functionality on simulated data. For the simulations, we have estimated branch lengths and HKY model parameters (equilibrium frequencies and transition/ transversion ratio) from the UCSC syntenic alignments (Karolchik et al., 2008) of human, chimp, and macaque on human chromosome 22.

We created 20 simulated gene clusters in each of the following two categories: slow evolving and fast evolving (duplication rate at 200 and 300 times substitutions per site, respectively). We have applied our algorithm to atomic segments derived from the simulation, with short ($<$ 500bp) atomic segments removed to emulate the increase in breakpoint reuse due to imperfect identification of alignment boundaries in real data sets (see Tab.1 for the data set overview). For each cluster, we ran two chains of the sampler from random starting points for up to 10,000 iterations each, discarding the first 2,500 samples as burn-in. The sampler seems to converge reasonably quickly (supplementary Fig.B1).

In the majority of cases, we predict the correct number of events (Tab.2; 14 out of 20 for slowly evolving, 15 out of 20 for fast evolving clusters). Note that in some cases the predicted number of events is lower than the actual number of events: this is likely due to events that become invisible in the extant sequences because of subsequent deletions. We have compared our results on human lineage to Zhang et al. (2008), and on the whole tree to Zhang et al. (2009). The performance has improved, especially in the case of fast evolving clusters. We also examined the correctness of distribution of events along the phylogeny (supplementary Tab.B1). Finally, we compared predicted and actual ancestral atomic sequences. To quantify the differences between the sequences, we have counted the number of breakpoints required to transform the predicted ancestral sequence to the actual ancestor (Fig.4). In the majority of cases (31 out of 40), the expected number of breakpoints is smaller than 0.5.

Beyond the simulated data, we have applied our algorithm to the following gene cluster sequences: PRAME (human-macaque phylogeny), AMY (human-macaque

Table 1. Overview of simulated and real data sets

	slow rate			**fast rate**			PRAME	AMY	UGT1A
	min	max	mean	min	max	mean			
Seq. len (kb)	91	295	172	120	387	219	1000, 200	221, 170	210, 210, 250
No. atom types	15	53	36	39	57	48	39	44	55
No. duplications	5	24	15	18	29	23	34.9 ± 0.8	23.4 ± 0.8	22.9 ± 0.8
No. deletions	0	3	0.8	0	3	1.1	9.4 ± 1.9	15.2 ± 1.9	20.2 ± 1.3
Species		H,C,R			H,C,R		H,R	H,R	H,C,O

Table 2. Performance evaluation. The table shows the histogram of differences between the real number of events and the predicted number of events along the human lineage and on the whole tree on the 40 simulated data sets (20 with slow duplication rate 200, 20 with fast duplication rate 300). MCMC: rounded expected number of events from all samples. ML: highest likelihood sample. We compare to results of Zhang et al. (2008) on single species (Z2008) and Zhang et al. (2009) on the whole tree (Z2009). Note that the results of the two programs are not directly comparable, since our program was run on correct atomization with short atoms filtered out (giving Z2008 and Z2009 advantage of smaller amount of breakpoint reuse in the data), while Z2008 and Z2009 used their own built-in atomization method (giving advantage to our program, since the results of their atomization may be potentially incorrect).

	human lineage										**whole tree**									
	slow rate					**fast rate**					**slow rate**					**fast rate**				
Method	<0	0	1	2	>2	<0	0	1	2	>2	<0	0	1	2	>2	<0	0	1	2	>2
MCMC	1	15	1	3	0	1	16	1	2	0	3	14	2	0	1	2	15	2	1	0
ML	1	15	1	2	1	1	16	0	3	0	3	14	1	1	1	2	13	2	3	0
Z2008	3	14	2	1	0	1	6	4	3	6										
Z2009											0	8	5	2	5	0	0	3	2	15

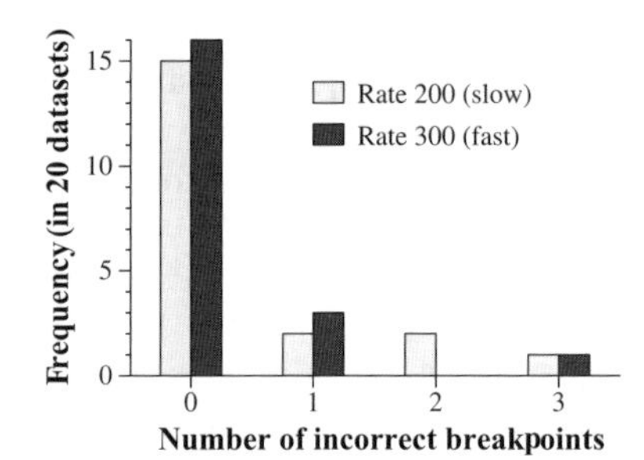

Fig. 4. Histogram of expected number of incorrect breakpoints on the 40 simulated data sets. The number of breakpoints required to transform predicted sequences to actual sequences is computed over all MCMC samples and the average is rounded to the closest integer.

phylogeny), and UGT1A (human-chimp-orang phylogeny). PRAME cluster (preferentially expressed antigen in melanoma) is one of the most active gene clusters in the human genome, and shows strong evidence of positive selection (Birtle et al., 2005; Gibbs et al., 2007). AMY cluster contains five amylase genes that are responsible for digestion of starch. It appears to have expanded much faster in humans than in other primates, according to aCGH experiments (Dumas et al., 2007). The UGT1A cluster consists of multiple isoforms of a single gene, instrumental in transforming small molecules into water-soluble and excretable metabolites. This gene has at least thirteen unique alternate first exons resulting from duplications at various stages of mammalian evolution. UGT1A provides an unusual opportunity for studying promoter evolution.

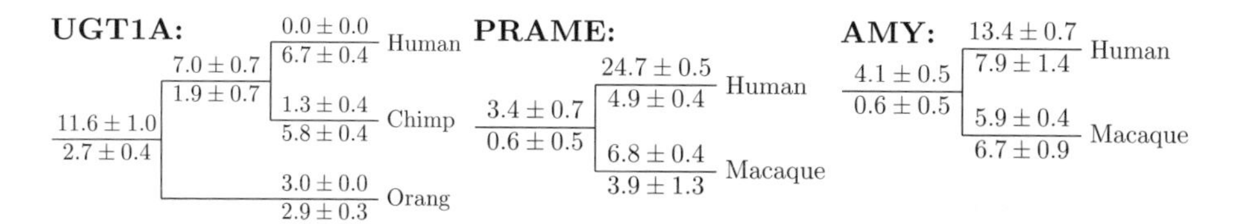

Fig. 5. Estimated numbers of events. For each cluster, we show the posterior mean and standard deviation of the number of duplications (above the branch) and deletions (below the branch) as assessed by MCMC sampling. The root branch shows events up until 90% sequence similarity cutoff.

Recently duplicated clusters are grossly missassembled in shotgun based genomes (Green, 2001; Zhang et al., 2008). To prepare our data sets, we have first screened sequenced BACs from chimp, orangutan, and macaque for similarity with the corresponding human sequences, assembled BACs into longer contigs, and selected subregions whose ends showed clear homology with upstream and downstream sequences of the human cluster. To identify atomic segments, we divide the sequences into equally sized 500bp windows, and for each window we find approximate copies in all available sequences at 90% identity cutoff. The atoms are assigned in a greedy way (starting from the windows with the largest number of copies), and windows overlapping already assigned atoms are discarded. Finally, atoms that always occur in pairs are merged into longer atoms. Table 1 shows an overview of the resulting sequences and atoms.

For each cluster, we ran five chains from different starting points for 5,000–10,000 samples, discarding the first 2,500 samples. We have estimated the number of duplications and deletions overall (Table 1), and on individual branches of the phylogeny (Fig.5). The estimated numbers of duplications for PRAME and AMY are comparable to those of Zhang et al. (2008). With UGT1A, we obtain higher estimates possibly due to differences in our atomization procedure or effects of the additional species in the analysis.

Figure 6a shows the highest likelihood reconstruction of the history of the UGT1A cluster. The cluster contains several isoforms of the same five-exon gene. Exons 2-5 are shared among all the isoforms, while exon 1 is alternatively spliced. The reconstruction shows division of the first exons into three distinct groups, and their ortholog/paralog relationships in human, chimp, and orangutan.

While the duplication history of the UGT1A cluster consists of mostly ancient events, the PRAME cluster (Fig.6b) shows recent large-scale duplications, especially in the human lineage. Figure 6 shows such events by several co-linear bifurcations at the same level of the tube tree. The reconstruction of the history by traditional methods (gene tree/species tree reconciliation) is complicated by the presence of recent duplications (99% similarity), and chimeric genes (Gibbs et al., 2007). We address these issues by considering multiple guide trees for each atom as well as spatial configuration of atoms in multiple species. However, the predicted history is by no means perfect. Rhesus sequence exhibits large regions that apparently arose by a single event, yet we split this event due to mistakes

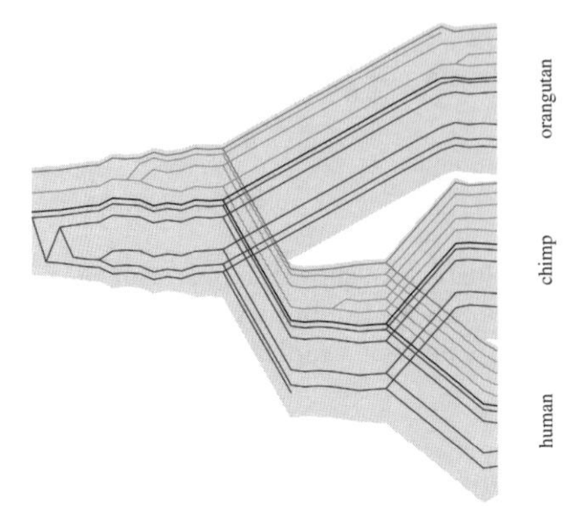

(a) UGT1A cluster consists of several five exon isoforms of the same gene. Exons 2-5 (red) are shared among all the isoforms, the first exons (blue, black, green) are alternatively spliced.

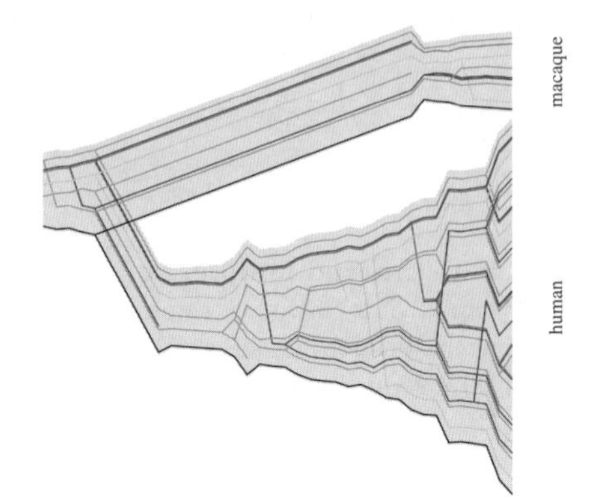

(b) PRAME cluster consists of multiple copies of the PRAME gene. A typical copy has three coding exons, the highlighted atoms overlap exon 2. Some of the genes were pseudogenized.

Fig. 6. Highest likelihood reconstruction of duplication histories. The branch lengths in the figures do not correspond to the actual branch lengths. The atoms are ordered in their order along the genomic sequences (extant and ancestral).

in atomic representation. We expect that improved procedure for segmenting sequence into atoms will address this problem.

5 Discussion

In this paper, we have introduced a new model of evolution of gene clusters and designed an algorithm to reconstruct high probability evolutionary histories of these clusters. We have tested our method on both simulated and real data. Comparative genomics methods traditionally concentrate on sequences where 1:1 orthology can be established. In case of gene clusters, this is rarely the case. Our efforts in reconstruction of gene cluster histories will support further development of comparative genomic tools to analyze these complex regions.

Gene clusters should not be seen only as a confounding factor. The number of orthologous sequences, their divergence, and phylogenetic relationships greatly impact the accuracy of comparative genomic studies. For example, Kosiol et al. (2008) has shown that the sensitivity of positive selection scans is improved by considering sequences from a complex phylogeny. Studies based on orthologous regions between species can at present use a phylogeny of up to 10 orthologous copies of a particular mammalian gene from genomes sequenced at reasonable quality. On the other hand, some clusters contain many more copies with significantly more complex phylogeny even within a single species (for example, the PRAME cluster contains more than 30 copies in the human genome alone).

Thus, the gene clusters provide an opportunity for refined look at evolution of genes and genomes. Multiple sources of evidence suggest that many interesting developments in genomes happen within the boundaries of gene clusters, which further increases interest in their study. Multiple efforts are currently under way to BAC sequence selected gene clusters in multiple species and in multiple populations (Zhang et al., 2008; Zody et al., 2008). Accurate methods and models for reconstruction of duplication histories of these clusters are essential in understanding the evolution, function, and biomedical implications of these regions.

The general framework of our method allows future developments. One limitation of our sampler is its low sample acceptance ratio, indicating low level of mixing in the Markov chain. We plan to devise a systematic way for tuning the parameters in the proposal distribution towards better acceptance ratios. We also plan to improve the underlying probabilistic model. Currently the branch lengths in segment trees are chosen independently of the duplication history. Instead, we plan to consistently date duplication events on each branch, and use a scaling parameter for each atom type so that we can accurately model correlation between branch lengths of individual atom types and at the same type allow rate variation in different parts of the sequence. We use a simple HKY substitution model with variance in rates allowed between individual atomic segments. In future work, it will be possible to employ more complex models of sequence evolution, such as variable rate site models and models of codon evolution, within the same framework. Such extensions will allow us to identify sites and branches under selection in gene clusters in a principled way, and contribute towards better functional characterization of these important genomic regions. An interesting alternative approach might be to use combinatorial optimization instead of sampling to find maximum likelihood history in the above model.

Acknowledgements. We would like to thank Devin Locke and LaDeana Hillier at Washington University, St. Louis for providing us with the BAC sequences from chimp, orangutan, and macaque. We would also like to thank Webb Miller and Yu Zhang for helpful discussions on this problem. Research of TV and BB is funded by European Community FP7 grants IRG-224885 and IRG-231025.

References

An, P., et al.: APOBEC3G genetic variants and their influence on the progression to AIDS. J. Virol. 78(20), 11070–11076 (2004)

Becker, T.S., Lenhard, B.: The random versus fragile breakage models of chromosome evolution: a matter of resolution. Mol. Genet. Genomics 278(5), 487–491 (2007)

Bertrand, D., Lajoie, M., El-Mabrouk, N.: Inferring ancestral gene orders for a family of tandemly arrayed genes. J. Comput. Biol. 15(8), 1063–1067 (2008)

Birtle, Z., Goodstadt, L., Ponting, C.: Duplication and positive selection among hominin-specific PRAME genes. BMC Genomics 6, 120 (2005)

Bitti, P.P., et al.: Association between the ancestral haplotype HLA A30B18DR3 and multiple sclerosis in central Sardinia. Genet. Epidemiol. 20(2), 271–273 (2001)

Degenhardt, J.D., et al.: Copy number variation of CCL3-like genes affects rate of progression to simian-AIDS in Rhesus Macaques (Macaca mulatta). PLoS Genet. 5(1), e1000346 (2009)

Dumas, L., Kim, Y.H., Karimpour-Fard, A., Cox, M., Hopkins, J., Pollack, J.R., Sikela, J.M.: Gene copy number variation spanning 60 million years of human and primate evolution. Genome Res. 17(9), 1266–1267 (2007)

Edgar, R.C.: MUSCLE: a multiple sequence alignment method with reduced time and space complexity. BMC Bioinformatics 5, 113 (2004)

Elemento, O., Gascuel, O., Lefranc, M.P.: Reconstructing the duplication history of tandemly repeated genes. Mol. Biol. Evol. 19(3), 278 (2002)

Finckh, U., et al.: Genetic association of a cystatin C gene polymorphism with late-onset Alzheimer disease. Arch. Neurol. 57(11), 1579–1583 (2000)

Gibbs, R., et al.: Evolutionary and biomedical insights from the rhesus macaque genome. Science 316(5822), 222–224 (2007)

Green, E.D.: Strategies for the systematic sequencing of complex genomes. Nat. Rev. Genet. 2(8), 573 (2001)

Hajdinjak, T., Zagradisnik, B.: Prostate cancer and polymorphism D85Y in gene for dihydrotestosterone degrading enzyme UGT2B15: Frequency of DD homozygotes increases with Gleason Score. Prostate 59(4), 436–439 (2004)

Hastings, W.K.: Monte Carlo sampling methods using Markov chains and their applications. Biometrika 57, 97–109 (1970)

Holmes, I., Bruno, W.J.: Evolutionary HMMs: a Bayesian approach to multiple alignment. Bioinformatics 17(9), 803–810 (2001)

Karolchik, D., et al.: The UCSC Genome Browser Database: 2008 update. Nucleic Acids Res. 36(Database issue), D773–D779 (2008)

Kosiol, C., Vinar, T., da Fonseca, R.R., Hubisz, M.J., Bustamante, C.D., Nielsen, R., Siepel, A.: Patterns of positive selection in six Mammalian genomes. PLoS Genet. 4(8), e1000144 (2008)

Lajoie, M., Bertrand, D., El-Mabrouk, N., Gascuel, O.: Duplication and inversion history of a tandemly repeated genes family. J. Comput. Biol. 14(4), 462–468 (2007)

Lander, E.S., et al.: Initial sequencing and analysis of the human genome. Nature 409(6822), 860–921 (2001)

Ma, J., Ratan, A., Raney, B.J., Suh, B.B., Miller, W., Haussler, D.: The infinite sites model of genome evolution. Proc. Natl. Acad. Sci. USA 105(38), 14254–14261 (2008)

Nadeau, J.H., Taylor, B.A.: Lengths of chromosomal segments conserved since divergence of man and mouse. Proc. Natl. Acad. Sci. USA 81(3), 814–818 (1984)

Ohno, S.: Evolution by Gene Dupplication. Springer, Berlin (1970)

Peng, Q., Pevzner, P.A., Tesler, G.: The fragile breakage versus random breakage models of chromosome evolution. PLoS Comput. Biol. 2(2), e14 (2006)

Ronquist, F., Huelsenbeck, J.P.: MrBayes 3: Bayesian phylogenetic inference under mixed models. Bioinformatics 19(12), 1572–1574 (2003)

Schwartz, S., Kent, W.J., Smit, A., Zhang, Z., Baertsch, R., Hardison, R.C., Haussler, D., Miller, W.: Human-mouse alignments with BLASTZ. Genome Res. 13(1), 103–107 (2003)

Tang, K.S., Chiu, H.F., Chen, H.H., Eng, H.L., Tsai, C.J., Teng, H.C., Huang, C.S.: Link between colorectal cancer and polymorphisms in the uridine-diphosphoglucuronosyltransferase 1A7 and 1A1 genes. World J. Gastroenterol 11(21), 3250–3254 (2005)

Wapinski, I., Pfeffer, A., Friedman, N., Regev, A.: Automatic genome-wide reconstruction of phylogenetic gene trees. Bioinformatics 23(13), i549–i558 (2007)

Zhang, Y., Song, G., Hsu, C.-H., Miller, W.: Simultaneous History Reconstruction for Complex Gene Clusters in Multiple Species. In: Pacific Symposium on Biocomputing (PSB), vol. 14, pp. 162–173 (2009)

Zhang, Y., Song, G., Vinar, T., Green, E.D., Siepel, A., Miller, W.: Reconstructing the Evolutionary History of Complex Human Gene Clusters. In: Vingron, M., Wong, L. (eds.) RECOMB 2008. LNCS (LNBI), vol. 4955, pp. 29–49. Springer, Heidelberg (2008)

Zody, M.C., et al.: Evolutionary toggling of the MAPT 17q21.31 inversion region. Nat. Genet. 40, 1076–1083 (2008)

Co-evolutionary Models for Reconstructing Ancestral Genomic Sequences: Computational Issues and Biological Examples

Tamir Tuller[1,2,3,4,⋆], Hadas Birin[1,⋆], Martin Kupiec[2], and Eytan Ruppin[1,3]

[1] School of Computer Science
[2] Department of Molecular Microbiology and Biotechnology
[3] School of Medicine, Tel Aviv University
{tamirtul,birinhad,martin,ruppin}@post.tau.ac.il
[4] Faculty of Mathematics and Computer Science,
Weizmann Institute of Science

Abstract. The inference of ancestral genomes is a fundamental problem in molecular evolution. Due to the statistical nature of this problem, the most likely or the most parsimonious ancestral genomes usually include considerable error rates. In general, these errors cannot be abolished by utilizing more exhaustive computational approaches, by using longer genomic sequences, or by analyzing more taxa. In recent studies we showed that co-evolution is an important force that can be used for significantly improving the inference of ancestral genome content.

In this work we formally define a computational problem for the inference of ancestral genome content by co-evolution. We show that this problem is NP-hard and present both a Fixed Parameter Tractable (FPT) algorithm, and heuristic approximation algorithms for solving it. The running time of these algorithms on simulated inputs with hundreds of protein families and hundreds of co-evolutionary relations was fast (up to four minutes) and it achieved an approximation ratio < 1.3.

We use our approach to study the ancestral genome content of the Fungi. To this end, we implement our approach on a dataset of 33,931 protein families and 20,317 co-evolutionary relations. Our algorithm added and removed hundreds of proteins from the ancestral genomes inferred by maximum likelihood (ML) or maximum parsimony (MP) while slightly affecting the likelihood/parsimony score of the results. A biological analysis revealed various pieces of evidence that support the biological plausibility of the new solutions.

Keywords: Co-evolution, reconstruction of ancestral genomes, maximum parsimony, maximum likelihood.

1 Introduction

The problem of reconstructing ancestral genomic sequences is as old as the field of molecular evolution. The first approach for inferring ancestral genomic sequences

⋆ TT and HB contributed equally to this work.

F.D. Ciccarelli and I. Miklós (Eds.): RECOMB-CG 2009, LNBI 5817, pp. 164–180, 2009.

was suggested by Fitch around 40 years ago [18]. This first algorithm assumed a binary alphabet, and was based on the Maximum Parsimony (MP) criteria, *i.e.* find the labels to the internal nodes of a tree that minimize the number of mutations along the tree edges. Over the years this basic algorithm was generalized in many ways. Sankoff showed how to efficiently solve versions of the maximum parsimony problem with a non-binary alphabet and with multiple edge weights [43]. Similar algorithms for inferring ancestral sequences based on maximum likelihood (ML; instead of maximum parsimony) were suggested more than 15 years later [1,40,8,17,30,38]. Recently, similar approaches were used to infer ancestral sequences in phylogenetic networks [25].

Dozens of biological studies have dealt with the reconstruction of ancestral genomic sequences and ancestral genomes. For example, reconstruction of ancestral sequences was used for understanding the origins of genes and proteins [54,46,45,24,22,21,4,3,37], and for aligning genomic sequences [23], and for inferring ancestral enzymes and genomes [51,29,41,34,7,6].

The main problem related to reconstructing ancestral sequences and genomes is that in practice many times the reconstructed sequences contain a large number of errors. A major source of this phenomenon is the existence of multiple local and/or global maxima in the solution space searched by both the maximum likelihood and the maximum parsimony approaches (see *e.g.* [5,48]). Thus, many times our confidence in the most likely or most parsimonious reconstructed ancestral state is not too high. As the above mentioned algorithms assume that different sites and different genes/proteins evolve independently, this problem cannot be solved by adding more samples or more taxa [32].

Several studies demonstrated that functionally and physically interacting proteins tend to co-evolve [26,44,12,49,16], and that co-evolutionary relations between proteins are quite ubiquitous [10,9,13,15,11]. Some of these previous investigations used the fact that interacting proteins have correlative evolution in order to successfully predict physical interactions (for example, see [26,44,12]).

Based on these results, we recently suggested a different approach, the Ancestral Co-Evolver (ACE), for improving the accuracy of reconstructed ancestral genomes [48]. Our approach was based on utilizing information embedded in the co-evolution of functionally/physically interacting proteins. We used this approach to study the genome content of the Last Universal Common Ancestor (LUCA). In this work we give a formal description of the ancestral co-evolution problem, we analyze its computational complexity and describe algorithms for solving it; the performances of these algorithms are demonstrated by a simulation study. Additionally, we use the ACE for studying a new biological example, the ancestral genome content of the Fungi.

2 Definitions and Preliminaries

For simplicity, we assume a binary alphabet. However, all the results here can be easily generalized to models with more than two characters. In this work we assume that in general, if they do not have co-evolutionary relation, neighbor sites in the input sequences evolve independently. Thus, the basic components in the model and algorithms are single characters.

Our goal is to reconstruct the ancestral states for a set of organisms $\mathcal{T}$ of size $|\mathcal{T}| = n$. A *phylogenetic tree* is a rooted binary tree $T = (\mathrm{V}(T), \mathrm{E}(T))$ together with a *leaf labeling* function λ, where $\mathrm{V}(T)$ is the set of vertices and $\mathrm{E}(T)$ the set of edges. In our context, a weight table is attributed to each edge $(u, v) = e \in \mathrm{E}(T)$. This *weight table* includes a weight (a positive real number) for each pair of labels of the two vertices $(u, v) = e$.

In this work, we assume that each node in a phylogenetic tree corresponds to a different organism. The leaves in a phylogenetic tree correspond to organisms that exist today ($\mathcal{T}$), while the internal nodes correspond to organisms that have become extinct ($\mathcal{T}'$). Thus, the *leaf labeling* function is a bijection between the leaf set $\mathrm{L}(T)$ and the set of organisms that exist today, $\mathcal{T}$.

In our binary case, each label is a binary sequence; all the sequences have the same length. In the case of conventional ML/MP, as we assume an i.i.d. case where different characters in a sequence evolve independently, we can describe the algorithm for sequences of lengths one: i.e. either $'1'$ or $'0'$. A *full labeling* of a phylogeny $\hat{\lambda}(T)$ is a labeling of all nodes of the tree such that the labels at the leaves are the same as in λ, i.e., for all $l \in \mathrm{L}(T)$ $\lambda(l) = \hat{\lambda}(l)$.

We can name each node after its corresponding organism. Let $O_T(\cdot)$ denote a function that returns the index of the organisms corresponding to each node in T, *i.e.* for every $v \in V(T)$, $O_T(v)$ is the index of the organism (from $\mathcal{T} \cup \mathcal{T}'$) corresponding to v.

A *co-evolving forest* $F = (S_F = \{T_1, T_2, ...\}, E_c(S_F))$ is a set of *phylogenetic trees,S_F*, with *identical* topology that correspond to the same organisms [*i.e.* each tree has the same $O(\cdot)$], and an additional set of edges, $E_c(S_F)$, that connect pairs of nodes in *different* trees. This set of edges and represent the co-evolutionary relations between pairs of protein families. Edges in $E_c(S_F)$ must connect pairs of nodes that correspond to the same organism (*i. e.* $(v, u) \in E_c(S_F), v \in V(T_1), u \in V(T_2) \Longrightarrow O_{T_1}(v) = O_{T_2}(u)$; Figure 1); we call such pairs of nodes *legal co-evolutionary pairs.*

The edges in $E_c(S_F)$ are named *co-evolution edges* while edges that are part of the evolutionary trees are named *tree edges.* For example, Figure 1 *A.* includes a *co-evolving forest* with two trees (the *co-evolution edges* are dashed with arrows while the *tree edges* are continuous). As co-evolutionary relations. In this work we assume that new *co-evolutionary edges* do not appear/dissapear during evolution. Namely, we assume that if there is a *co-evolutionary edge* between a *legal co-evolutionary pair* of nodes in two trees than all the *legal co-evolutionary pairs* of nodes in the two trees are connected by *co-evolutionary edge.*

A *full labeling* of a *co-evolving forest* $\hat{\lambda}(S_F)$ is a full labeling, $\{\hat{\lambda}(T_1), \hat{\lambda}(T_2), ...\}$, of all the nodes of the trees in S_F. The roots of a *co-evolving forest* are the set of roots of the *phylogenetic trees* in the *co-evolving forest.*

As mentioned, a *co-evolving forest* also includes a weight table for each *co-evolution edge* and each *tree edge.* These weight tables are cost functions that return a real positive number for each pairs of labels at the two ends of the edge.

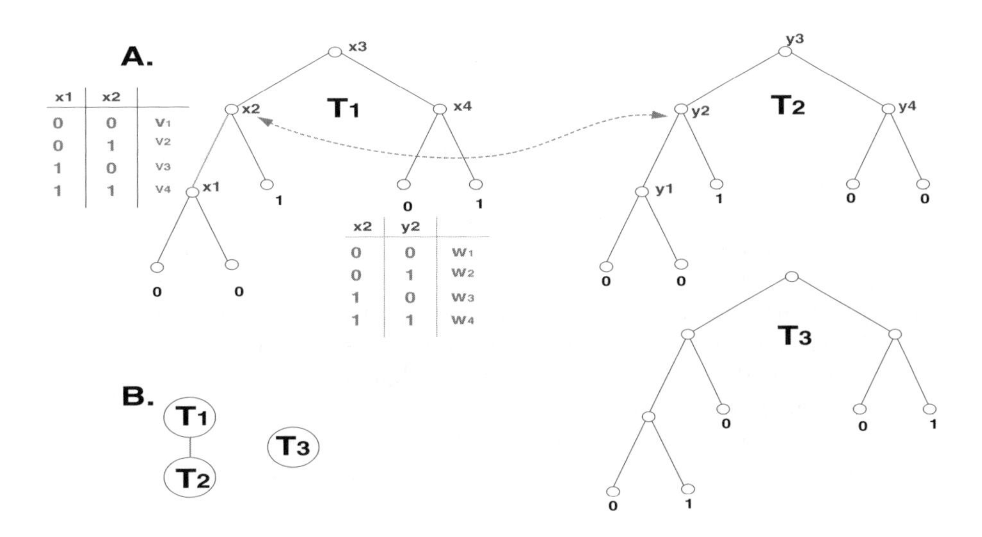

Fig. 1. *A.* A simple example of a *co-evolving forest* with three trees, and one *co-evolution edge* connecting node x_2 in tree T_1 and node y_2 in tree T_2; the weight table corresponding to this co-evolution edge is in red. The weight table corresponding to the tree edge (x_1, x_2) in T_1 is in green. *B.* The *co-evolutionary graph* corresponding to the *co-evolving forest* in *A*.

In the case of *tree edges*, these weights reflect the probability of a mutation along the edge. In the case of *co-evolution edges*, these weights reflect the distribution of mutual occurrence of the labels of the nodes at the ends of the edge.

This leads us to the formal definition of the problem we are concerned with, the *Ancestral co-evolution* problem:

Problem 1. Ancestral co-evolution ($Ancest - co - evol$)
Input: A *co-evolving forest*, $F = (S_F, E_c(S_F))$, and a real number, B.
Question: Are there labels for the internal nodes of all the trees in the *co-evolving forest* such that the sum of the corresponding weights along all the tree edges and the *co-evolution* edges is less than B?

The following example demonstrates the advantages of the ancestral co-evolver compared to the simple *i.i.d* parsimony approach.

Example 1. Consider the co-evolving forest that appears in Figure 1, and assume that all the weight tables of the tree edges are identical to the table that appears in the figure where $[v_1, v_2, v_3, v_4] = [0, 1, 1, 0]$. It is easy to see that there are two MP solutions for the labels of the internal nodes in the phylogenetic tree T_1: either $[x_1, x_2, x_3, x_4] = [0, 0, 0, 0]$ or $[x_1, x_2, x_3, x_4] = [0, 1, 1, 1]$ gives the same score (2). In the case of the tree T_2, it is easy to see that there is one MP solution: all the labels of the internal nodes are $'0'$. Now, suppose that in the weight table corresponding to the co-evolution edge (x_2, y_2), $w1$ is the smallest entry. Thus, by co-evolution study, the solution $[x_1, x_2, x_3, x_4] = [0, 0, 0, 0]$ is more plausible and the ambiguity in the labels of T_1 is solved.

Note that in general it is not necessarily required that the solution of each tree *separately* will be the most parsimonious (see the next sections). The minimal sum of edge weights corresponding to a *co-evolving forest*, F (problem 1) is named the *cost* of F. A *co-evolutionary graph* is an undirected graph that describes the *co-evolution* edges in the *co-evolving forest.* In such a graph, each node corresponds to a tree in the *co-evolving forest*, and two nodes are connected by an edge if there is at least one co-evolution edge between their corresponding trees. A connected component in the *co-evolving forest* is a sub-set of trees whose corresponding nodes in the co-evolutionary graph induce a connected component (see an example in Figure 1 B.)

We finish this section with an observation that will be used in the next sections (the proof is omitted due to lack in space and will appear in the full version of the paper).

Observation 1. *The optimization problem of inferring the ancestral states of a phylogenetic tree when the optimization criteria is maximum likelihood (see, for example, [40]) under i.i.d models such as Jukes Cantor (JC) [27], Neyman [36], or the model of Yang* et al. *[51] can be formalized as a maximum parsimony problem for non-binary alphabet and with multiple edge weights [43].*

Observation 1 teaches us that the *Ancestral co-evolution* problem without *co-evolution edges* ($|E_c(S_F)| = 0$) can describe a Maximum Likelihood (ML) problem. In this work, indeed the weights of the *tree edges* correspond to the probabilities to gain/lose proteins and thus we describe and solve a generalization of both the ML and the MP problems on trees.

3 Hardness Issues

In this section we show that the *Ancestral co-evolution* problem is NP-hard. We will show it by reduction from the $max-2-sat$ problem which is known to be NP-hard [20] (the reduction appears in Figure 2).

Problem 2. Maximum 2-Satisfiability ($max - 2 - sat$)
Input: Set U of variables, collection C of clauses over U such that each clause $c \in C$ has $|c| = 2$, and a positive integer $K \leq |C|$.
Question: Is there a truth assignment for U that simultaneously satisfies at least K of the clauses in C?

Theorem 1. *The* $Ancest - co - evol$ *problem is NP-hard.*

Proof. By reduction from $max - 2 - sat$. Given an input $< U, C, K >$ to the $max-2-sat$ problem reconstruct the following input to the $Ancest-co-evol$ problem (See Figure 2): $|S_F| = |U|$, and each tree in S_F corresponds to one variable in U; each tree has the same structure and leaf labels as described in

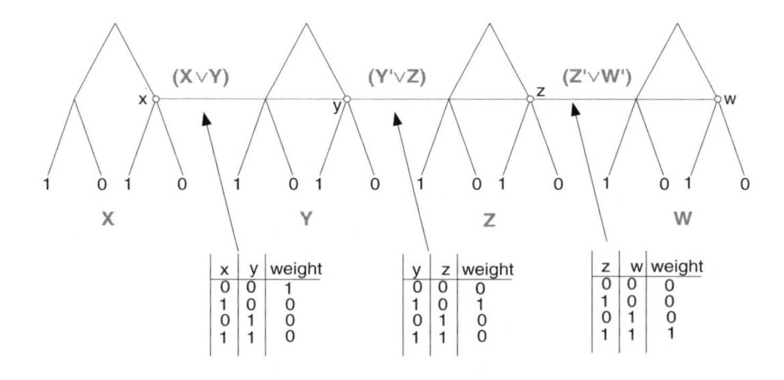

x	y	weight
0	0	1
1	0	0
0	1	0
1	1	0

y	z	weight
0	0	0
1	0	1
0	1	0
1	1	0

z	w	weight
0	0	0
1	0	0
0	1	0
1	1	1

Fig. 2. Reduction from $max - 2 - sat$ to $Ancest - co - evol$

Figure 2. Each *co-evolution* edge in E_C corresponds to one clause in C, and connects the right-most internal nodes in two trees (see Figure 2). The translation of the four types of possible clauses (true-true, true-false, false-false) to the weight matrix of its corresponding *co-evolution* edge appears in Figure 2. Finally, choose $B = 2 \cdot |U| + |C| - |K|$.

It is easy to see that the optimal parsimony score for each tree in the $Ancest - co - evol$ problem (excluding the *co-evolution* edges) is 2; either a solution that labels all the internal nodes with '0' or a solution that labels all the internal nodes with '1' gives this score.

$\Longrightarrow$ Suppose the answer to the $max - 2 - sat$ problem is YES (*i.e.* there is a truth assignment for U that simultaneously satisfies at least K of the clauses in C). In this case, we choose the labeling of all the internal nodes of each tree to be identical to the assignment of its corresponding variable in U. Thus, the contribution of all the tree edges to the parsimony score (*i.e.* excluding the *co-evolution* edges) is $2 \cdot |U|$.

By the construction of the *weight tables* (Figure 2), the contribution to the parsimony score due to *co-evolution* edges that correspond to one of the K satisfied clauses is 0, and the contribution to the parsimony score due to each of the other *co-evolution* edges is at most 1. Thus, the contribution of all the *co-evolution* edges to the parsimony score is at most $|C| - |K|$, and the answer to the $Ancest - co - evol$ problem is YES.

$\Longleftarrow$ Suppose the answer to the $Ancest - co - evol$ problem is YES (*i.e.* the parsimony score along all the edges is no more than $2 \cdot |U| + |C| - |K|$). As mentioned earlier, for optimal parsimony score, in each tree all the internal nodes should be labeled with the same label as that of the internal node that is connected by *co-evolution* edges. Thus, the contribution of all the *tree edges* to the parsimony score is $2 \cdot |U|$, and the contribution of all *co-evolution* edges to the parsimony score is at most $|C| - |K|$. Thus, the contribution to the parsimony score for K of the *co-evolution* edges is 0. By the reconstruction of the *co-evolution* edges, they correspond to K clauses in C that are be satisfied when the assignment to each variable in U is identical to the labeling of the internal nodes of its corresponding tree. Thus, the answer to the $max - 2 - sat$ problem is YES.

4 Methods

4.1 An FPT Algorithm and Approximation Heuristics

As we have shown in the previous section, the *Ancestral co-evolution* problem is NP-hard. In this section, we describe an FPT algorithm and an approximation algorithm for the *Ancestral co-evolution* problem. The approximation algorithm is described in Figure 3. It has three main steps: 1) The input *co-evolving forest* is partitioned into smaller *co-evolving forests*; 2) optimal labels are assigned to the internal nodes of each of these *co-evolving forests* by an FPT algorithm; in total, these labels are an approximation of the solution for the input *co-evolving forest*; 3) finally, the solution is further improved greedily.

We will start by describing an FPT algorithm that finds the optimal solution for the *Ancestral co-evolution* problem in time complexity that is exponential with the number of trees in S_F but polynomial with the other properties of the input. This algorithm is used in step 2) of the approximation algorithm where it is implemented on subsets of S_F. Similarly to many algorithms for computing the labels of internal tree nodes [19,43], our algorithm has two phases (in the first phase, it traverses the *co-evolving forest* from the leaves to the root; in the second phase, it traverses the *co-evolving forest* from the root to the leaves). However, our algorithm is performed jointly for all the trees in each connected component.

Let $e = (i, j)$ denote a *co-evolution edge*. Let $W^c_{e;(a,b)} = W^c_{(i,j);(a,b)}$ denote the cost of assigning a to node i, and b to node j where (i, j) is a co-evolution edge. Similarly, if (i, j) is a *tree edge* we use $W^b_{e;(a,b)} = W^b_{(i,j);(a,b)}$ to denote the cost of assigning a and b to the two nodes of e respectively. Let $S_{\bar{t}}(\bar{v})$ denote the cost of a sub-forest in the *co-evolving forest* whose roots are $\bar{t}$, when assigning $\bar{v}$ to these roots.

Let t_j denote the j-th node in the vector of nodes $\bar{t}$. Let $\bar{k}$ and $\bar{l}$ denote the corresponding vectors of children of $\bar{t}$ in the *co-evolving forest*. In the first phase, all $S_{\bar{t}}(\bar{v})$ are computed by the following dynamic programming formula (see Figure 3 C.):
$S_{\bar{t}}(\bar{v}) = min_{\overline{u}}\{\sum_j W^b_{(t_j,l_j);(v_j,u_j)} + S_{\bar{l}}(\overline{u})\} + min_{\overline{w}}\{\sum_j W^b_{(t_j,k_j);(v_j,w_j)} + S_{\overline{k}}(\overline{w})\} + \sum_{j_1,j_2:(j_1,j_2)\in E_c} W^c_{(t_{j_1},t_{j_2});(v_{j_1},v_{j_2})}$

In the second phase, the algorithm traverses the sub-forest from the roots to the leaves, and optimal values are assigned to the internal nodes of the *co-evolving forest* by the following dynamic programming formula (see Figure 3 C.):

For the roots of the *co-evolving forest*:
$\bar{v^*} = argmin_{\overline{v}}\{S_{\bar{t}}(\overline{v})\}$
For a general vector of internal nodes $\bar{k}$ corresponding to the same organism in the *co-evolving forest* (after the values $\bar{t^*}$ were assigned to its parents, $\bar{t}$):
$\bar{v^*} = argmin_{\overline{v}}\{\sum_j W^b_{(t_j k_j);(t^*_j,v_j)} + S_{\overline{k}}(\overline{v})\}$

The running time of this algorithm on a *co-evolving forest* with m' trees of size n is $O(n \cdot 2^{2*m'})$ since, in each of the $O(n)$ it checks all the $2^{m'} \times 2^{m'}$ possible simultaneous labels to all nodes corresponding to the direct descendant of all nodes corresponding to the same organism.

As the running time of the algorithm described above is exponential with the size of *co-evolving forest*, the general algorithm (see Figure 3 A.) has an initial stage (stage 1) where the input graph is partitioned into small enough connected components.

As the running time of the algorithm described above is exponential with the size of the largest connected component in the *co-evolutionary graph*, the general algorithm (see Figure 3 A.) has an initial stage where the input graph is partitioned into small enough connected components. The input to the general algorithm (ACE) includes the maximal size of a connected component after this stage. Let K denote this parameter.

This step is described in Figure 3 B. and is performed by an algorithm that recursively implements k-means [35] on the *co-evolutionary graph*. On the first iteration, the number of clusters is $|V|/K$ where $|V|$ is the number of vertices in the co-evolutionary graph. If the size of some cluster is larger than K, the algorithm is executed recursively on this cluster to further partite it to smaller connected components. The algorithm stops when all parts of the graph (connected components) are smaller than K. Though the problem of clustering is NP-hard, in practice, and as reported in the next section, the k-means algorithm is very fast.

The input to this step is a weighted graph whose edges correspond to the edges in the *co-evolutionary graph*. The weights of the graph edges can be any measure that represents the strength of the co-evolution between the corresponding trees (for example, the correlation between the phyletic pattern of the corresponding proteins).

The final step of the *Ancestral-Co-Evolver* algorithm is a greedy stage (3 *D.*). In each iteration, the greedy algorithm searches for an edge and labels its ends in a way that improves the cost of the *co-evolving forest*. As demonstrated in the simulations in section 5.2, this algorithm converge to a local optimum faster than the running time of the dynamic programming stage with $K = 7$. Note that the greedy algorithm can be stopped after a certain number of iteration if it does not converges to a local optimum. The greedy stage can be used as an independent algorithm when running it from various initial points (*e.g.*, one of the initial points can be the ML solution).

Let $\hat{\lambda}(S_F)$ denote the labels found by the *Ancestral-Co-Evolver* algorithm. Let $MP(\hat{\lambda}(S_F))$ denote the parsimony score induced by these labels. Let $MP^{-}(\hat{\lambda}(S_F))$ denote the parsimony score induced by these labels when not considering the co-evolution edges *between* the connected components found by the *Partite* algorithm; let E^{-} denote this set of co-evolution edges. An upper bound on the approximation ratio of the general algorithm is given in the following observation (the proof is omitted due to lack in space and will appear in the full version of the paper):

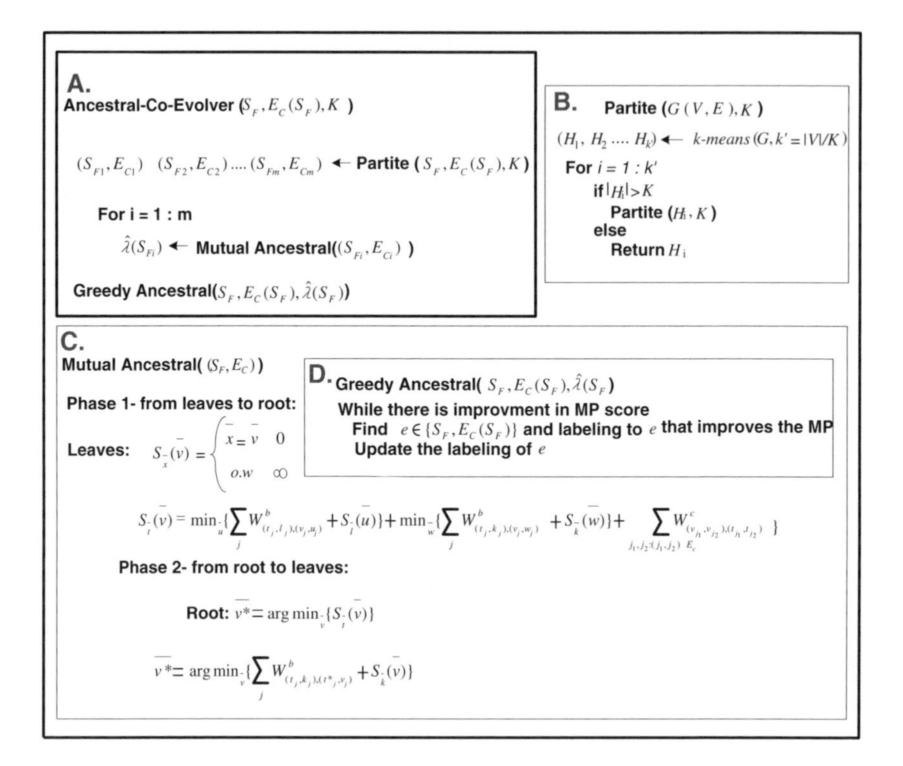

Fig. 3. A. The general algorithm for inferring ancestral states under co-evolution. B. The algorithm for partitioning the co-evolutionary graph. C. The algorithm for finding the ancestral states of a connected component (output of the Partite algorithm). D. The Greedy stage of the algorithm.

Observation 2. *The approximation ratio of the* Ancestral-Co-Evolver *algorithm is* $\leq \frac{MP(\hat{\lambda}(S_F))}{MP^{-}(\hat{\lambda}(S_F))}$.

We used Observation 2 in order to estimate the approximation ratios of the different algorithms in the simulations.

One important property of the algorithm is that it enables a trade-off between accuracy and speed. A larger K (see Figure 3) increases the running time exponentially but at the same time increases the accuracy of the solution; $K = |S_F|$ will give the optimal solution for the *Ancestral co-evolution* problem.

Finally, it is important to note that by weighting the *co-evolution edges* relatively to the *tree edges* we can control the relative influence of these two sources of information (co-evolution *vs.* the evolutionary tree) on the resulting labels. Thus, for example, it is easy to see that (and a very similar proof can be outlined for the case where *tree edges* are rational numbers):

Let $\hat{\lambda}(S_F/E_C)$ denote the set of the optimal labels of S_F when not considering the co-evolution edges (*i.e.* $E_C = 0$).

Observation 3. *If the weight tables of the* tree edges *are natural numbers and all the entries in the weight tables of the* co-evolution edges $< \frac{1}{|E_c(S_F)|}$ *then the* optimal *labeling of the* co-evolving forest *is* one *of the optimal labels in* $\hat{\lambda}(S_F/E_C)$.

Due to lack in space the proof of Observation 3 is omitted and will appear in the full version of this paper. By Observation 1 the *ancestral co-evolution* problem without *co-evolution edges* describes a conventional maximum likelihood problem. Thus, by Observation 3, if we choose small enough weights for the *co-evolution edges* our method can be used for choosing *one* of the optima (or a point very close to *one* of the optima) of the maximum likelihood function – the one that is supported by co-evolutionary relations.

4.2 Weighting of the Co-evolution Edges

We tested several values for the weights of the *co-evolution edges* compared to the *tree edges*. At one extreme, the entries of the tree edges weight tables are multiplied by a very large constant. In this case, the tree edge weight tables are dominant compared to the weights of the *co-evolution edges* (the solution is one of the ML/MP solutions). At the second extreme, the fifth weighting, the *co-evolution edge* weights are dominant compared to the tree edge weights. In this case, the entries of the tree edges weight tables are multiplied by a very large constant.

Let $MP^b(S_F, W)$ denote the parsimony score when solving the *ancestral co-evolution* problem with weighting W and when considering only *tree edges*. Let $MP^c(S_F, W)$ denote the parsimony score when solving the *ancestral co-evolution* problem with weighting W and when considering only *co-evolution edges*. In this work, we used the weighting, W^*, that optimizes the sum of the two sources of information (co-evolution, and the evolutionary trees); *i.e.* we used $W^* = argmin_W(\frac{MP^c(S_F,W)}{min_W(MP^c(S_F,W))} + \frac{MP^b(S_F,W)}{min_W(MP^b(S_F,W))})$.

5 Experimental Results

5.1 Simulated Evolution

To analyze the performances of the algorithms described in the previous section we generate a probabilistic process that describes the evolution of a *co-evolving forest*. In the simulation, each character evolves along the branches of the evolutionary trees, but also has correlations with the other characters that interact with it. The simulation was performed as in [48]; due to lack in space, all the details about the simulations are deferred to the full version of this paper.

5.2 Simulation Results

We compared the performances of the following algorithms: 1) The Partitioning algorithm (Figure 3 *B*.) with the Dynamic Programming algorithm (Figure 3

C.). 2) The greedy algorithm (Figure 3 D.) with a few initial points (one of them is the the ML solution). 3) The ACE algorithm (all the stages in Figure 3). 4) The ML and MP algorithms that do not consider the *co-evolution edges*. Let DPX denotes a Dynamic Programming algorithm with $K = X$; let $ACEX$ denotes an ACE algorithm with $K = X$.

A summary of the simulation results appears in Figure 4; sub-figures $A - C$ depict the running times (log scale) and sub-figures $D - F$ describe the approximation ratios as functions of the size of the *evolutionary trees*, the number of *evolutionary trees*, the number of co-evolution edges per node in the *co-evolutionary graph.* All the running finished in less than a four minutes. As can be seen, the running time increases exponentially with K (see $ACE7$, $DP7$ *vs.* $ACE2$, $DP2$ sub-figures $A - C$) while the running time of the greedy algorithm alone is larger than $DP2$ but exponentially smaller than $DP7$. As can be seen, in the case of running time, the most influential parameter is the number of *evolutions trees* in the input.

In the case of the approximation ratio (the upper bound from Observation 2), the most influential parameter is number of co-evolution edges per node in the *co-evolutionary graph.* As can be seen, $ACE7$ performs better than all the other algorithms and always has approximation ratio < 1.3. Interestingly, the greedy algorithm is only a few percentages worse. These results support using the greedy algorithm if running time matters. The fact that the *upper bound* of the $ACE7 < 1.3$ demonstrates that our approach can find solutions that are very close to the optimal ones. As we used here an *upper bound* on the approximation ratio the actual ratio can be significantly lower.

In the simulation, as in the case of biological data (see the next section), the ML solutions (when ignoring co-evolution) are relatively similar to the solutions found by our approach. Thus, it is not surprising that approximation ratio of ML is bound to be < 1.7. On the other hand, the margin between the approximation ratio of the ML and that of the ACE is significant: up to 30%. This result demonstrates the essentiality of our approach.

5.3 A Biological example: Reconstruction of the Ancestral Genome Content the Fungi

Using the method outlined above we set to reconstruct the ancestral genome content of 17 Fungi whose evolutionary tree appears in Figure 5. The input included 33,931 families of Fungi orthologs (downloaded from [11]) and a total of 20,317 co-evolution edges.

We represented each of the 17 genomes by a binary string of length 33931 where $'1'$ in the x position of a string means that there is a gene/protein from the x group of orthologs in this genome, and $'0'$ means that there is no gene/protein from the x group of orthologs in this genome. We used Neyman's two state model [36], a version of Jukes Cantor (JC) model [27] for inferring the edge lengths of the tree by maximum likelihood. This was done by PAML [50]. These edge lengths correspond to the probabilities that a protein will appear/vanish along the corresponding lineage.

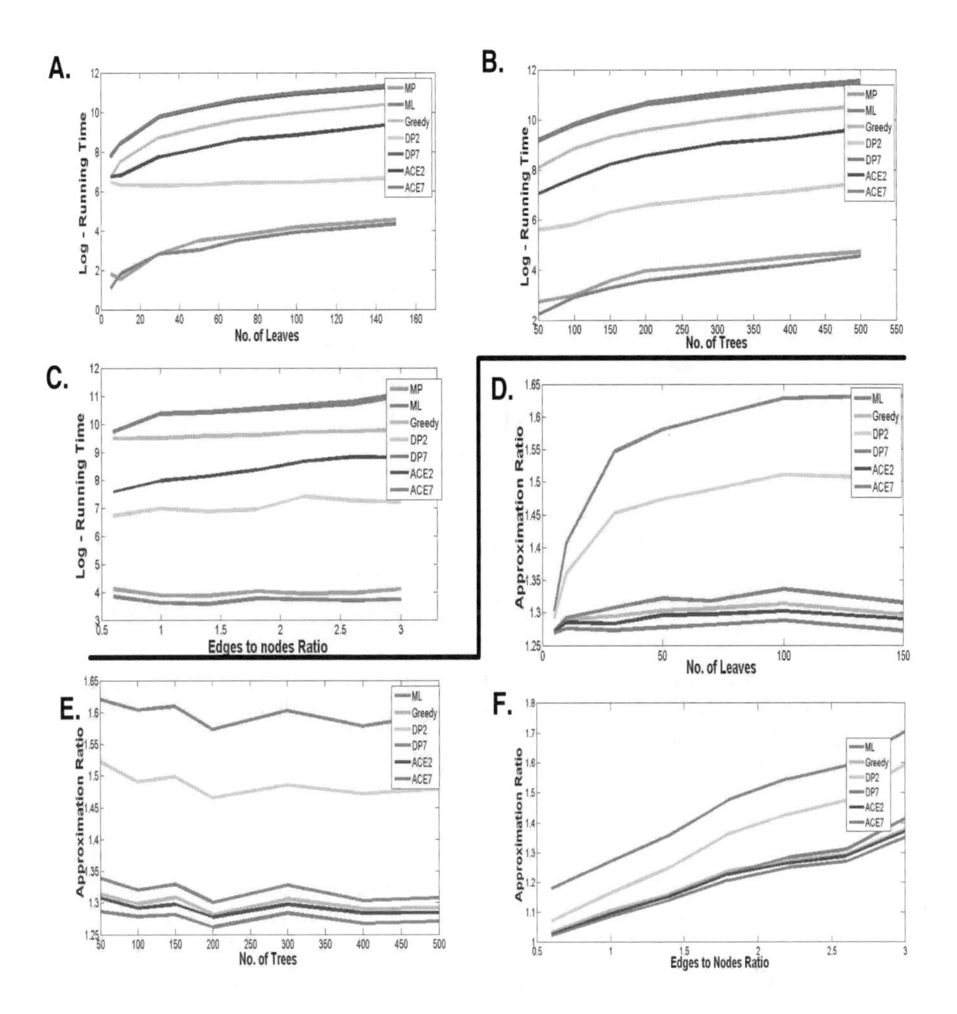

Fig. 4. Running time and accuracy of the partitioning algorithm with the Dynamic Programming algorithm, the greedy algorithm, the ACE, and the ML/MP algorithms. *A.–C.*: The *log* running time (in ms) of the algorithms as function of the size of the *evolutionary trees* (*A.*; 200 trees and 2.5 *co-evolution edges* per node in the *co-evolutionary graph*), number of *co-evolutionary trees* (*B.*; 2.5 *co-evolution edges* per node in the *co-evolutionary graph*, each *evolutionary tree* with 70 leaves), and the number of *co-evolution edges* per node in the *co-evolutionary graph* (*C.*; 200 *evolutionary trees*, each with 70 leaves). *D. – F.*: The upper bound on the approximation ratio (Observation 2) of the solution found by each of the algorithms as function of the number of leaves in each *evolutionary tree* (*D.*), the number of *evolutionary trees* (*E.*), and the number of *co-evolution edges* per node in the *co-evolutionary graph* (*F.*).

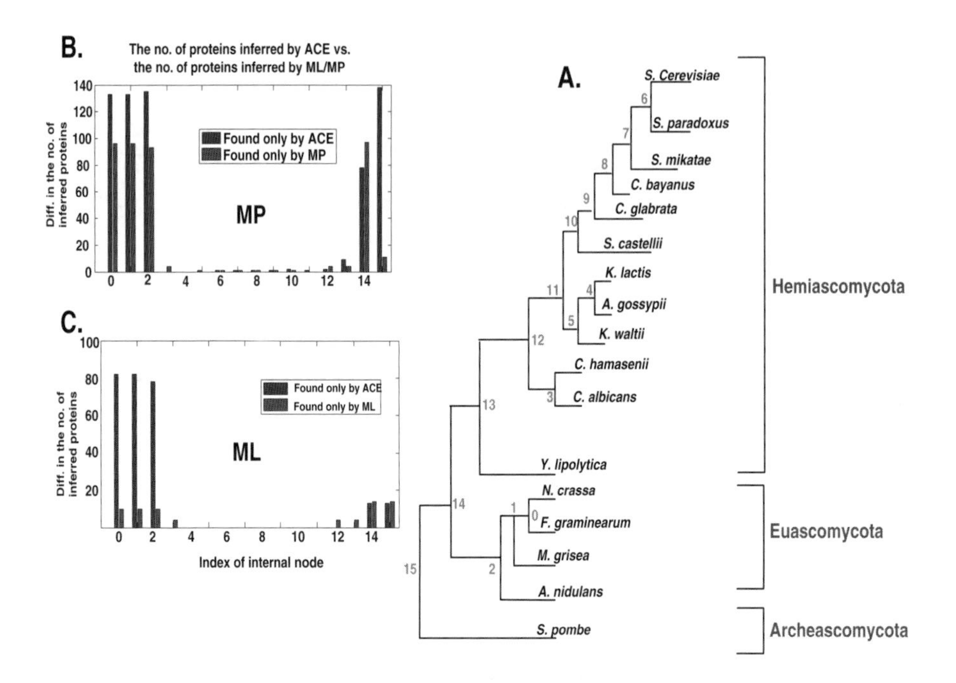

Fig. 5. *A.* The evolutionary tree of the Fungi that was analyzed by the ACE. *B.* Summary of the results for the MP case; the number of proteins inferred by ACE and not by MP and vice versa. *C.* Summary of the results for the ML case; the number of proteins inferred by ACE and not by ML and vice versa.

As co-evolution edges we use various physical and functional interactions that were downloaded from String [14] (http://string.embl.de/; for example, [15,49] reported the relation between co-evolution and similar functionality). We filtered co-evolutionary edges whose their ratio between the highest and lowest probability in the co-occurrence distribution table was less than 4.25. The weights in the tables were computed according to the co-occurrence probabilities of the corresponding pairs of orthologs. We used the weighting whose corresponding solution optimizes both the score of the *co-evolution edges* and the score of the *tree edges* (see more details in subsection 4.2). The annotation of ancestral proteins was based on the GO annotations of *S. cerevisiae.*

We compared our results of the ACE to those obtained by using only *tree edges* (by using MP and ML).

The total running time of the ACE algorithm on this large biological dataset was about 1.5 hours on a conventional PC. Figure 5 summarizes the results of the analysis by ACE. As can be seen, ACE removed/added hundreds of proteins to ML/MP labels of the internal nodes of the evolutionary trees. A major fraction of the discrepancies between the results of the ACE algorithm and those obtained with the conventional methods (ML and MP) appears in the *Euascomycota* subtree (internal nodes $0-2$ in the Fungi tree; see Figure 5 *A.*). In general, ACE mainly added proteins to these nodes, implying that ML/MP underestimated the size of these ancestral genomes. Additionally, both in the case of MP and ML, the ACE added/removed many proteins from internal nodes 14 and 15. It is important to note that the likelihood score of the ACE solution was only 0.25% lower than the ML score, demonstrating that this solution was *very* close to the ML point. Similarly, when the ACE was implemented with MP the parsimony score of its solution was only 2.2% higher than that of the MP solution. These solutions, however, are supported by the co-evolutionary information and thus are more biologically plausible.

We further analyzed the proteins added by the ACE to the ML solution for the ancestors of the *Euascomycota* (internal nodes $0-2$): the nodes with the largest number of discrepancies between ML and ACE. The groups of proteins added to each of these nodes were very similar (around 95% similarity); thus we report only the results for node 2, the ancestor of the *Euascomycota*.

ACE added 89 proteins to the ML solution of internal node 2. Various pieces of evidence support the biological plausibility of the addition of these proteins by ACE: First, the group of proteins added to this node was enriched with proteins that take part in basic and essential metabolic processes. Specifically, it was enriched with the cellular process: *protein amino acid phosphorylation* (p-value = 0.00054), *amine transport* (p-value = 0.00153), and *amino acid transport* (p-value = 0.00518); all p-values passed the False Discovery Rate (FDR) control for multiple hypothesis testing [2]. Second, all these proteins have orthologs in most of the analyzed Fungi (on average in 76% of the Fungi); this fact also supports the essentiality of these proteins. Third, in *S. cerevisiae*, many of these proteins are part of the same complex with proteins inferred by the standard ML (note that co-evolutionary relations used by ACE do not necessarily imply association in the same complex).

The following is a typical example that demonstrates the three points mentioned above:

Example 1: Three of the proteins added by the ACE are orthologs of *S. cerevisiae TPK1*, *TPK2*, and *TPK3*. The presence of at least one of these genes is required for normal growth in *S. cerevisiae* [47]. These genes are part of the complex *cAMP-dependent protein kinase complex* that includes another protein (*BCY1*), which was inferred by ML.

6 Concluding Remarks

In this study we formally described a new computational approach for reconstructing ancestral genomic sequences using information about co-evolution. Our model captures co-evolutionary dependencies between different proteins and uses this information to disambiguate the labels of the reconstructed ancestral genomes.

In the future we intend to generalize this work in several ways. First, currently, our approach is presented in the setup of ancestral genome reconstruction, due to the importance of this problem and because co-evolutionary information can be readily obtained on the gene/protein level. However, it is important to note that the potential scope of our approach goes beyond the ancestral genome reconstruction problem, to tackle the more general problem of ancestral sequence reconstruction: that is, the reconstruction of different sites or domains in proteins, and even (provided that sufficient information is available) the reconstruction of single sites in DNA or RNA sequences (see [52,53,33,39,28,42]). In this setup, the success of such future applications depends on the existence of reliable co-evolutionary information on the individual site level. Second, it is also clear that our approach can be generalized in the future to more complex reconstruction models (for example, using non-binary alphabets, dependency between close sites, and various versions of maximum likelihood). Third, we intend to design algorithms that may compete with those described in this work. Specifically, we intend to check algorithms that are based on the belief propagation [31] approach. Finally, we believe that a generalization of the approach described in this work can be used for *joint* inference of ancestral genomes and protein interactions or for *joint* inference of ancestral genomes and metabolic networks of ancestral and extant organisms.

Acknowledgment. T.T. is a Koshland Scholar at Weizmann Institute. MK and ER were supported by grants from the Israel Science Foundation.

References

1. Barry, D., Hartigan, J.: Statistical analysis of humanoid molecular evolution. Stat. Sci. 2, 191–210 (1987)
2. Benjamini, Y., Hochberg, Y.: Controlling the false discovery rate: a practical and powerful approach to multiple testing. J. Roy. Statistical Society 57(1), 289–300 (1995)
3. Blanchette, M., et al.: Reconstructing large regions of an ancestral mammalian genome in silico. Genome Res. 14, 2412–2423 (2004)
4. Cai, W., et al.: Reconstruction of ancestral protein sequences and its applications. BMC Evolutionary Biology 4, e33 (2004)
5. Chor, B., et al.: Multiple maxima of likelihood in phylogenetic trees: An analytic approach. Mol. Biol. Evol. 17(10), 1529–1541 (2000)
6. Cohen, O., et al.: A likelihood framework to analyse phyletic patterns. Philos. Trans. R. Soc. Lond. B. Biol. Sci. 363(1512), 3903–3911 (2008)
7. Csurös, M., Miklós, I.: Streamlining and large ancestral genomes in archaea inferred with a phylogenetic birth-and-death model. Mol. Biol. Evol. (2009)
8. Elias, I., Tuller, T.: Reconstruction of ancestral genomic sequences using likelihood. J. Comput. Biol. 14(2), 216–237 (2007)
9. Barker, D., et al.: Constrained models of evolution lead to improved prediction of functional linkage from correlated gain and loss of genes. Bioinformatics 23(1), 14–20 (2007)
10. Pazos, F., et al.: Correlated mutations contain information about protein-protein interaction. J. Mol. Biol. 271, 511–523 (1997)
11. Wapinski, I., et al.: Natural history and evolutionary principles of gene duplication in fungi. Nature 449, 54–65 (2007)

12. Wu, J., et al.: Identification of functional links between genes using phylogenetic profiles. Bioinformatics 19, 1524–1530 (2003)
13. Marino-Ramirez, L., et al.: Co-evolutionary rates of functionally related yeast genes. Evol. Bioinformatics, 2295–2300 (2006)
14. Jensen, L.J., et al.: String 8–a global view on proteins and their functional interactions in 630 organisms. Nucleic Acids Res. 37, D412–D416 (2009)
15. Chena, Y., et al.: The coordinated evolution of yeast proteins is constrained by functional modularity. Trends Genet. 22(8), 416–419 (2006)
16. Felder, Y., Tuller, T.: Discovering local patterns of co-evolution. In: Nelson, C.E., Vialette, S. (eds.) RECOMB-CG 2008. LNCS (LNBI), vol. 5267, pp. 55–71. Springer, Heidelberg (2008)
17. Felsenstein, J.: Phylip (phylogeny inference package) version 3.5c. Technical report, Department of Genetics, University of Washington, Seattle (1993)
18. Fitch, W.: Toward defining the course of evolution: minimum change for a specified tree topology. Syst. Z. 20, 406–416 (1971)
19. Fitch, W.M., Margoliash, E.: Construction of phylogenetic trees. Science 155, 279–284 (1967)
20. Garey, M.R., Johnson, D.S.: Computer and Intractability. Bell Telephone Laboratories, incorporated (1979)
21. Gaucher, E.A., et al.: Inferring the palaeoenvironment of ancient bacteria on the basis of resurrected proteins. Nature 425, 285–288 (2003)
22. Hillis, D.M., et al.: Application and accuracy of molecular phylogenies. Science 264, 671–677 (1994)
23. Hudek, A.K., Brown, D.G.: Ancestral sequence alignment under optimal conditions. BMC Bioinformatics (2005)
24. Jermann, T.M., et al.: Reconstructing the evolutionary history of the artiodactyl ribonuclease superfamily. Nature 374, 57–59 (1995)
25. Jin, G., et al.: Maximum likelihood of phylogenetic networks. Bioinformatics 22(21), 2604–2611 (2006)
26. Juan, D., et al.: High-confidence prediction of global interactomes based on genome-wide coevolutionary networks. Proc. Natl. Acad. Sci. USA 105(3), 934–939 (2008)
27. Jukes, T.H., Cantor, C.R.: Evolution of protein molecules. In: Munro, H.N. (ed.) Mammalian protein metabolism, pp. 21–123. Academic Press, New York (1969)
28. Knudsen, B., Hein, J.: Rna secondary structure prediction using stochastic context-free grammars and evolutionary history. Bioinformatics 15, 446–454 (1999)
29. Koshi, M., Goldstein, R.: Probabilistic reconstruction of ancestral protein seuences. JME 42, 313–320 (1996)
30. Krishnan, N.M., et al.: Ancestral sequence reconstruction in primate mitochondrial dna: Compositional bias and effect on functional inference. MBE 21(10), 1871–1883 (2004)
31. Kschischang, F.R., et al.: Factor graphs and the sum-product algorithm. IEEE Transactions on Information Theory 47(2), 498–519 (2001)
32. Li, G., et al.: More taxa are not necessarily better for the reconstruction of ancestral character states. Systematic Biology 57(4), 647–653 (2008)
33. Lockless, S.W., Ranganathan, R.: Evolutionarily conserved pathways of energetic connectivity in protein families. Science 286(5438), 295–299 (1999)
34. Ma, J., et al.: Reconstructing contiguous regions of an ancestral genome. Genome Res. 16(12), 1557–1565 (2006)
35. MacQueen, J.B.: Some methods for classification and analysis of multivariate observations. In: Proc. of 5-th Berkeley Symposium on Mathematical Statistics and Probability, vol. 1, pp. 281–297 (1967)

36. Neyman, J.: Molecular studies of evolution: A source of novel statistical problems. In: Gupta, S., Jackel, Y. (eds.) Statistical Decision Theory and Related Topics, p. 127. Academic Press, New York (1971)
37. Ouzounis, C.A., et al.: A minimal estimate for the gene content of the last universal common ancestor–exobiology from a terrestrial perspective. Res. Microbiol. 157(1), 57–68 (2006)
38. Pagel, M.: The maximum likelihood approach to reconstructing ancestral character states of discerete characters on phylogenies. Systematic Biology 48(3), 612–622 (1999)
39. Pedersen, J.S., et al.: Identification and classification of conserved rna secondary structures in the human genome. PLoS. Comp. Bio. 2, e33 (2006)
40. Pupko, T., et al.: A fast algorithm for joint reconstruction of ancestral amino acid sequences. Mol. Biol. Evol. 17(6), 890–896 (2000)
41. Rascola, V.L., et al.: Ancestral animal genomes reconstruction. Current Opinion in Immunology 19(5), 542–546 (2007)
42. Rzhetsky, A.: Estimating substitution rates in ribosomal rna genes. Genetics 141, 771–783 (1995)
43. Sankoff, D.: Minimal mutation trees of sequences. SIAM Journal on Applied Mathematics 28, 35–42 (1975)
44. Sato, T., et al.: The inference of proteinprotein interactions by co-evolutionary analysis is improved by excluding the information about the phylogenetic relationships. Bioinformatics 21(17), 3482–3489 (2005)
45. Tauberberger, J.K., et al.: Characterization of the 1918 influenza virus polymerase genes. Nature 437, 889–893 (2005)
46. Thornton, J.W., et al.: Resurrecting the ancestral steroid receptor: ancient origin of estrogen signaling. Science 301, 1714–1717 (2003)
47. Toda, T., et al.: Three different genes in s. cerevisiae encode the catalytic subunits of the camp-dependent protein kinase. Cell 50(2), 277–287 (1987)
48. Tuller, T., et al.: Reconstructing ancestral gene content by co-evolution (submitted 2009)
49. Tuller, T., et al.: Co-evolutionary networks of genes and cellular processes across fungal species. Genome Biol. 10 (2009)
50. Yang, Z.: Paml: a program package for phylogenetic analysis by maximum likelihood. Computer Applications in BioSciences 13, 555–556 (1997)
51. Yang, Z., et al.: A new method of inference of ancestral nucleotide - and amino acid sequences. Genetics 141, 1641–1650 (1995)
52. Yeang, C.H., et al.: Detecting the coevolution of biosequences–an example of rna interaction prediction. Mol. Biol. Evol. 24(9), 2119–2131 (2007)
53. Yeang, C.H., Haussler, D.: Detecting coevolution in and among protein domains. PLoS Comput. Biol. 3(11), e211 (2007)
54. Zhang, J., Rosenberg, H.F.: Complementary advantageous substitutions in the evolution of an antiviral rnase of higher primates. Proc. Natl. Acad. Sci. USA 99, 5486–5491 (2002)

Whole-Genome Analysis of Gene Conversion Events

Chih-Hao Hsu, Yu Zhang, Ross Hardison, and Webb Miller

Center for Comparative Genomics and Bioinformatics
Pennsylvania State University, University Park, PA 16802 USA
chih-hao@bx.psu.edu

Abstract. Gene conversion events are often overlooked in analyses of genome evolution. In a conversion event, an interval of DNA sequence (not necessarily containing a gene) overwrites a highly similar sequence. The event creates relationships among genomic intervals that can confound attempts to identify orthologs and to transfer functional annotation between genomes. Here we examine 1,112,202 paralogous pairs of human genomic intervals, and detect conversion events in about 13.5% of them. Properties of the putative gene conversions are analyzed, such as the lengths of the paralogous pairs and the spacing between their sources and targets. Our approach is illustrated using conversion events in the beta-globin gene cluster.

1 Introduction

Several classes of evolutionary operations have sculpted the human genome. Nucleotide substitutions have been studied in great detail for years, and much attention is now focused on large-scale events such as insertions, deletions, inversions, and duplications. Frequently overlooked are gene conversion events (reviewed by [1], [2]), in which one region is copied over the location of a highly similar region; before the operation there are two genomic intervals, say A and B with 95% identity, and afterwards there are two identical copies of A, one in the position formerly occupied by B.

Conversion events need to be accounted for when attempting to understand the human genome based on identification of orthologous regions in other species. To take a hypothetical example, suppose human genes A and B are related by a duplication event that pre-dated the separation of humans and Old World monkeys, so that rhesus macaques also have genes A and B. A conversion event in a human ancestor that overwrote some of A with sequence from B could cause all or part of the amino-acid sequence of A to be more closely related to the rhesus B protein than to the rhesus A protein, even though the regulatory regions of A might remain intact. Successful design and interpretation of experiments in rhesus to understand gene A might well require knowledge of these evolutionary relationships.

Gene conversion events have been studied in a variety of species, including the following investigations. [3] characterized conversions within 192 yeast gene

F.D. Ciccarelli and I. Miklós (Eds.): RECOMB-CG 2009, LNBI 5817, pp. 181–192, 2009.

families; [4] detected conversion events in 7,397 Caenorhabditis elegans genes; [5] studied 2,641 gene quartets, each consisting of two pairs of orthologous genes in mouse and rat, and found that 488 (18%) appear to have undergone gene conversion; and [6] detected 377 gene conversion events within 626 multigene families in the rice genome. However, all of these studies detect gene conversion events only between pairs of protein-coding genes, although conversion can occur between any pair of highly similar regions [2]. Besides, some analyses of gene conversion are done in the human genome [7], [8], [9]. While none of the datasets used in these previous studies compare to the size of the one analyzed here. In this paper, we cover more than one million paralogous pairs of regions, requiring a more efficient method to deal with such a large dataset.

A number of statistical tests have been proposed to detect gene conversions. However, most of these tests are only efficient for small data sets, e.g. individual gene clusters. [10] nicely summarize computational methods available for detecting mosaic structure in sequences, and propose a new method that is particularly economical in terms of computer execution time for large data sets. One drawback is that their algorithm requires large amounts of computer memory. However, we show here that this method can be reformulated so that the memory requirements are no longer a limiting factor, which allows us to conduct a comprehensive scan for gene-conversion events in the human genome, starting with 1,112,202 pairs of paralogous human intervals. For each pair of paralogous intervals, say H_1 and H_2, we choose a sequence from another species, say C_1, that is believed to be orthologous to H_1. These triplets of sequences are examined to find cases where part of H_1 is more similar to H_2 than to C_1, while another part is more similar to C_1. In such cases, the interval of high H_1 - H_2 similarity is inferred to have resulted from a conversion event.

2 Methods

Boni [10] developed a time-efficient method for identifying conversions and other recombination events, using the H_1 - H_2 and the H_1 - C_1 alignments to identify informative positions in H_1, such that either H_1 and H_2 have one nucleotide while C_1 has another (score -1), or H_1 and C_1 have one nucleotide while H_2 has another (score $+1$). The cumulative sum of these scores along H_1 constitutes what is called a hypergeometric random walk (HGRW; [11]) under the assumption that the relationships of H_1 to H_2 and C_1 are invariant across the interval; see Fig. 1. Conversions are detected using the test statistic $x_{m,n,k}$, which is the probability of a *maximum descent* of k occurring by chance for a triplet (H_1, H_2, C_1) with m +1s and n −1s. The maximum descent is the maximum decrease of scores across the interval (in one direction only), e.g. the regions of blue and red rectangles in Fig. 1. [10] give a dynamic programming algorithm for computing $x_{m,n,k}$, which creates a table that can be consulted for an arbitrary number of triplets.

To apply Boni's method to the entire human genome, for each given pair H_1 and H_2, we needed to find a sequence, C_1, from a species at an appropriate

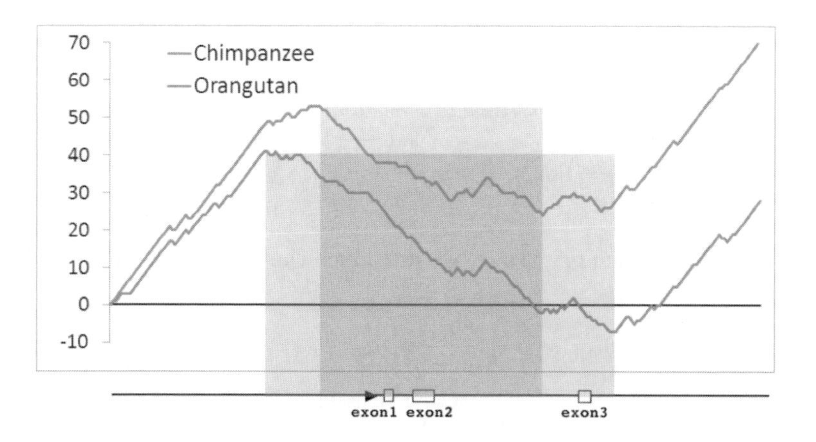

Fig. 1. Maximum descent of the hypergeometric random walks for an alignment between the two human gamma-globin genes. Using the orangutan sequence as C_1 identifies a wider converted interval than does using chimpanzee, possibly because of nested conversion events at different times during the evolution of the human lineage.

evolutionary distance, i.e., that split from the human lineage somewhat after the duplication event and before the most recent conversion. Thus, we tried a gamut of available mammalian genome sequences: chimpanzee, orangutan, rhesus, marmoset, dog, and opossum. Each of these species could be used to detect gene conversion events in a particular period of evolution along the lineage leading to human. Because the orthologs of H_1 and H_2 often differ, up to 12 triplets were used to look for gene conversion in a given human paralogous pair.

2.1 Space-Efficient Modifications

The original formulation by [10] requires an amount of computer time and memory that is proportional to B^4, where B is an upper bound for m, n, and k. For a triplet with 400 informative sites, this approach would use 6.4 GB of computer memory, allowing the method to work only with relatively short sequences. We modified that method to need only space proportional to $mn + n^2 + SP$, as we now describe.

In the notation of [10], the test statistic $x_{m,n,k}$ is defined as $P(md\ H_{m,n} = k)$ and can be calculated using the equation:

$$x_{m,n,k} = \sum_{j=0}^{k} y_{m,n,k,j} \tag{1}$$

where:

$$y_{m,n,k,j} = P(md\ H_{m,n} = k \bigcap min\ H_{m,n} = -j) \tag{2}$$

The probabilities y can be obtained by dynamic programming based on the following recursive relationships.

$$y_{m,n,k,j} = \begin{cases} (\frac{m}{m+n})[y_{m-1,n,k,1} + y_{m-1,n,k,0}] & \text{if } j = 0 \\ (\frac{m}{m+n})y_{m-1,n,k,j+1} + (\frac{n}{m+n})y_{m,n-1,k,j-1} & \text{if } k > j > 0 \\ (\frac{n}{m+n})[y_{m,n-1,k-1,j-1} + y_{m,n-1,k,j-1}] & \text{if } j = k > 0 \\ 0 & \text{if } j > k \geq 0 \end{cases} \quad (3)$$

In order to reduce memory usage, we introduce the additional variable $A_{m,n,k}$, defined as:

$$A_{m,n,k} = y_{m,n,k,k} = (\frac{n}{m+n})[A_{m,n-1,k-1} + y_{m,n-1,k,k-1}] \quad (4)$$

Then,

$$x_{m,n,k} = (\frac{m}{m+n})x_{m-1,n,k} + (\frac{n}{m+n})[x_{m,n-1,k} - A_{m,n-1,k} + A_{m,n-1,k-1}] \quad (5)$$

The key observation is that for fixed k, the only component of (3) that depends on $k-1$ is when $j = k > 0$, and in that case the required value is $A_{m,n-1,k-1}$. (On the other hand, the initialization of $y_{m,n,k,j}$ for $m = 0$ does depend on k.) Consequently, provided that we record the 3-dimensional array of values $A_{m,n,k}$, we can store the values of y for a fixed k in another 3-dimensional array that we call $y_{m,n,j}$ and overwrite them with the values corresponding to $k+1$ as the computation proceeds. The resulting algorithm uses only two arrays of size mn^2 (x and y can be stored in the same array). It can handle triplets with 2000 informative sites on a mid-sized workstation.

Furthermore, since the value of x depends only on the values in the same loop, e.g. $x_{m,n-1,k}$, and in the previous loop, e.g. $x_{m-1,n,k}$ (when using m as the outer loop), an $O(mn+n^2+SP)$ space method (where S = number of outgroup species and P = number of pairs of sequences) is possible. First, the values of m, n, and k for the triplets of all pairs of sequences are determined and stored in a three-dimensional linked list, which consumes $O(SP)$ space. Then the value of x is calculated and summed to the relevant triplets. Since only those values that are necessary for further calculation are kept, the maximum table size required for the calculation of x is $O(mn+n^2)$.

Although the space requirement is thus reduced, the time complexity is still quartic (exponent 4). Also, the longest interval in our data is 251,067 base pairs. In order to deal with long alignments, those with length greater than 5000 are divided into several sub-alignments with 1000 sites overlapped. The p-value for each sub-alignment is then calculated, and a multiple-comparison correction method [12] is used to determine if the set of sub-alignments supports an assertion that the whole alignment shows significant signs of a conversion.

2.2 Extension to Quadruplet Testing

It is not uncommon that we have a pair of paralogs in the other species, say C_1 and C_2 in chimpanzee, that are orthologs for H_1 and H_2 in human, respectively.

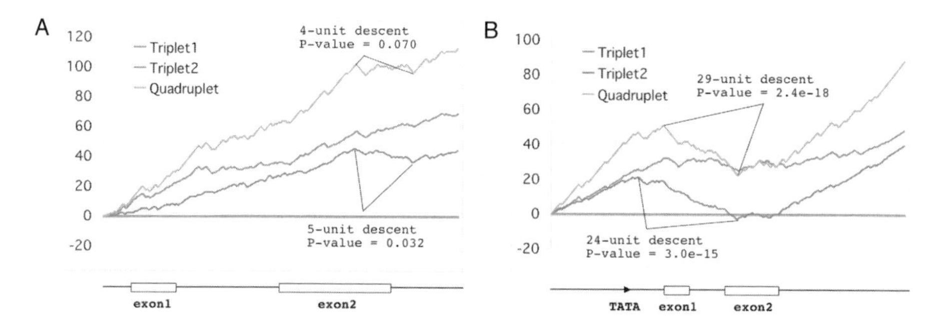

Fig. 2. Comparisons between quadruplet testing and triplet testing. (A) the HBD and HBBP1 paralog pair; (B) the HBB and HBD paralog pair.

In a fashion similar to the triplet testing for gene conversions, we can perform quadruplet testing (H_1, H_2, C_1, C_2) that is the summation of the hypergeometric random walks of two triplets, i.e. (H_1, H_2, C_1) and (H_1, H_2, C_2), as shown in Fig. 2. Quadruplet testing may have higher specificity and sensitivity than triplet testing for detecting conversions. For example, in Fig. 2A, a weakly significant (0.032) conversion event was detected between the HBD and HBBP1 paralog pair in one triplet testing, which is inconsistent with a previous study [13] in the β-globin gene cluster. This could be due to a faster evolutionary rate in HBBP1, which is a pseudo-gene. However, quadruplet testing did not show any evidence of conversion in this region. This suggests that the effect of one triplet can be neutralized by that of another triplet when there is no conversion between a paralog pair. On the other hand, when applying quadruplet testing for a conversion region, we can get more significant result, as shown in Fig. 2B (2.4e-18). Therefore, whenever orthologs for both H_1 and H_2 are available in a particular outgroup species, we combine the results of the two triplets to perform quadruplet testing, and use the same formula as triplet testing, i.e. equation (5), to get p-values.

2.3 Multiple-Comparison Correction

When several statistical tests are performed simultaneously, a multiple-comparison correction should be applied. In our study, the Bonferroni correction [12] is used; we multiply the smallest p-value for each paralogous human pair by the number of tests (up to 6) as the adjusted p-value.

Multiple-comparison correction is also applied to the tests for all pairs of paralogous sequences. For the 1,112,202 pairs that were analyzed, we used a multiple-comparison correction method that controls the false discovery rate (FDR), proposed by [14]. The cutoff threshold for p-values can be found by the following algorithm:

```
CutOff(α, p-values)
1  sort p-values in ascending order
2  for i ← 1 to number of p-values do
3     if p_i > (i ÷ number of p-values) × α
4        return (i ÷ number of p-values) × α
```

In our study, α is 0.05 and the cutoff threshold for p-values is 0.006818. This means that only a test whose adjusted p-value is less than 0.006818 is considered as significant for gene conversion.

2.4 Directionality of Gene Conversion

We attempt to determine the source and target of a conversion event as follows. As shown in Fig. 3B, let us suppose that a conversion event happened z years ago, with $x > y > z$, and consider a converted position. Regardless of the direction of the conversion (from H_1 to H_2, or vice versa), in the converted region, H_1 and H_2 are separated by $2z$ total years. If H_1 converted H_2 (i.e. part of H_1 overwrote part of H_2), then the separation of H_1 and C_1 is $2y$ but the separation of H_2 and its ortholog, C_2, is $2x > 2y$. This observation serves as a basis for determining the conversion direction. Fig. 4 shows an example of determining the source and target of a conversion from HBB to HBD.

Specifically, assume (m_1, n_1) with maximum descent k_1 in the first triplet (H_1, H_2, and C_1), and (m_2, n_2) with maximum descent k_2 in the second triplet (H_1, H_2, and C_2). Note that m_i and n_i here are not the m and n in (1); rather, they are the numbers of ups and downs within the common maximum descent region of the two triplets (union). The probabilities of going down in the common maximum descent regions of two triplets are:

$$p_1 = n_1 \div (m_1 + n_1) \tag{6}$$

$$p_2 = n_2 \div (m_2 + n_2) \tag{7}$$

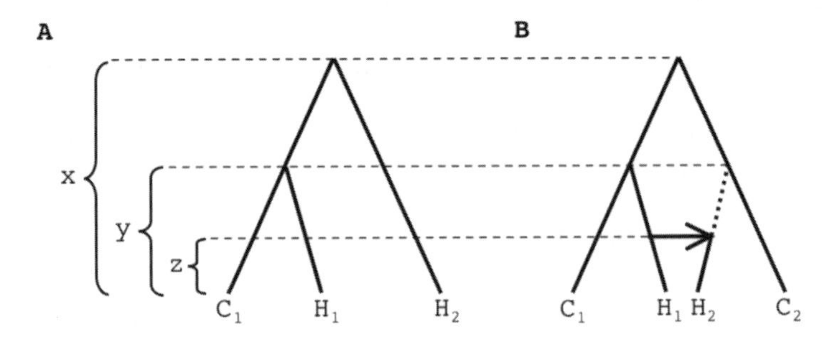

Fig. 3. Timing of evolutionary events. The assumed duplication, speciation, and conversion events occurred respectively x, y, and z years ago. See text for further explanation.

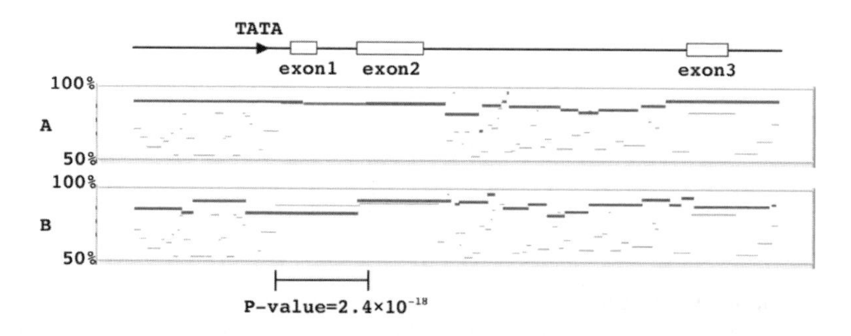

Fig. 4. Evidence that the β-globin gene (HBB) converted the δ-globin gene (HBD). Percent identity plots for (A) HBB and (B) HBD showing alignments to the human paralog in red, and alignments to the putative marmoset ortholog in blue. In the converted region, the human-marmoset alignments have 92% identity for β-globin and 85% identity for δ-globin.

When combining these data, there are a total of $(m_1 + m_2)$ ups and $(n_1 + n_2)$ downs, and the possibility of going down in the combined data is:

$$p = (n_1 + n_2) \div (m_1 + m_2 + n_1 + n_2) \tag{8}$$

As shown in Fig. 3B, if H_1 converted H_2, the separation of H_1 and C_1 is closer than the separation of H_2 and C_2 in the converted region. Thus, p_1 should be smaller than p_2. Our objective function (O) is therefore to determine how significant the difference of $p_1 - p_2$ is, based on the binomial distribution:

$$O = ((p_1 - p_2) - E(p_1\prime - p_2\prime)) \div sqrt(V(p_1\prime - p_2\prime)) \tag{9}$$

where:

$$E(p_1\prime - p_2\prime) = 0 \tag{10}$$

$$V(p_1\prime - p_2\prime) = (\frac{1}{m_1 + n_1} + \frac{1}{m_2 + n_2}) \times p \times (1 - p) \tag{11}$$

In this paper, 6 outgroup species are used to detect gene conversions. We use the outgroup species that shows the most significant difference of $p_1 - p_2$ to determine the directionality of conversion for a given paralogous pair. However, there are several reasons why the direction of a conversion might not be clear, even when using several outgroup species, including conversions in the outgroup species and missing outgroup data. Our approach indicates a direction for 65.4% of the putative conversions.

3 Results

3.1 Highly Conserved Pairs of Sequences

We aligned each pair of human chromosomes, including self-alignments, using BLASTZ [15] with T=2 and default values for the other parameters. Chaining

Table 1. Distribution of intra- and inter-chromosomal gene conversions

	intra-chromosome	inter-chromosome	total
gene conversion	42,927 (17.8%)	106,872 (12.3%)	149,799 (13.5%)
no gene conversion	195,180 (80.9%)	760,486 (87.3%)	955,666 (85.9%)
unknown	3,034 (1.3%)	3,703 (0.4%)	6,737 (0.6%)
total	241,141	871,061	1,112,202

of the human-human alignments was performed using the method of [16]. For alignments between human intervals and their putative orthologs in other species, we used the pairwise alignment nets [17] downloaded from the UCSC Genome Browser website [18]. We applied the modified Boni's method described in this paper and recorded information about the inferred conversion events in Table 1.

Of the 1,112,202 analyzed pairs of human sequences, 149,799 (13.5%) indicated a gene conversion event. The occurrence of a gene conversion for 6,737 (0.6%) pairs could not be tested due to a lack of available orthologous sequence in the other species used in this study. These results are consistent with those of [5], where about 13% of the mouse sequence pairs show signs of gene conversion. Among our putative gene conversion events, the fraction of intra-chromosomal pairs indicating a conversion (17.8%) is significantly higher than for inter-chromosomal pairs (12.3%).

3.2 Association with Gene Conversion

To study the correlation between various genomic features and gene conversion, we used a logistic regression model [19] to characterize gene conversions based on the following factors.

$strand$: binary; strand of the second paralog relative to the first one
seq_len: paralog size in basepairs
$pair\_dist$: distance between the paralogs
seq_sim: percent identity between the two paralogs
gc: percentage of G+C in both paralogs combined
$gc1$: percentage of G+C in the first paralog
$gc2$: percentage of G+C in the second paralog
$coding1$: binary; whether or not the first paralog contains coding regions
$coding2$: binary; whether or not the second paralog contains coding regions

For ease of interpretation, we only included the main effects of the variables, and carried out the analysis separately for inter-chromosome and intra-chromosome pairs, as shown below under (1) and (2), respectively. For a logistic regression model, the response is binary. In our case, it is gene conversion (indicated by 1) or not (indicated by 0). The logistic regression relates the logit of the probability of gene conversion to a function of predictors, where the logit function is $logit(x) = log(x \div (1 - x))$. The actual gene conversion event is regarded as a binary outcome with the probability given by the regression model.

(1) For inter-chromosome paralog pairs, the model is:
$logit(conversion\_rate) \sim seq\_len + seq\_sim + gc + coding1 + coding2$
Coefficients:

	Estimate	Std. Error	z Value	Pr(>|z|)
(Intercept)	-3.196198	0.011088	-288.254	< 2e-16 ***
seq_len	0.594905	0.002778	214.143	< 2e-16 ***
seq_sim	0.004164	0.002862	1.455	0.146
gc	0.077565	0.002507	30.939	<2e-16 ***
*coding*1	0.105050	0.009025	11.640	< 2e-16 ***
*coding*2	0.211697	0.009364	22.608	< 2e-16 ***

If the estimated coefficient of a variable is positive, the variable increases the gene conversion probability (or rate), while a negative value indicates a decrease.

(2) For intra-chromosome paralog pairs, the model is:
$logit(conversion\_rate) \sim strand + seq\_len + seq\_sim + gc + coding1 + coding2 + pair\_dist$
Coefficients:

	Estimate	Std. Error	z Value	Pr(>|z|)
(Intercept)	-2.931604	0.026803	-109.374	< 2e-16 ***
strand	0.037740	0.011775	3.205	0.00135 **
seq_len	0.660065	0.004254	155.150	< 2e-16 ***
seq_sim	0.094515	0.004684	20.178	< 2e-16 ***
gc	0.169814	0.004113	41.283	< 2e-16 ***
*coding*1	0.082249	0.014652	5.613	1.98e-08 ***
*coding*2	0.127370	0.014548	8.755	< 2e-16 ***
pair_dist	-0.117358	0.004307	-27.251	< 2e-16 ***

Based on the results of (1) and (2), we see that:

- The conversion rate is higher when H_1 and H_2 are on the same chromosome; this can be seen from the larger intercept of (2) than of (1).
- Strand has little effect on gene conversion. From (2) we see that strand effect is positive (0.03774) and weakly significant.
- Conversion rate increases as paralog size increases; this can be seen from the positive coefficient of *seq_len* in both (1) and (2).
- Similarity of sequences has a significant contribution to conversion rate for intra-chromosome pairs; however, it has less effect on conversion rate, given other variables in the model, for inter-chromosome pairs.
- Both GC-content and the presence of coding sequences contribute positively to the conversion rate for inter- and intra-chromosome pairs.
- Conversion rate decreases as distance between two paralogs increases; this can be seen from (2) where *pair_dist* is negative.

The model used here is simplified; it does not account for interactions among factors. We did not include interactions in the models because (a) interactions are more complicated to interpret, and (b) they require much more computer memory, considering that over a million paralog pairs are being tested. Instead, we performed small-scale studies using subsets of the data, and we observed that although some interactions are significantly related to gene conversion rate, the magnitude of their contributions is relatively small compared to the factors direct effects. Furthermore, we did not account for differences across chromosomes, although conversion rates do vary significantly depending on the chromosome.

3.3 Directionality of Gene Conversion

To obtain a logistic regression model for the conversion direction (in the cases where it could be determined), we used the discrete variable *con_direction*, set to 1 if H_2 converts H_1, and 0 if H_1 converts H_2.

(1) For inter-chromosome paralog pairs, the model is:
$logit(con\_direction) \sim seq\_len + seq\_sim + gc1 + gc2 + coding1 + coding2$
Coefficients:

	Estimate	Std. Error	z Value	Pr(>|z|)
(Intercept)	0.066604	0.026105	2.551	0.0107 *
seq_len	-0.016826	0.006711	-2.507	0.0122 *
seq_sim	-0.013648	0.006294	-2.168	0.0301 *
gc1	-0.004214	0.009325	-0.452	0.6514
gc2	0.010791	0.009225	1.170	0.2421
coding1	-0.559428	0.020543	-27.233	< 2e-16 ***
coding2	0.506014	0.021184	23.886	< 2e-16 ***

(2) For intra-chromosome paralog pairs, the model is:
$logit(conversion\_rate) \sim strand + seq\_len + seq\_sim + gc1 + gc2 + coding1 + coding2 + pair\_dist$
Coefficients:

	Estimate	Std. Error	z Value	Pr(>|z|)
(Intercept)	0.091931	0.059779	1.538	0.12408
strand	0.002046	0.024525	0.083	0.93351
seq_len	0.026179	0.009397	2.786	0.00534 **
seq_sim	-0.031367	0.009684	-3.239	0.00120 **
gc1	-0.011218	0.016916	-0.663	0.50725
gc2	-0.005629	0.017142	-0.328	0.74263
coding1	-0.354766	0.030856	-11.497	< 2e-16 ***
coding2	0.391011	0.030674	12.747	< 2e-16 ***
pair_dist	-0.014240	0.009266	-1.537	0.12436

Based on the results of (1) and (2), we see that:

- Strand, GC-content, and pair-distance are not significantly associated with conversion direction.
- Sequence length and similarity have slight effects on conversion direction.
- Coding sequences significantly affect the direction for both inter- and intra-chromosome pairs. Negative *coding1* and positive *coding2* mean that the

conversion direction tends to be from coding sequences to non-coding sequences.

4 Discussion

For much of the half-century since multigene families were discovered, it has been known that copies of the repeated genes within a species are more similar than would be expected from their interspecies divergence. The processes generating this sequence homogeneity in repeated DNA are mechanisms of concerted evolution. Gene conversion is one of those processes, and while its impact on disease genes is appreciated [2], the extent of its impact on the evolution of the human genome has not been fully investigated in previous studies. Our work documents about one hundred and fifty thousand conversions (13.5%) between duplicated DNA segments in humans. Similarly large fractions of conversion events among duplicated segments have been reported in whole-genome studies of yeast [3], Drosophila melanogaster [20] and rodents [5], even though the total number of observed gene conversions is much higher in our study. The genome-wide identification of DNA segments undergoing concerted evolution via gene conversions will make the application of comparative genomics to functional annotation considerably more accurate. This resource will allow the conversion process to be factored into functional inference based on sequence similarity to other species; for example it could flag potential false positives for inferred positive or negative selection.

Some of the genomic features associated with gene conversion are expected, given that the conversions often result from one pathway of resolving intermediates in homologous recombination. In particular, we find that the length of paralogous segments has a strong positive effect on conversion events for both inter- and intra-chromosomal paralog pairs, as expected for a process requiring homologous pairing. Also, two features indicate that closer proximity between the homologous pairs increases the likelihood of a conversion event: the conversion frequency is higher for intra-chromosomal pairs than for inter-chromosomal ones, and it is also higher for paralog pairs that are closer together on a chromosome. The closer proximity may be expected to increase the frequency of homologous pairing in recombination. Furthermore, the effects of coding sequences are very interesting. Positive correlation with coding sequences could result from higher similarity. More frequent conversion direction from coding sequences to non-coding sequences could be a consequence of selection. However, some of the associations were surprising; for instance, there is less correlation with sequence similarity for inter-chromosomal paralog pairs than for intra-chromosomal paralog pairs. This, plus the curious effects of GC-content, point to aspects of the gene conversion mechanism(s) that need further investigation.

Acknowledgements. This work was supported by grant HG02238 from the National Human Genome Research Institute and grant DK065806 from the National Institute of Diabetes, Digestive and Kidney Diseases. We thank Maciek Boni for explaining his strategies for reducing the space required by his method.

References

1. Hurles, M.: Gene duplication: the genomic trade in spare parts. PLoS Biology 2, 900–904 (2004)
2. Chen, J.M., Cooper, D.N., Chuzhanova, N., Ferec, C., Patrinos, G.P.: Gene conversion: mechanisms, evolution and human disease. Nat. Rev. Genet. 8, 762–775 (2007)
3. Drouin, G.: Characterization of the gene conversions between the multigene family members of the yeast genome. J. Mol. Evol. 55, 14–23 (2002)
4. Semple, C., Wolfe, K.H.: Gene duplication and gene conversion in the Caenorhabditis elegans genome. J. Mol. Evol. 48, 555–564 (1999)
5. Ezawa, K., Oota, S., Saitou, N.: Genome-wide search of gene conversions in duplicated genes of mouse and rat. Mol. Biol. Evol. 23, 927–940 (2006)
6. Xu, S., Clark, T., Zheng, H., Vang, S., Li, R., Wong, G.K., Wang, J., Zheng, X.: Gene conversion in the rice genome. BMC Genomics 9, 93–100 (2008)
7. Jackson, M.S., et al.: Evidence for widespread reticulate evolution within human duplicons. Am. J. Hum. Genet. 77, 824–840 (2005)
8. McGrath, C.L., Casola, C., Hahn, M.W.: Minimal Effect of Ectopic Gene Conversion Among Recent Duplicates in Four Mammalian Genomes. Genetics 182, 615–622 (2009)
9. Benovoy, D., Drouin, G.: Ectopic gene conversions in the human genome. Genomics 93, 27–32 (2009)
10. Boni, M.F., Posada, D., Feldman, M.W.: An exact nonparametric method for inferring mosaic structure in sequence triplets. Genetics 176, 1035–1047 (2007)
11. Feller, W.: An Introduction to Probability Theory and Its Application, vol. I. John Wiley & Sons, New York (1957)
12. Holm, S.: A simple sequential rejective multiple test procedure. Scand. J. Statistics 6, 65–70 (1979)
13. Papadakis, M.N., Patrinos, G.P.: Contribution of gene conversion in the evolution of the human β-like globin gene family. Human Genetics 104, 117–125 (1999)
14. Benjamini, Y., Hochberg, T.: Controlling the false discovery rate: a practical and powerful approach to multiple testing. J. Royal Stat. Soc. B 85, 289–300 (1995)
15. Schwartz, S., Kent, W.J., Smit, A., Zhang, Z., Baertsch, R., Hardison, R.C., Haussler, D., Miller, W.: Human-mouse alignments with BLASTZ. Genome Res 13, 103–107 (2003)
16. Zhang, Z., Raghavachari, B., Hardison, R.C., Miller, W.: Chaining multiple-alignment blocks. J. Comp. Biol. 1, 217–226 (1994)
17. Kent, W.J., Baertsch, R., Hinrichs, A., Miller, W., Haussler, D.: Evolutions cauldron: duplication, deletion, and rearrangement in the mouse and human genomes. Proc. Natl. Acad. Sci. USA 100, 11484–11489 (2003)
18. Kent, W.J., et al.: The human genome browser at UCSC. Genome Res. 12, 996–1006 (2002)
19. Agresti, A.: Categorical Data Analysis. Wiley-Interscience, New York (2002)
20. Osada, N., Innan, H.: Duplication and gene conversion in the Drosophila melanogaster genome. PLoS Genet. 4, e1000305 (2008)

A Statistically Fair Comparison of Ancestral Genome Reconstructions, Based on Breakpoint and Rearrangement Distances

Zaky Adam[1] and David Sankoff[2]

[1] School of Information Technology and Engineering, University of Ottawa
[2] Department of Mathematics and Statistics, University of Ottawa

Abstract. We introduce a way of evaluating two mathematically independent optimization approaches to the same problem, namely how good or bad each is with respect to the other's criterion. We illustrate this in a comparison of breakpoint and rearrangement distances between the endpoints of a branch, wehre total branch-length is minimized in reconstructing ancestral genomes at the nodes of a given phylogeny. We apply this to mammalian genome evolution and simulations under various hypotheses about breakpoint re-use. Reconstructions based on rearrangement distance are superior in terms of branch length and dispersion of the multiple optimal reconstructions, but simulations show that both sets of reconstructions are equally close to the simulated ancestors.

1 Introduction

Breakpoint distance and rearrangement distances provide alternative ways of evaluating phylogenetic trees and reconstructing ancestral genomes based on whole genome data. When applied to a set of sufficiently diverse genomes, these approaches will generally lead to different results. Ancestral genomes optimal under the breakpoint criterion will not minimize the total rearrangement distance on the edges of a given tree, and optimality according to rearrangement distance will not minimize the total number of breakpoints.

Can we say that one of these methods is superior to the other? Lacking any widely accepted probability model for genomes evolving through rearrangements, we can have no analytic framework for the statistical properties of reconstructions, in particular the accuracy and reliability of these reconstructions. Even assessment through simulations, though informative [7], is highly dependent on the assumptions necessary for generating the data, assumptions that are either highly simplified such as uniform weights on a small repertoire of rearrangement events or highly parametrized models pertinent to limited phylogenetic domains.

Is there any sense, then, in which we could affirm that one objective function on reconstructions is better than the other? In this paper we introduce a way of evaluating two mathematically independent[1] optimization approaches relative to

[1] Of course, breakpoints and rearrangement distances provide upper and lower bounds for each other, but between these bounds they are not mutually constrained.

F.D. Ciccarelli and I. Miklós (Eds.): RECOMB-CG 2009, LNBI 5817, pp. 193–204, 2009.

each other, namely how good or bad each is with respect to the other's criterion. The idea is that *the approach that comes the closest to satisfying the other's criterion as well as its own, is more desirable.*

We will illustrate this method on two data sets on mammalian evolution, as well as three different simulations modelling each of these data sets in a different way, eight data sets in all. We use a given, well-accepted phylogeny for each data set and each simulation. We reconstruct all the ancestral genomes, once minimizing the total breakpoints over all tree branches (using the polynomial-time median method in [12]), and once with the minimum rearrangement distance [1]. Because optimal reconstructions are not unique, we sample five different reconstructions in each case.

We then apply our "fair" method to four aspects of these reconstructions. First we assess all the branch lengths in each reconstruction and then compare, in two different ways, the dispersion (or compactness), for each tree node, of the five different reconstructions. Finally, for the simulated data sets, we measure the distances of the reconstructed ancestors from the simulated ancestors.

For the analysis of the branch lengths and the dispersion of reconstructed ancestors, the results show a clear and systematic advantage of rearrangement distance over breakpoint distance. Despite this, the two methods prove to be equally good at reconstructing the known simulated ancestor genomes.

In Section 2, we formalize our proposal for a fair comparison of metrics, illustrating with four aspect of the "small" phylogeny problem, branch lengths, node dispersion (looked at two ways) among optimal solutions, and distance between reconstructed and true ancestral genomes in simulations. In Section 3 we formalize the breakpoint distance and the rearrangement distance, and sketch how they are used in solving the small phylogeny problem. In Section 4, we briefly describe the two real data sets on mammalian evolution as well as the simulations carried out under various "breakpoint re-use" conditions. The results are presented in Section 5 and commented in the Discussion.

2 A Fair Comparison

Let $\mathcal{P}$ be a phylogeny where each of the M terminal nodes is labelled by a known genome, and let d_A and d_B be two metrics on the set of genomes. Each branch of $\mathcal{P}$ may be incident to at most one terminal node and at least one of the N ancestral nodes. Consider reconstructions $R_A = (G_1^A \dots, G_N^A)$ and $R_B = (G_1^B, \dots, G_N^B)$ of the set of genomes to label the ancestral nodes such that

$$L_A(R_A) = \sum_{\text{branch } XY \in \mathcal{P}} d_A(X^A Y^A) \quad (1)$$

is minimized, and

$$L_B(R_B) = \sum_{\text{branch } XY \in \mathcal{P}} d_B(X^B Y^B) \quad (2)$$

is also minimized. Without taking into account any additional, external criteria, there is no justification for saying one of d_A or d_B is better as a criterion for reconstructing the ancestral genomes. In general, we can expect that

$$L_A(R_A) < L_A(R_B), \tag{3}$$

i.e., strict inequality holds and

$$L_B(R_B) < L_B(R_A). \tag{4}$$

2.1 Branch Lengths

Inequalities (1) and (2) are not necessarily inherited by the individual terms in the sums, e.g., $d_A(X^AY^A)$ may not always be less than $d_A(X^BY^B)$. We call $e_A(X^BY^B) = d_A(X^B, Y^B) - d_A(X^A, Y^A)$ the *excess length* of branch X^BY^B with respect to distance d_A, though it may sometimes be less than zero.

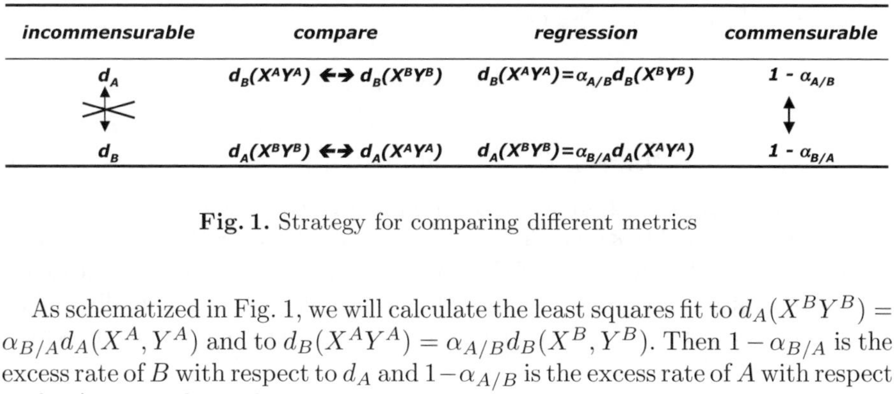

incommensurable	***compare***	***regression***	***commensurable***
d_A	$d_B(X^AY^A) \leftrightarrow d_B(X^BY^B)$	$d_B(X^AY^A)=\alpha_{A/B}d_B(X^BY^B)$	$1 - \alpha_{A/B}$
d_B	$d_A(X^BY^B) \leftrightarrow d_A(X^AY^A)$	$d_A(X^BY^B)=\alpha_{B/A}d_A(X^AY^A)$	$1 - \alpha_{B/A}$

Fig. 1. Strategy for comparing different metrics

As schematized in Fig. 1, we will calculate the least squares fit to $d_A(X^BY^B) = \alpha_{B/A}d_A(X^A, Y^A)$ and to $d_B(X^AY^A) = \alpha_{A/B}d_B(X^B, Y^B)$. Then $1 - \alpha_{B/A}$ is the excess rate of B with respect to d_A and $1-\alpha_{A/B}$ is the excess rate of A with respect to d_B. A metric that induces a reconstruction with a lower excess rate with respect to the other metric may be considered superior. I.e., if the excess rate of A with respect to d_B is less than the excess rate of B with respect to d_A, then d_A is better in the sense of being more universal or less "parochial": the branches in the A reconstruction are closer to optimal length according to d_B than the branches in the B reconstruction are according to d_A.

2.2 Node Dispersion

Among the N_s different optimal reconstructions under d_A, let $G_i^A = \{G_{i1}^A, \ldots, G_{iN_s}^A\}$ be set of reconstructions of ancestral genome G_i. Then

$$V^A(G_i^A) = \max_{0<j<k\leq N_s} d_A(G_{ij}^A, G_{ik}^A) \tag{5}$$

is the dispersion of the N_s reconstructions. Since the reconstructions may be expected to cluster around the "true" value of G_i as measured by d_A, we may also expect that

$$V^A(G_i^A) \leq V^A(G_i^B), \tag{6}$$

where

$$V^A(G_i^B) = \max_{0<j<k\leq N_s} d_A(G_{ij}^B, G_{ik}^B) \tag{7}$$

since the genomes in G_i^B are not necessarily close to the true G_i as measured by d_A, and similarly

$$V^B(G_i^B) \leq V^B(G_i^A). \tag{8}$$

In both (6) and (8), the inequality would normally be strict.

As in Section 2.1, we will calculate the least squares fit to $V^A(G_i^B) = \alpha_{B/A} V^A(G_i^A)$ and to $V^B(G_i^A) = \alpha_{A/B} V^B(G_i^B)$ over all ancestral nodes. Then $1 - \alpha_{B/A}$ is the excess rate of B with respect to d_A and $1 - \alpha_{A/B}$ is the excess rate of A with respect to d_B. As with the branch lengths, we can see whether one of the two metrics has a systematically lower excess rate.

2.3 Distance to True Genome

In contrast to the real data sets, with the simulated data sets, we actually know the ancestral genomes G_i. We can expect

$$d_A(G_i, G_i^A) \leq d_A(G_i, G_i^B), \tag{9}$$

$$d_B(G_i, G_i^B) \leq d_B(G_i, G_i^A). \tag{10}$$

As in Sections 2.1 and 2.2, we will calculate the least squares fit to $d_A(G_i, G_i^B) = \alpha_{B/A} d_A(G_i, G_i^A)$ and to $d_B(G_i, G_i^A) = \alpha_{A/B} d_B(Gi, G_i^B)$ over all ancestral nodes. Then $1-\alpha_{B/A}$ is the excess rate of B with respect to d_A and $1-\alpha_{A/B}$ is the excess rate of A with respect to d_B. As with the branch lengths and node dispersion, we can see whether one of the two metrics has a systematically lower excess rate.

3 Breakpoints and Rearrangements

Given genomes $g_1, \ldots, g_M$ associated with the terminal nodes of $\mathcal{P}$, the small phylogeny problem is to construct a set of genomes $G_1, \ldots, G_N$ to associate with the non-terminal nodes of $\mathcal{P}$, such that the phylogenetic tree length L is minimal, as in Section 2. We consider the simplest structure for $\mathcal{P}$, namely an unrooted, binary-branching tree. All nodes are of degree one (terminal) or three (non-terminal). Our algorithms for searching for a minimum L depend on median algorithms as shown in the pseudo-code presented as Algorithm 1 here.

Simply stated, the median problem is: considering three genomes H, J, K as points in some metric space (E, d), find another genome $C \in E$ such that $d(C, H) + d(C, J) + d(C, K)$ is minimal.

There is a large literature on median problems in comparative genomics [12], with all studied versions except one proving to be NP-hard. For our purposes we take E to be the set of oriented multichromosomal or unichromosomal genomes on the same set of n elements or genes. The genomes in this set may have circular as well as linear chromosomes, a property which has little consequence for the

Algorithm 1. Outline of phylogenetic reconstruction based on medians

Algorithm for Small Phylogeny Using Metric d
input $\mathcal{P}, g_1, \ldots, g_M$
set $L = \infty$
initialize $G_1, \ldots, G_N$
calculate $L' = L(G_1, \ldots, G_N)$
while $L' < L$
  set $L = L'$
  for each $G_1, \ldots, G_N$, with neighbours H,J,K
    $G =$ **Median Algorithm for** $d(H, J, K)$
  $L' = L(G_1, \ldots, G_N)$
end while
output

numerical results of the median problem, but has computational advantages for both the two metrics we study; the double cut and join (DCJ) distance [13,2], where we have implemented highly accurate code [1] and the breakpoint distance, which gives rise to the only known version of the median problem that is of polynomial complexity [12].

The breakpoint distance between two genomes is defined to be

$$d_{BP} = n - \text{ the number of common gene adjacencies in the genomes} \\ + \frac{1}{2} \text{ the number of common chromosomal endpoints (telomeres)} \quad (11)$$

where gene adjacency requires conservation of their relative orientation (strandedness). We have carried out the first implementation of the Tannier *et al.* algorithm. The median problem turns out to be directly transformable to a version of the maximum weight perfect matching problem. We made use of the code in [5] for this purpose. Although the maximum weight perfect matching algorithm is polynomial, the execution time is not negligible, being at least $O(n^3)$.

The DCJ metric d_{DCJ} counts the minimal number of operations necessary to transform one genome into another, where the repertoire of operations includes inversions, reciprocal translocations, chromosome fissions and fusions, as well as block interchanges, which count as two operations. Transposition of chromosomal segments from one site to another are a special case of block interchange. Our version of the median solver in [1] is based on the MGR algorithm [3], with some differences due to the different rearrangement distances used, but also because of extensive use of additional routines to escape from local minima.

In the breakpoint method, the ancestral gene orders converge after 4-6 iterations within Algorithm 1, and each iteration takes 3-5 minutes for the data we consider in the next section. Our implementation of the DCJ method requires extensive computing time, taking more than 20 iterations to converge on the same data, where each iteration takes many hours to finish.

4 The Empirical Studies

The Data. The first data set, drawn from [9], consists of the placental mammalian genomes from human, rat, mouse, cat, dog, pig and cow. Each genome consists of 307 HSB (homologous synteny blocks). The second data set includes the marsupial opossum along with the placental mammalian genomes human, rat, mouse and dog. These data are from the supplementary information for reference [6]. Each genome consists of 603 HSB. The given phylogenies are shown in Fig. 2. Although these data sets are obviously not independent, the differences in the species involved, and the large number of extra HSB induced by the presence of the opossum genome, assures that these problems are rather different from the computational point of view.

The Sample of Optimal Reconstructions and Breakpoint Re-use. For each of the two data sets, we ran the breakpoint phylogeny algorithm 40 times with different random initializations of $G_1, \ldots, G_N$. We retained the runs that gave the five minimum total tree lengths, usually the same value. For each branch of the tree and for each of the five results, we computed the corresponding DCJ distance between the two breakpoint-inferred genomes determining that branch, and then computed the breakpoint re-use [3,10] quotient $r = 2d_{DCJ}/d_{BP}$.

We ran the DCJ phylogeny algorithm five times only with different random initializations of $G_1, \ldots, G_N$. For each branch of the tree and for each of the five results, we computed the corresponding breakpoint distance between the two DCJ-inferred genomes determining that branch, and then computed the quotient $r = 2d_{DCJ}/d_{BP}$.

In the simulations we make use of the average $\bar{r}(X,Y)$ of the ten values of the re-use statistic calculated for each branch XY in these two ways, as well as the average of the ten branch lengths $\bar{d}_{DCJ}(X,Y)$.

Simulated Data Experiments. For each of the two data sets, we generated a random genome at an arbitrarily chosen "root", distributing the HSB over

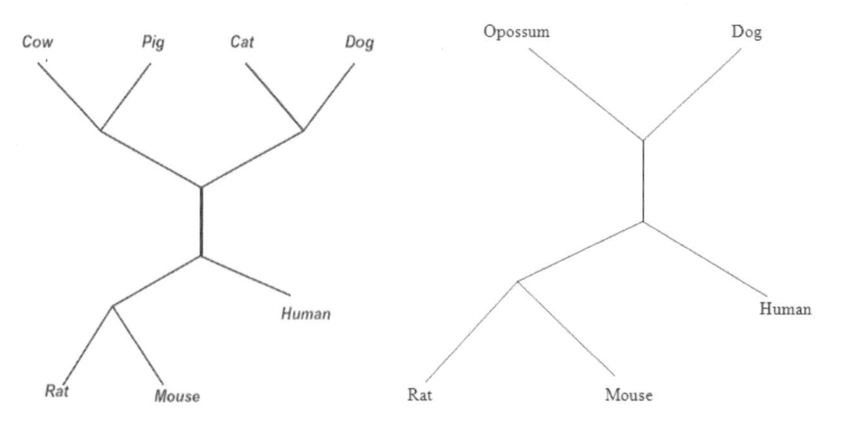

Fig. 2. Phylogenies for two mammalian data sets

$\chi = \text{mean}_i^M \chi(g_i)$ chromosomes and then generated its descendants along the tree branches using 90% inversions and 10% reciprocal translocations for the $\bar{d}_{DCJ}(X, Y)$ random rearrangements on branch XY. We ran three separate simulations, constraining the rearrangements on a branch XY in three different ways: once assuring each operation on XY used two new breakpoints so that $r = 1$, once assuring each successive rearrangement on a branch used one new breakpoint and re-used one existing breakpoint so that $r = 2 - 1/n$, and once assuring $r = \bar{r}(X, Y)$. We then implemented the breakpoint and DCJ phylogeny algorithms exactly as for the real data, obtaining five reconstructions for each criterion.

The Measurements. For each of the $8 = 2$ criteria $\times$ (1 real+3 simulated) data sets, we calculated the following aggregate statistics over the reconstructions, in each case measured by *both* criteria, the one used to infer the reconstructions and the opposing one:

- average branch length $d_A(X^A, Y^A)$ and $d_B(X^A, Y^A)$, for each branch XY,
- maximum intranode distance, $\max_{jk} d_A(G_{ij}^A, G_{ik}^A)$ and $\max_{jk} d_B(G_{ij}^A, G_{ik}^A)$, for each G_i,
- maximum intranode, intercriteria, distance, $\max_{jk} d_A(G_{ij}^A, G_{ik}^B)$, for each G_i,
- (simulations) average distance between reconstruction and true ancestor, $d_A(G_i, G_i^A)$ and $d_B(G_i, G_i^A)$, for each G_i.

5 Results

Before examining the results, we reiterate that the key to this methodology is that it is fair, i.e., not inherently biased either towards BP or DCJ. Each comparison is made according to a single criterion; we do not compare BP scores with DCJ scores. The BP measurements are slightly worse on the DCJ reconstructions and the DCJ measurements are slightly worse on the BP reconstructions. How bad they are is measured in normalized terms – slope of a least squares line anchored at (0,0). The excess of this slope over 1.0 we call the excess explanatory rate (of using the ancestral genomes reconstructed under criterion A instead of those reconstructed under criterion B, when B is used to make the measurements). The smaller this cost, the closer the A ancestors are to being solutions, not only under criterion A but also B. Thus we see that is the real data, the excess rate of the DCJ reconstruction is systematically lower than that of the BP reconstruction, for both real data sets and for all the simulations.

Average Branch Length. In Table 1, we see that the excess rate for DCJ (in boldface), which measures how far the DCJ reconstructions are from being BP-optimal, is only of the order of half the excess rate of the BP reconstructions. This is true in both real data sets and in all the simulations, regardless of re-use rate.

To illustrate the derivation of the excess rates as detailed in Section 2.1 and Fig. 1, we present scattergrams of competing criterion versus reconstructing criterion branch lengths in Fig. 3 for the real data sets, corresponding to the top row in Table 1.

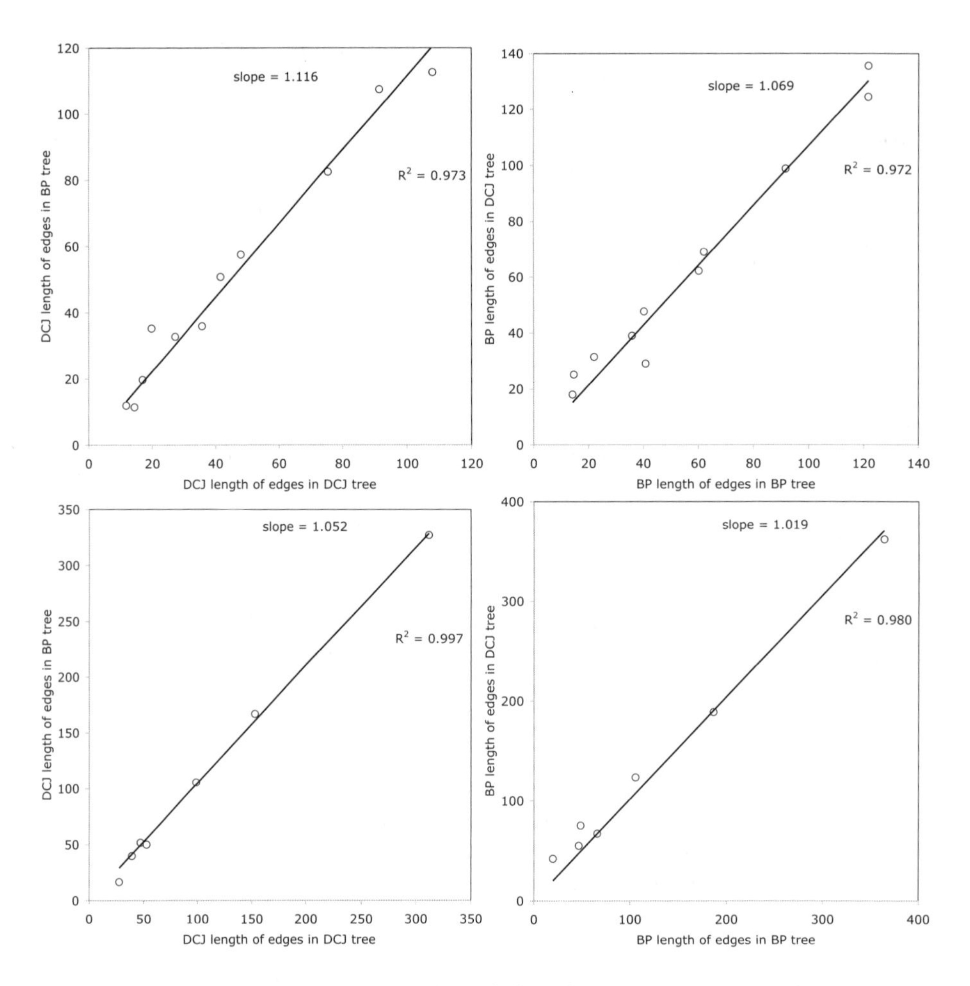

Fig. 3. Competing criterion average branch lengths versus reconstructing criterion. Top: seven placentals data. Bottom: placentals plus opossum data.

Maximum Intranode Distance. In calculating the intranode distances, no patterns could be discerned for the excess rates within the second data set, the one containing the opossum genome. In particular, there were only three points on each scattergram, making inference very sensitive to statistical fluctuation in any of them. The correlations were very poor, and sometimes even negative. We tracked this latter problem down to two very different DCJ reconstructions of the node closest to opossum, occurring in at least two of the sets of simulations. Generally local minima for these problems tend to be relatively close together and this is what justifies our using summary statistics for them. Exploring the large-scale structure of the set of optimal solutions is beyond the scope of this paper, however, so we confined the study of intranode distances to the seven placentals data set. Here, Table 2 shows that in seven out of eight comparisons,

Table 1. Excess rate, in %, for each reconstruction, measured by competing criterion. Average branch length results.

dataset	seven placentals		marsupial, placentals	
reconstruction	BP	DCJ	BP	DCJ
	real data			
	5.23	**1.94**	11.55	**6.92**
	simulated			
no re-use	6.31	**3.05**	7.22	**5.56**
re-use 2	8.55	**2.02**	8.65	**0.44**
actual re-use	6.06	**3.85**	8.69	**3.30**

Table 2. Excess rate, in %, for each reconstruction, measured by competing criterion. Maximum intranode distance data.

dataset	intranode		intranode, intercriteria	
reconstruction	BP	DCJ	BP	DCJ
	real data			
	34.1	**-27.8**	11.55	**6.92**
	simulated			
no re-use	40.0	**-35.9**	59.2	**14.3**
re-use 2	8.09	**5.46**	**13.75**	26.4
actual re-use	5.96	**-3.26**	62.5	**16.6**

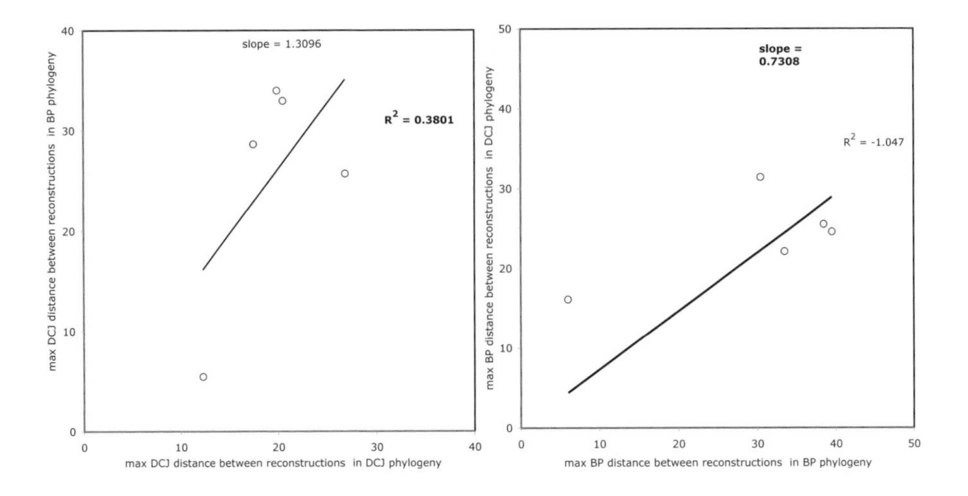

Fig. 4. Competing criterion maximum intranode distance *vs.* reconstructing criterion

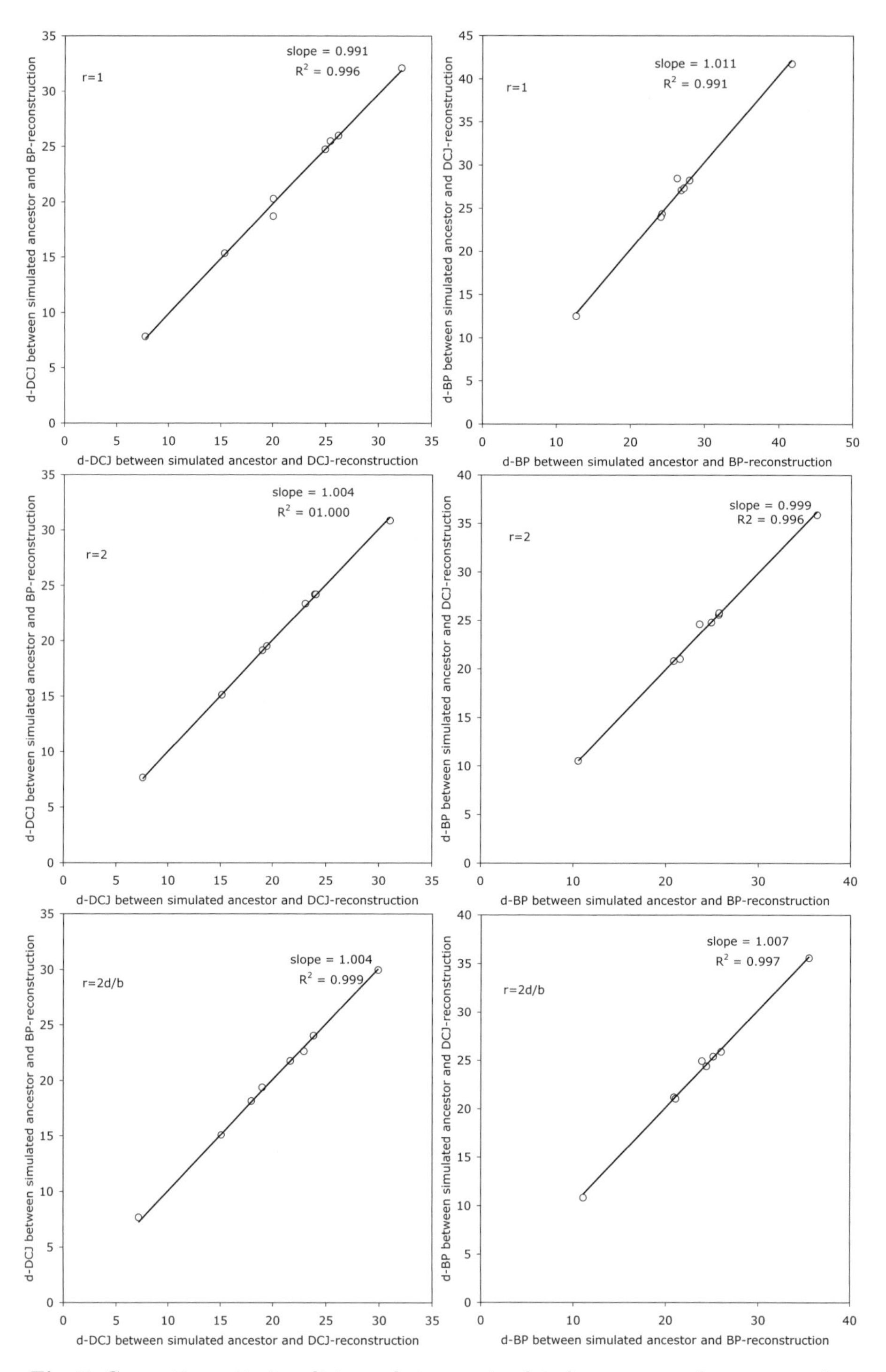

Fig. 5. Competing criterion distance between simulated ancestor and reconstruction

the smaller excess rate (in boldface) is that for the DCJ reconstructions. These rates are often negative, showing that the DCJ reconstructions are often more closely clustered in terms of d_{BP} than the BP reconstructions are.

Fig. 4 illustrates the derivation of the excess rates for the real data. Note the smaller correlations compared with the average branch length comparisons.

Average Distance between Reconstruction and True Ancestor. In the case of the simulated data sets, we actually know the ancestral genomes. Thus as in Section 2.3, we can assess the relative performance of d_{BP} and d_{DCJ} with respect to how close the reconstructed genomes are to the true genomes. Table 3 shows the results of combining the results of both data sets to estimate the $\alpha_{DCJ/BP}$ and the $\alpha_{BP/DCJ}$. Fig. 5 illustrates the calculations of these values. We note the extremely small excess rates, surely within the noise level when we compare with Tables 1 and 2, so that it appears that the reconstructed genomes are equally close to the true simulated ancestors no matter which method was used to infer them, or used to measure the distance.

Table 3. Excess rate, in %, for each reconstruction, measured by competing criterion. Average distance between reconstructed and simulated ancestor results.

dataset	combined data	
reconstruction	BP	DCJ
	simulated	
no re-use	**-0.9**	1.1
re-use 2	0.4	**-0.1**
actual re-use	**0.4**	0.7

6 Discussion

It is often taken for granted that rearrangement distance is "better" than breakpoint distance in phylogenetic inference because only the former is connected to a model of genome evolution. This reasoning is specious since (i) the rearrangements inferred in reconstructing the phylogeny are virtually always different and far fewer than those that actually generated the tree, (ii) the set of rearrangement operations in the evolutionary model cannot include all possible mechanisms, (iii) the uniform costs accorded to all operations is a weakness of the rearrangements approach (iv) breakpoint distance is in fact closely related to evolutionary models, especially those with relatively unconstrained rearrangement operations. In fact, there is no *a priori* biological or statistical reason to prefer one approach over the other. This is what motivates our search for the criterion than does least poorly as judged by the competing criterion.

The general picture that emerges from our analysis is that the ancestral genomes reconstructed according to the breakpoint criterion are more dispersed in the space of genomes than those reconstructed under the rearrangements criterion. This explains the dispersion analyses and by extension the branch-length results, all of which give the impression that the alternative optimal reconstructions under the DCJ criterion are relatively more compact in the set of genomes.

The larger dispersion does not entail that the breakpoint reconstructions are on the average farther from the true simulated ancestor. It might be that the high "dimensionality" of genome space allows the reconstructions to be far from each other, compared to the DCJ reconstructions, while still allowing them to be close to the real ancestor genome. Moreover, this dispersion is not dectectible using the coefficient of variation, which favours neither DCJ nor breakpoints, across data sets, across reuse levels or with respect to any reconstruction property.

References

1. Adam, Z., Sankoff, D.: The ABCs of MGR with DCJ. Evolutionary Bioinformatics Journal 4, 69–74 (2008)
2. Bergeron, A., Mixtacki, J., Stoye, J.: A unifying view of genome rearrangements. In: Bücher, P., Moret, B.M.E. (eds.) WABI 2006. LNCS (LNBI), vol. 4175, pp. 163–173. Springer, Heidelberg (2006)
3. Bourque, G., Pevzner, P.: Genome-scale evolution: reconstructing gene orders in the ancestral species. Genome Research 12, 26–36 (2002)
4. Cosner, M., Jansen, R., Moret, B.M.E., Raubeson, L., Wang, L., et al.: An empirical comparison of phylogenetic methods on chloroplast gene order data in Campanulaceae. In: Sankoff, D., Nadeau, J. (eds.) Comparative Genomics, pp. 99–121. Kluwer, Dordrecht (2001)
5. Lau, H.: A Java Library of Graph Algorithms and Optimization. Chapman, Boca Raton (2006)
6. Mikkelsen, T.S., Wakefield, M.J., Aken, B., Amemiya, C.T., Chang, J.L., et al.: Genome of the marsupial *Monodelphis domestica* reveals innovation in non-coding sequences. Nature 447, 167–177 (2007)
7. Moret, B.M.E., Siepel, A.C., Tang, J., Liu, T.: Inversion medians outperform breakpoint medians in phylogeny reconstruction from gene-order data. In: Guigó, R., Gusfield, D. (eds.) WABI 2002. LNCS, vol. 2452, pp. 521–536. Springer, Heidelberg (2002)
8. Moret, B.M.E., Wang, L., Warnow, T., Wyman, S.: New approaches for reconstructing phylogenies from gene order data. Bioinformatics 17(suppl.), 165–173 (2001)
9. Murphy, W.J., Larkin, D.M., Wind, A.E., Bourque, G., Tesler, G., et al.: Dynamics of mammalian chromosome evolution inferred from multispecies comparative maps. Science 309, 613–617 (2005)
10. Sankoff, D.: The signal in the genomes. PLOS 2, e35 (2006)
11. Swenson, K.M., Marron, M., Earnest-DeYoung, J.V., Moret, B.M.E.: Approximating the true evolutionary distance between two genomes. In: Proc. 7th Workshop on Algorithm Engineering and Experiments (ALENEX 2005), pp. 121–129. SIAM, Philadelphia (2005)
12. Tannier, E., Zheng, C., Sankoff, D.: Multichromosomal median and halving problems under different genomic distances. BMC Bioinformatics 10, 120 (2009)
13. Yancopoulos, S., Attie, O., Friedberg, R.: Efficient sorting of genomic permutations by translocation, inversion and block interchange. Bioinformatics 21, 3340–3346 (2005)

Comparative Genomics and Extensive Recombinations in Phage Communities

Guylaine Poisson[1], Mahdi Belcaid[1], and Anne Bergeron[2]

[1] Information and Computer Sciences, University of Hawaii
guylaine@hawaii.edu
[2] LaCIM, Université du Québec à Montréal, Canada

Abstract. Comparing the genomes of two closely related viruses often produces mosaics where nearly identical sequences alternate with sequences that are unique to each genome. When several closely related genomes are compared, the unique sequences are likely to be shared with third genomes, leading to virus mosaic communities. Here we present comparative analysis of sets of *Staphylococcus aureus* phages that share large identical sequences with up to three other genomes, and with different partners along their genomes. We introduce mosaic graphs to represent these complex recombination events, and use them to illustrate the breath and depth of sequence sharing: some genomes are almost completely made up of shared sequences, while genomes that share very large identical sequences can adopt alternate functional modules. Mosaic graphs also allow us to identify breakpoints that could eventually be used for the construction of recombination networks. These findings have several implications on phage metagenomics assembly, on the horizontal gene transfer paradigm, and more generally on the understanding of the composition and evolutionary dynamics of virus communities.

1 Introduction

Viruses that infect bacteria, known as phages, evolve by accumulating mutations, but also through recombination events in which they exchange genetic material with other phages. These events have been suggested to explain the mosaic structure that arises when the genomes of two phages are compared: nearly identical sequences alternate with sequences that are merely similar or even completely divergent. The first evidence of such exchanges in bacteriophages dates back to the early 90's and was obtained by heteroduplex mapping [1]. Since then, numerous mosaics have been identified by sequence comparison, and the mosaic structure of bacteriophages is now a well documented phenomenon, (see [2] for a recent review).

In this paper, we study *co-linear* phages that infect a common host. These phages have small genomes – around 44 000 bp – that often have conserved order of gene function, called *modules*, such as:

$$\text{head} \rightarrow \text{tail} \rightarrow \text{lysis} \rightarrow \text{integration} \rightarrow \text{DNA replication.}$$

F.D. Ciccarelli and I. Miklós (Eds.): RECOMB-CG 2009, LNBI 5817, pp. 205–216, 2009.

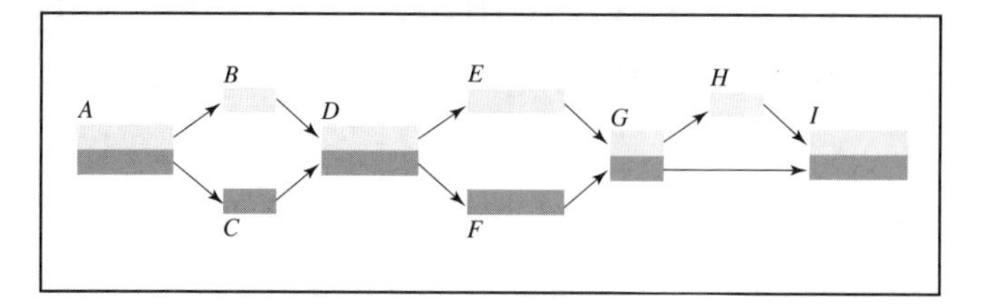

Fig. 1. Comparing two co-linear genomes with a directed graph. In this graph, the two colored bands represent the genomes. In block A both genomes are nearly identical, divergent in blocks B and C, identical in block D, and again divergent in blocks E and F. Block H do not appear in the second genome.

When two co-linear phages are compared along their genomes, the sequences coding for modules alternate between nearly identical sequences and more divergent ones. It may even be the case that two sequences coding for the same module do not have any recognizable homology. Graphs, such as the one in Figure 1, can be used to classify these events: nearly identical sequences are merged in a single block and arrows indicate transitions between consecutive intervals of the same genome.

Although most studies still rely solely on two by two comparisons of phages, the first hints of mosaic communities began to appear a few years ago. For example, in [3], a total of 27 *Staphylococcus aureus* phage genomes are compared, showing genomic regions having more than 98% identity over more than 50 bp shared between one phage and as much as 16 other phages. A recent study of 50 mycobacteriophages that infect a common host also reveals pervasive mosaicism [4]. Because only a tiny fraction of viruses are amenable to study using traditional cultivation methods, the wealth of genetic information is frequently accessed using metagenomic analysis. Viruses sampled in the same environment are in close contact, and recombination events observed in cultivated viruses are thus likely to happen in those communities. There is even evidence of long range propagation of recombination events: nearly identical sequences of phages have already been detected in multiple environments, with varying geographical locations, both in fresh and saline waters [5,6,7].

Understanding the structure and evolution of phage communities is a major challenge. Due to the presence of many recombination events, traditional tools such as phylogenetic trees [8] must be complemented by other methods. Gene *phamilies* [9], for example, compare the distinct evolutionary histories of genes that belong to a recombinant organism. Information on recombination events may also be incorporated in phylogenetic trees, such as in [10], or more general clustering approaches can be used to represent the evolutionary and functional relationships between phages in terms of shared genes [11]. *Recombination networks* have been extensively used in the context of population studies [12,13], modeling the evolution of fixed length sequences with point mutations and recombination events. Unfortunately, none of these approaches allows the representation and interpretation of the extensive recombination events observed in mosaic communities.

Even if numerous studies point out that recombination events are a major feature in phage evolution and organization, very little is known about the relationships between these events. Are sequences shared between more than two phages in closely related organisms? What types of transition may occur when similarity ends? Can all participating sequences become divergent from each other? Are the divergent sequences unique, or are they reused by third genomes? Do "parent" genomes survive along their recombinant descendants in a community? Are recombination breakpoints reused?

In this study, we report that sharing of large identical, or nearly identical, sequences (average 2723 base pairs with 99.98 % identity) is a frequent phenomenon among groups of phages that infect *Staphylococcus aureus*. Recombinations occur almost anywhere along the typical genome of 44 000 bp phages, often involving more than two species, and phages that are identical along 75 % of their genomes may diverge completely, gaining different partners in the divergent sequences. These interactions are complex, and we developed new concepts and representations in order to describe them: we introduce *mosaic graphs* as a complement to phylogenetic trees [8] and reticulate representations [9,10,11,13] in trying to understand phage communities.

2 Tiling ROSA

In order to assess the amount of "shared sequences" in phage communities, we conducted an exploratory experiment on the set of 27 phages described in [3], using a threshold of 98% identity over more than 500 bp (data not shown). This data turned out to be far more complex than we expected, but its analysis provided crucial observations.

A first unexpected finding was that the mismatches in alignments tended to be clustered in small regions, leaving between them large fragments that were identical. Using phage ROSA (AY954961) in BLAST [14] queries we identified 19 intervals of ROSA's genome of at least 400 bp that were 100 % identical to sequences in five other *S. aureus* phages. These intervals cover 35 501 bp of ROSA's 43 155 bp genome (82.26 %). Many pairs of intervals were at most one nucleotide apart, and we merged them to obtain 12 *tiles* of size ranging from 497 bp to 10390 bp, averaging 2958 bp. Table 1 shows the bounds of these intervals for ROSA and for the corresponding phages.

This data is perhaps best appreciated by looking at Figure 2 that shows how the five other genomes cover ROSA. Large identical, or nearly identical, sequences between phages are often attributed to recent recombination events [15]. If this is the case, then ROSA must have participated in – or is the product of – many recent recombination events.

Except for phage PhiNM4 (DQ530362), the coordinates of all phages of Table 1, are comparable, reflecting the co-linearity of the genomes. The modules of PhiNM4 form a cyclic permutation of the order of the other five. When genomes are linear,

Table 1. Covering phage ROSA with phages PhiNM4, 29, 53, 71, 80 and 88

		Start	End		Start	Size	Errors	% Identity
1	Rosa	1197	3212	Phage 71	1123	2016	1	99.95
2	Rosa	3271	6325	Phage 71	3197	3055	1	99.97
3	Rosa	6326	6863	PhiNM4	23734	538	0	100.00
4	Rosa	7874	18263	PhiNM4	24319	10390	0	100.00
5	Rosa	18265	22240	Phage 88	17226	3976	2	99.95
6	Rosa	22241	23491	Phage 53	22125	1251	0	100.00
7	Rosa	23492	28089	PhiNM4	39961	4598	1	99.98
8	Rosa	31047	36361	PhiNM4	4965	5314	1	99.98
9	Rosa	36462	36958	Phage 53	37544	497	0	100.00
10	Rosa	37062	39381	Phage 53	38144	2320	0	100.00
11	Rosa	39517	40064	Phage 53	40596	548	0	100.00
12	Rosa	42104	43101	Phage 29	41664	998	1	99.90

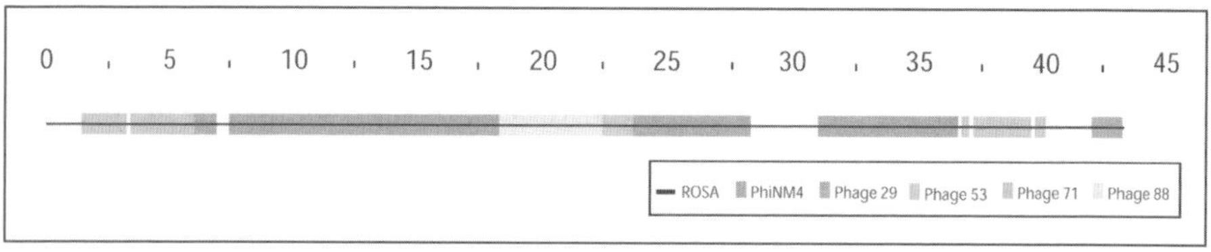

Fig. 2. Tiling ROSA. Each solid band of a single color, or *tile*, represents a sequence that is shared between phage ROSA and another phage. These 12 tiles cover 35 501 bp of ROSA's genome (82.26 %) with 7 errors, that is, 99.98 % identity on average.

such a cyclic permutation is due to a transposition in the genome. However, if the genomes are circular, a cyclic permutation merely accounts for a different start point in the assembly of the genome.

The fact that more than 80 % of ROSA's genome could be covered by large identical sequences from already sequenced genomes was a surprise, and a good one. But it became quickly evident that we did not have the representation tools to understand the relations between these genomes. These are developed in the next sections.

3 From Alignments to Multiple Alignments

The prevalence of large identical sequences in a set of genomes offers a rather unique opportunity in comparative genomics, in the sense that it is easy to construct multiple alignments with pairs of alignments. In this section, we will study multiple alignments that can be constructed from the comparison of the set of phages of the preceding section: phages ROSA, PhiNM4, 29 (AY954964), 53 (AY954952), 71 (AY954962) and 88 (AY954966). First we have:

Definition 1. *Let $i \leq k$, and two alignments of genome A, one with interval $A[i..j]$ with an interval of genome B, and one with interval $A[k..l]$ with an interval of genome C. The two alignment* overlap *if $k \leq j$. Two overlapping alignment induce a multiple alignment of genomes A, B and C in the interval $A[k..\min(j,l)]$.*

In general, multiple alignments induced by pairs of alignments can be of poor quality. However, starting with alignments of identical sequences, or sequences that have occasional single mismatches separated by a few hundred nucleotides, the result is pretty good, as long as the overlapping intervals have significant lengths.

Since we now work with a fixed set of genomes, we used the software REPuter [16] in order to identify identical sequences of at least 400 bp. This software is based on a very efficient algorithm to find repetitions in a genome, or, in our case, the concatenation of two genomes. As in the preceding section, intervals that were at most one nucleotide apart were merged in a single alignment. Table 2 shows 23 alignments between nearly identical sequences, whose size range from 497 to 12032 bp, averaging 2723 bp.

In order to show the induced multiple alignements, Figure 3 displays them using ROSA coordinates (in 1000 bp) for most of the genomes, and phage PhiNM4 coordinates when ROSA is absent. In the first line, between positions 0 and 12 000, only pairs of genomes align well. On the second line, overlapping alignments start to appear: for example, the alignment of ROSA and PhiNM4 recruits phage 88 around position 15 000. On the third line, there are three examples of

Table 2. Shared sequences between phages Rosa, PhiNM4, 29, 53, 71 and 88

		Start	End		Start	End	Size	Errors	% Identity
1	Rosa	1197	3212	Phage 71	1123	3138	2016	1	99.95
2	Rosa	3271	6325	Phage 71	3197	6251	3055	1	99.97
3	Rosa	6319	6863	PhiNM4	23734	24271	545	0	100.00
4	Rosa	7874	19905	PhiNM4	24319	36350	12032	2	99.98
5	Rosa	15319	22240	Phage 88	14280	21201	6922	3	99.96
6	Rosa	22192	26454	Phage 53	22125	26387	4263	1	99.98
7	Rosa	22570	28089	PhiNM4	39038	1368	5520	1	99.98
8	Phage 71	25735	26922	Phage 88	25412	26599	1188	0	100.00
9	Phage 29	27408	28500	Phage 88	26777	27872	1096	0	100.00
10	PhiNM4	**1370**	2312	Phage 29	28596	29538	943	0	100.00
11	PhiNM4	**1370**	2521	Phage 53	29098	30249	1152	0	100.00
12	PhiNM4	2624	5291	Phage 71	29166	31833	2668	1	99.96
13	Rosa	31047	**36361**	PhiNM4	4965	10279	5315	1	99.98
14	Rosa	**32642**	**34064**	Phage 29	33988	35410	1423	0	100.00
15	Rosa	**32642**	**34064**	Phage 71	33671	35093	1423	3	99.79
16	Phage 53	33996	35065	Phage 88	32454	33523	1070	0	100.00
17	Rosa	**34637**	36238	Phage 88	34723	36322	1602	1	99.94
18	Rosa	**34637**	**36361**	Phage 53	35719	37443	1725	0	100.00
19	PhiNM4	8554	13285	Phage 53	35719	40450	4732	0	100.00
20	Rosa	36462	36958	Phage 53	37544	38040	497	0	100.00
21	Rosa	37062	39381	Phage 53	38144	40463	2320	0	100.00
22	Rosa	39517	40064	Phage 53	40596	41143	548	0	100.00
23	Rosa	42104	43101	Phage 29	41664	42661	998	1	99.90

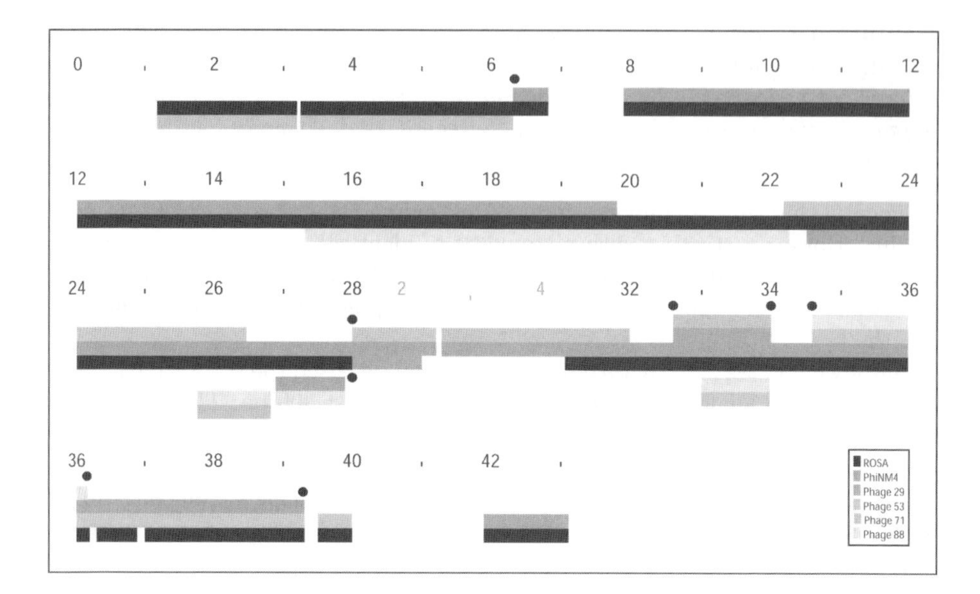

Fig. 3. Shared sequences between phages ROSA, PhiNM4, 29, 53, 71 and 88. The 23 alignments of Table 2 are drawn along phage ROSA coordinates (in 100 bp), except at marker 28 where coordinates are temporarily switched to phage PhiNM4 coordinates. Each solid line of color represents a phage genome. When two sequences are identical, the corresponding lines are stacked, creating multiple alignments when alignments overlap. Complex breakpoints, where at least two sequences are dropped and/or recruited simultaneously, are indicated by back dots.

parallel alignments, meaning that two distinct alignments exist at approximately the same positions along the genomes. Line 3 contains two examples of multiple alignments of four sequences: phages ROSA, PhiNM4, 29 and 71 are equal on a length of 1423 bp except for 3 mismatches; and phages ROSA, PhiNM4, 53 and 88 are equal on a length of 1602 bp except for 1 mismatch.

We define *breakpoints* between multiple alignments when a new sequence is recruited into the alignment, or when a sequence is dropped from the alignment. A breakpoint is *complex* if at least two sequences are recruited and/or dropped almost simultaneously. Black dots in Figure 3 mark the 8 complex breakpoints in which sequences are dropped or recruited into an alignment within 15 bp. For example, phage ROSA recruits phage PhiNM4 at position 6319, and drops phage 71 at position 6325, a difference of 6 bp. A more complex example occurs around position 28 000: phage PhiNM4 drops phage ROSA at position 1368, and synchronously recruits phages 29 and 53 at position 1370. It is a striking feature of this dataset that 5 of the 8 complex breakpoints are synchronous. The positions of these synchronous events are bolded in Table 2.

4 Mosaic Graphs

The representation of Section 3 can give a quite accurate description of the relations between phages, as long as there is one genome – such as ROSA – that can be used almost always as a reference. In general, this will not be the case, and we need a more general representation that is independent from a particular genome.

In order to develop this representation, we must make a certain number of assumptions on the relations between compared genomes. In the preceding sections, we used the term "shared sequences" rather informally, but we were able to establish that there are ample evidence of sequence sharing in phage communities in the following sense:

Definition 2. *A* shared sequence *between genomes A and B is a subsequence that appears exactly once in each genome, and that is of maximal length.*

When a genome is compared to k other genomes, its sequence can be decomposed into alternating intervals of overlapping shared sequences and *unique* intervals. These unique intervals can be as short as one nucleotide. Using Definition 2, the multiple alignments induced by shared sequences are trivial: every column contain the same nucleotide. Such multiple alignments can thus be described by the bounds of the intervals of each of its participating genome. We next introduce the concept of *mosaic graphs*:

Definition 3. *Given a set of genomes $\mathcal{G}$, and a collection of shared sequences between pairs of genomes in S. A* block *is defined as a maximal induced multiple alignment, or as a unique interval of a genome. The* mosaic graph *of $\mathcal{G}$ is a directed graph whose vertices are blocks, and in which block S is connected to block T, represented as $S \to T$, if S contains interval $[i..j]$ of a genome in $\mathcal{G}$, and T contains the interval $[j + 1..k]$ of the same genome.*

Mosaic graphs are meant to capture the relations between sequences that evolved by recombination events, but they have the advantage of being uniquely defined by the initial collection of shared sequence. In practice, two blocks separated by a single point mutation can be merged, and small blocks can be omitted from a graphical representation in order to yield a better visual representation. We next discuss one such example.

Our preliminary experiment showed that phages 88 and 92 (AY954967) shared their genomes along the initial half, but then each phage took a different path: on those parallel paths there was a pair of distinct multiple alignments of unusual depth. In these alignments, phage 88 associates with phages 29 and 187 (AY954950), and phage 92 with phages 53 and 85 (AY954953).

The mosaic graph of these 6 phages – with partial sequences[1] for phages 29, 53, 85 and 187 – uses the blocks of Table 3. There are 8 alignments of two sequences, and 2 multiple alignments of three sequences. The error count for the

[1] Phage 29 (AY954964[26198..30755]), phage 53 (AY954952[30257..32909]), phage 85 (AY954953[25417..31888]), and phage 187 (AY954950[20618..23998]).

Table 3. Blocks of the mosaic graph of phages 88, 92, 29 53, 85 and 187

		Start	End		Start	Size	Errors	% Matches
A	Phage 88	4	22484	Phage 92	1	22481	4	99.98
D	Phage 88	24403	25404	Phage 85	26417	1002	0	100.00
E	Phage 92	25305	26155	Phage 85	27658	851	1	99.88
F	Phage 88	26566	27946	Phage 29	27198	1381		
	Phage 88	26566	27946	Phage 187	21618	1381	11	99.73
	Phage 29	27198	28577	Phage 187	21618	1380		
G	Phage 92	26156	27688	Phage 85	28509	1533		
	Phage 92	26156	27688	Phage 53	27552	1533	3	99.93
	Phage 53	27552	29084	Phage 85	28509	1533		
I	Phage 53	29085	30256	Phage 29	28583	1172	1	99.91
H	Phage 92	27689	28534	Phage 85	30042	846	0	100.00
J	Phage 92	28547	29859	Phage 53	30597	1313	4	99.70
M	Phage 88	34186	39299	Phage 92	33383	5114	0	100.00
N	Phage 88	39790	43231	Phage 92	38987	3442	0	100.00

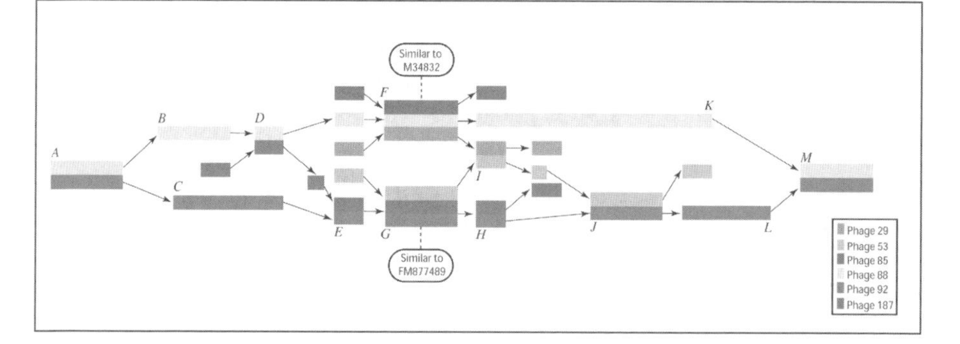

Fig. 4. The partial mosaic graph of phages 29, 53, 85, 88, 92 and 187. The vertices of the graph are the multiple alignments of nearly identical sequences of Table 3, together with sequences that are unique to each genome. Two vertices S and T are connected if one of the genomes has consecutive intervals that belong to S and T. The two multiple alignments in the center have no recognizable homology between them but both contain sub-sequences highly similar to integrase genes. [Blocks are not necessarily to scale.]

multiple alignments was obtained by counting the number of nucleotides that differ from the majority in a column of the alignment. The last column shows the percentage of correct nucleotides in the alignment. Figure 4 displays the resulting graph in the interval ranging from approximate positions 21 000 to 35 000 of phage 88.

The two blocks with multiple alignments of three sequences, F and G, have no recognizable homology between them. However both contain a sub-sequence highly similar to annotated integrase genes: phage Phi-11 (M34832) for sequences that clusters with phage 88, and phage Phi6390 (FM877489) for sequences that clusters with phage 92.

This region displayed also contains 3 complex breakpoints. One breakpoint is synchronous, $G \rightarrow I$, and the two others occur within 12 bp: 5 extra bp in $F \rightarrow I$, and 12 extra bp in $H \rightarrow J$. Two of these complex breakpoints occur at the same 'position' in the genomes, begging the following question: is the transition $F \rightarrow H$, missing for good reasons, or is there a phage out there waiting to be sequenced and that would make the required transition? In order to test the second possibility, we created a chimera composed of the last 495 bp of phage 88 in block F, followed by the 5 extra bp after block F, followed by the first 500 bp of phage 92 in block G. This composite sequence has an alignment with phage PhiMR25 (AB370205) with 996 identities over 1000 nucleotides. Thus we confirm that the desired transition is already sequenced.

4.1 From Mosaic Graphs to Recombination Networks

Mosaic graphs of related genomes and multiple alignments of similar sequences both give representations of evolutionary events, one focussing on recombination events and the other on point mutations. These representations do not specify the nature and order of the evolutionary events: this is done by phylogenetic inference, and the resulting constructions are subject to evaluation by parsimony or likelihood criteria.

Based on multiple alignments, phylogenetic trees are constructed under the assumption of point mutations modifying a common ancestral sequence, allowing for occasional horizontal gene transfers [10]. Recombination networks used in population studies [12,13] recognize recombination events and point mutations on an equal footing [12,13], but still rely on the common ancestor assumption.

Untangling the evolution history of phages that underwent frequent recent recombination events can certainly use the framework of recombination networks. In this case, they would be defined as networks that describe the nature and order of recombination events explaining a given mosaic graph. However, the construction of these networks cannot rest on the assumptions that held in population studies: gene sequences that code for analog functions often have no recognizable homology, ruling out the possibility of a fairly recent common ancestor; and breakpoints are given by the mosaic graph instead of being inferred. This opens up some exciting combinatorial problems:

Problem 1. Given a set of co-linear genomes that evolved by recombination events, and possible extinctions, from divergent ancestors, when is it possible to reconstruct its evolution history?

Problem 2. How can the phylogenetic information available by alignments of nearly identical sequences be used to guide the reconstruction?

To the best of our knowledge, very few results seem to exist for Problem 1. In Figure 5, an instance of the problem is given with its mosaic graph. Each genome has four modules, and equal labels indicate shared sequences. We have the following two results that, unfortunately, are only sufficient conditions. The

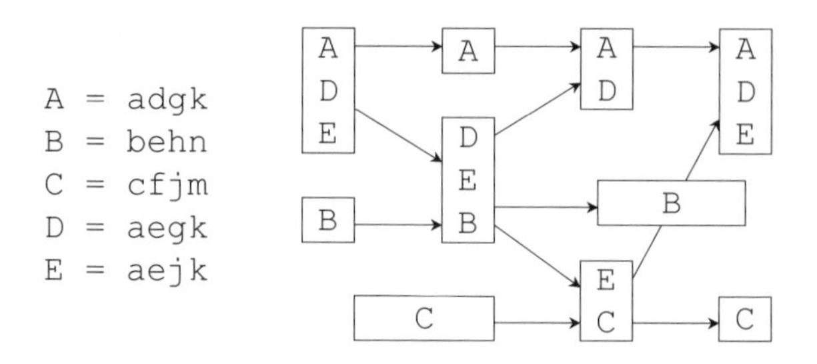

Fig. 5. An instance of the reconstruction problem and its mosaic graph

first one is immediate since any recombinant share each part of its sequence with at least one parent.

Proposition 1. *If there are no extinction events, then any genome that has a unique sequence in the graph is an ancestor.*

The second one is almost as obvious, but can be used to resolve the example in Figure 5.

Proposition 2. *If removing genome G disconnects the graph, then G has a parent in each remaining component.*

In the example, genome E is the only genome that disconnects the graph. Of the two remaining components, one is trivial containing only genome C, and removing genome D from the other further disconnects the graph. Thus there exist a reconstruction with two recombination events, with genomes A and B as parents of genome D, and genomes D and C as parents of genome E. Note that if, for example, B became extinct, either A or D could be ancestral in an optimal reconstruction.

5 Conclusion

We showed that recombination events involving identical or nearly identical sequences are numerous in groups of phages that infect *Staphylococcus aureus*, and we currently have strong evidence of the same behavior in at least three other families of phages that infect other bacteria. If this phenomenon is widespread, it has an immediate impact on the environmental metagenomics sequencing strategies. Next generation sequencing, such as 454 [17] or Illumina [18], allows deeper sequencing of short reads, but these are often single reads. While such technologies are suitable for re-sequencing, mate-paired reads are necessary for *de novo* sequencing of genomes with long repeats. Even if phages seldom have long repeats, shared sequences introduce artificial repeats that can fool assembly software: when two genomes share a large sequence S such as in

$$A \to S \to B \text{ and } C \to S \to D,$$

an assembler working with single reads cannot distinguish between

$$A \to S \to B, C \to S \to D, A \to S \to D \text{ and } C \to S \to B.$$

The output of the assembler would thus be, at best, the mosaic graph itself.

A second consequence of our findings is that it sheds a new light on the concept of gene transfers between phages. Conventional wisdom recognizes transfers as *from* an organism *to* another, with recently transfered sequences more similar than sequences resulting from older transfers. In this context, what could be the meaning of long identical sequences in three or more different phages? Since phage ROSA is mostly made up of sequences identical to at least 5 other phages, how many events are necessary to construct ROSA, even assuming that transfers between two phages are not necessarily contiguous? Or is ROSA a donor?

Such questions are not easy to answer since the precise mechanisms that lead to mosaic structures are not entirely elucidated [19]. One interesting suggestion [20] is random recombination on a large scale followed by selection of the fittest. Clearly, some newly created genomes would lack vital parts, but if the recombinations happen at the junctions of the mosaic graph, then the resulting chimera could very well be able to reproduce itself. In this sense, a mosaic graph would be a better representation of a community of phages than a collection of individual genomes. We saw that junctions between blocks of the mosaic graph are complex and varied, and that many of the possible paths in the graph are indeed followed by particular phages. Are the paths that were unobserved in our datasets mostly non-existent, or just waiting to be sequenced?

Acknowledgments

AB was supported by NSERC Grant no. 121768. MB and GP are supported by NIH Grant no. P20 RR-16467 from the National Center for Research Resources. The paper's contents are solely the responsibility of the authors and do not necessarily represent the official views of the NIH.

References

1. Highton, P.J., Chang, Y., Myers, R.J.: Evidence for the exchange of segments between genomes during the evolution of lambdoid bacteriophages. Mol. Microbiol. 4, 1329–1340 (1990)
2. Hatfull, G.F.: Bacteriophage genomics. Curr. Opin. Microbiol. 11, 447–453 (2008)
3. Kwan, T., Liu, J., DuBow, M., Gros, P., Pelletier, J.: The complete genomes and proteomes of 27 Staphylococcus aureus bacteriophages. Proc. Natl. Acad. Sci. USA 102, 5174–5179 (2005)
4. Hatfull, G.F., Cresawn, S.G., Hendrix, R.W.: Comparative genomics of the mycobacteriophages: insights into bacteriophage evolution. Res. Microbiol. 159, 332–339 (2008)
5. Breitbart, M., Miyake, J.H., Rohwer, F.: Global distribution of nearly identical phage-encoded DNA sequences. FEMS Microbiol. Lett. 236, 249–256 (2004)
6. Short, C.M., Suttle, C.A.: Nearly identical bacteriophage structural gene sequences are widely distributed in both marine and freshwater environments. Appl. Environ. Microbiol. 71, 480–486 (2005)

7. Bryan, M.J., Burroughs, N.J., Spence, E.M., Clokie, M.R., Mann, N.H., et al.: Evidence for the intense exchange of MazG in marine cyanophages by horizontal gene transfer. PLoS ONE 3, e2048 (2008)
8. Rohwer, F., Edwards, R.: The Phage Proteomic Tree: a genome-based taxonomy for phage. J. Bacteriol. 184, 4529–4535 (2002)
9. Hatfull, G.F., Pedulla, M.L., Jacobs-Sera, D., Cichon, P.M., Foley, A., et al.: Exploring the mycobacteriophage metaproteome: phage genomics as an educational platform. PLoS Genet. 2, e92 (2006)
10. Glazko, G., Makarenkov, V., Liu, J., Mushegian, A.: Evolutionary history of bacteriophages with double-stranded DNA genomes. Biol. Direct 2, 36 (2007)
11. Lima-Mendez, G., Van Helden, J., Toussaint, A., Leplae, R.: Reticulate representation of evolutionary and functional relationships between phage genomes. Mol. Biol. Evol. 25, 762–777 (2008)
12. Gusfield, D., Bansal, V.: A fundamental decomposition theory for phylogenetic networks and incompatible characters. In: Miyano, S., Mesirov, J., Kasif, S., Istrail, S., Pevzner, P.A., Waterman, M. (eds.) RECOMB 2005. LNCS (LNBI), vol. 3500, pp. 217–232. Springer, Heidelberg (2005)
13. Huson, D.H., Bryant, D.: Application of phylogenetic networks in evolutionary studies. Mol. Biol. Evol. 23, 254–267 (2006)
14. Altschul, S.F., Gish, W., Miller, W., Myers, E.W., Lipman, D.J.: Basic local alignment search tool. J. Mol. Biol. 215, 403–410 (1990)
15. Nesbo, C.L., Dlutek, M., Ford Dolittle, W.: Recombination in Thermotoga: Implications for Species Concepts and Biogeography. Genetics 172, 759–769 (2006)
16. Kurtz, S., Choudhuri, J., Ohlebusch, E., Schleiermacher, C., Stoye, J., et al.: REPuter: The Manifold Applications of Repeat Analysis on a Genomic Scale. Nucleic Acids Res. 29, 4633–4642 (2001)
17. Margulies, M., Egholm, M., Altman, W.E., Attiya, S., Bader, J.S., et al.: Genome sequencing in microfabricated high-density picolitre reactors. Nature 437, 376–380 (2005)
18. Bentley, D.R.: Whole-genome re-sequencing. Curr. Opin. Genet. Dev. 16, 545–552 (2006)
19. Martinsohn, J.T., Radman, M., Petit, M.A.: The lambda red proteins promote efficient recombination between diverged sequences: implications for bacteriophage genome mosaicism. PLoS Genet. 4, e1000065 (2008)
20. Hendrix, R.W.: Bacteriophages: evolution of the majority. Theor. Popul. Biol. 61, 471–480 (2002)

Properties of Sequence Conservation in Upstream Regulatory and Protein Coding Sequences among Paralogs in *Arabidopsis thaliana*

Dale N. Richardson and Thomas Wiehe*

University of Cologne, Institute for Genetics, Zülpicher Str. 47,
50674 Cologne, Germany
dalesan@gmail.com, twiehe@uni-koeln.de

Abstract. Whole genome duplication (WGD) has catalyzed the formation of new species, genes with novel functions, altered expression patterns, complexified signaling pathways and has provided organisms a level of genetic robustness. We studied the long-term evolution and interrelationships of 5' upstream regulatory sequences (URSs), protein coding sequences (CDSs) and expression correlations (EC) of duplicated gene pairs in Arabidopsis. Three distinct methods revealed significant evolutionary conservation between paralogous URSs and were highly correlated with microarray-based expression correlation of the respective gene pairs. Positional information on exact matches between sequences unveiled the contribution of micro-chromosomal rearrangements on expression divergence. A three-way rank analysis of URS similarity, CDS divergence and EC uncovered specific gene functional biases. Transcription factor activity was associated with gene pairs exhibiting conserved URSs and divergent CDSs, whereas a broad array of metabolic enzymes was found to be associated with gene pairs showing diverged URSs but conserved CDSs.

Keywords: Upstream, regulation, promoter, expression correlation, duplication, paralog, Arabidopsis, evolution, similarity.

1 Introduction

Whole genome duplication (WGD) is a powerful force that has shaped the evolution of many, if not all, eukaryotic genomes. WGD is especially prevalent in the flowering plants, with duplications occurring multiple times throughout multiple lineages [1, 2]. WGD has had an important role in the origin and diversification of flowering plants [1] and it is estimated that at least 70% of flowering plants have polyploidy in their history [3]. Arabidopsis has experienced at least three WGD events [4, 5], with the most recent WGD event having occurred between 20-60 million years ago [6, 7].

Previous studies in Arabidopsis have focused primarily on how duplicated genes diverged in protein coding sequence (CDS) and expression divergence since the time of duplication. Blanc and Wolfe (2004) showed that gene pairs are not lost randomly

* Corresponding author.

F.D. Ciccarelli and I. Miklós (Eds.): RECOMB-CG 2009, LNBI 5817, pp. 217–228, 2009.

over time, but lost according to gene functional biases, where transcription factors and signal transduction proteins were preferentially retained, whereas genes involved in DNA repair were preferentially lost. Furthermore, they reported an asymmetric rate of sequence evolution in the CDSs of more than 20% of the analyzed pairs [8]. Haberer et al. (2004) reported that expression divergence occurs frequently between duplicated gene pairs and may be the primary mechanism behind preserving redundant genes [9]. They also revealed a moderate but significant correlation between promoter sequence similarity and expression divergence for polyploidy derived pairs. Ganko et al. (2007) examined levels of expression divergence between duplicated genes in Arabidopsis and reported that the strength of purifying selection acting on CDSs is coupled to the corresponding pair's expression pattern [10].

Most studies thus far have focused primarily on the properties of coding sequence evolution after duplication, measured in terms of non-synonymous (K_a), synonymous (K_s), or the ratio of non-synonymous to synonymous substitutions (K_a/K_s). However, the evolution of upstream regulatory sequences (URSs) of duplicated genes has received less attention in Arabidopsis. This is partly due to the inherent difficulty of assessing sequence similarity in non-coding DNA, where the number, order and spacing of shared sequence elements may confound traditional, alignment-based approaches. Moreover, the limited number of known or computationally predicted transcription factor binding sites (TFBS) for Arabidopsis makes it difficult to recover a meaningful signal from background noise.

Previously, Haberer et al. (2004) looked at similarities/dissimilarities in duplicated Arabidopsis promoters using a simple alignment-based approach [9]. Here, by using the Shared Motif Method [11], DIALIGN-TX (an improved version of the algorithm used in [9]) [12, 13] and an alignment-free measure of word repetitiveness, we were able to characterize aspects of Arabidopsis URS similarity at a more detailed level. Furthermore, the incorporation of positional information on exact matches between paralogous URSs provided insights into URS sequence evolution and expression divergence that raw similarity values simply cannot provide. An evolutionary analysis of the protein coding sequences of the WGD-derived paralogs revealed distinct functional classifications of the duplicates, dependent on whether the URS or CDS is more conserved. Moreover, the joint consideration of URSs and CDSs revealed that different components of a gene experience different selective pressures following gene duplication.

2 Methods

2.1 Arabidopsis Duplicate Gene Pair Sequences

A list of accession numbers for whole genome derived duplicate genes was obtained from [8]. Genes were considered tandem duplicates and were excluded from this analysis if their protein alignments had a blast E-value $\leq$ 1x10-20 and the corresponding sequences resided less than 15 genes apart on the same chromosome [7]. The TAIR accession numbers were used to obtain the 5' upstream regulatory sequences (URSs) and corresponding protein coding sequences (CDSs) for 2,584 gene pairs assumed to originate from the most recent whole genome duplication event in Arabidopsis (20-60 mya) [8].

To reduce ambiguity regarding 5' transcription start site (TSS) annotation, we kept only those sequences that had annotated 5' untranslated regions (UTRs) supported by cDNA evidence from TAIR (http://www.arabidopsis.org). We ensured that URSs were at least 600 bp long and did not interrupt other upstream, annotated genes. We discarded any pairs showing evidence for alternative 5' TSSs according to blast searches against a database of Arabidopsis ESTs downloaded from GenBank.

Gene pairs that met the above criteria were queried against the ATTED-II [14] database to obtain their Pearson (r) correlation coefficients of co-expression. ATTED-II contains Pearson correlation coefficients of robust multi array (RMA) normalized gene expression levels for 22,263 genes spanning 1,388 samples taken from the AtGenExpress project at TAIR, which cover a variety of experimental conditions: i.e., different developmental stages, biotic, abiotic, nutrient, hormone and chemical treatments.

Application of the above criteria resulted in a total of 815 duplicate gene pairs, each assumed to be polyploidy derived, unobtrusive to other genetic elements, clearly demarcated with a single TSS, replete with expression information and have available protein coding sequences.

2.2 Upstream Regulatory Sequence Analysis

Two data sets were constructed based on the URSs of the 815 Arabidopsis duplicate pairs. One URS data set consisted of six fixed-length sequence intervals, all anchored at the 5' TSS. Each interval increases by 100 bp until the maximum of 600 bp is reached. The second URS data set consisted of nine sliding window intervals, each with a window size of 200 bp, and step size of 50 bp. For each data set and for each window interval, URS conservation was measured using three distinct applications.

The first application is the Shared Motif Method (SMM). The SMM was previously used in ab initio sequence divergence analysis of cis-regulatory DNA [11]. It was used as an index of cis-regulatory divergence in promoter sequences, specifically for organisms with poorly annotated transcription factor binding site (TFBS) motifs, such as *C. elegans*. The SMM depends upon a parameter, L, which dictates alignment sensitivity (for details see [11]). We determined L for each window size by randomly shuffling our input sequences such that on average, these randomly shuffled sequences showed at least 90% sequence divergence. We applied the SMM with empirically derived parameter choices for each window interval within each URS data set. As the SMM outputs a score (d_{SM}) reflective of the percentage of two sequences that do not share similar fragments, we report 1- d_{SM} so that values would reflect similarity and be comparable to programs that output values in terms of similarity.

The second application is a modified version [15] of the previously described Index of Repetitiveness [16]. In this version a distinction is made between query and subject sequences. The algorithm for calculating this query/subject IR essentially determines for each position in the query the longest exact match in the subject. The resulting values are summed and the sum is divided by its expectation, assuming unrelatedness of query and subject. Finally, the logarithm of this ratio is taken to yield the IR [16]. Therefore, the IR may take values between $-\infty$ and +1, with an expectation of 0 for unrelated sequences.

For each Arabidopsis duplicate pair, we calculated the IR as the arithmetic mean between the two IR values for the two possible query/subject configurations (e.g., gene 1 vs. gene 2 and gene 2 vs. gene 1). Generally, sequence fragments that are identical between subject and query, and are longer than expected by chance, contribute to positive values of IR and have an upper bound of 1; however, random shuffles of the input sequences yield IR values ~ 0. IR was written in ANSI C and is freely available for download (http://guanine.evolbio.mpg.de/ir).

The third application is DIALIGN-TX [12, 13], an updated version of the segment-based alignment software used in [9]. We calculated alignment similarity between URSs in the same fashion as in [9].

2.3 Coding Sequence Analysis

For each of the 815 duplicate gene pairs, the synonymous (K_s), non-synonymous (K_a) and the ratio of non-synonymous to synonymous substitution rates (K_a/K_s) were calculated using MutationsHunter (http://aurelien.mazurie.oenone.net/index.php?p=research/tools/mutationshunter&l=en). Coding sequences were aligned according to their amino acid alignments and substitution rates were estimated using PAML, according to the Yang-Nielsen method [17].

3 Results

3.1 Similarity Profiles of Arabidopsis URSs

First, we confirmed that the IR and SMM values for the real data (Fig. 1A and 1B, black boxes) are significantly different from the randomized data (Fig. 1A and 1B, red boxes). A noticeable property of the IR was that its level of variance depended more heavily on the sequence lengths of analyzed URSs than the SMM. The inter-quartile range (IQR) for the IR values for sequences of length 600 bp was only 20% of the IQR for sequences of length 100 bp. For the SMM, the IQR was almost independent of sequence length. Furthermore, there are many more outliers and extreme values in the real data measured by IR (Fig. 1A, black boxes) than in the SMM (Fig. 1B, black boxes), whereas the converse is true for the random data (Fig. 1A and 1B, red boxes). However, in general, median values of conservation for duplicated URSs decrease as the size of the TSS-anchored window increases.

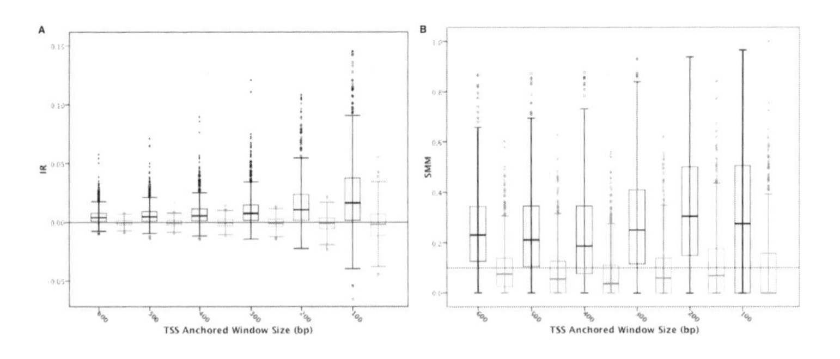

Fig. 1. Anchored window analysis. Observed and randomized data (black and red box plots, respectively) for the IR (Panel A) and the SMM (Panel B).

Next, we tried to localize the regions within the URSs that harbor the highest conservation signals. We performed a sliding window analysis with the window size fixed at 200 bp and moving away from the 5' TSS in 50 bp steps (Fig. 2). Again, similar trends were observed with respect to the behavior of the two methods: all real data values were significantly different from random data (Wilcoxon Signed Ranks test, $p << 10^{-5}$) and the pattern of outliers was reflective of what was observed in Figure 1. Furthermore, note that the variability pattern in the IR and SMM for random data, exhibited fewer outliers for IR and a relatively constant level of variance (Fig. 2A and 2B, red boxes). Nevertheless, despite the qualitative differences in both programs, a distinct pattern emerged with respect to the sliding window analysis. Both programs showed a nearly monotonic decrease as distance from the 5' TSS increased. Such a result suggests that most of the conservation signal (sequence similarity) is within the first 300 bp upstream of the TSS.

3.2 Inter-relationships between URSs, CDSs and Expression Correlation

Haberer et al. (2004) found only a marginal correlation ($r = 0.12$, $0.01 < p < 0.05$) between URS similarity and expression correlation in their study of WGD-derived duplicate pairs in Arabidopsis [9]. Therefore, we correlated the IR and SMM data compiled in the window analyses (Fig. 1 and 2) with relative levels of gene expression between each gene pair (Table 1). We observed highly significant Spearman rank correlations between URS similarity and expression correlation (range 0.159 – 0.277, $p << 0.01$); in some cases, the correlation was more than twice that previously reported. The IR yielded higher correlations than the SMM, except for the first 100 bp TSS-anchored window. A peak in correlation was observed in the 300 bp window for IR (0.277), whereas a bimodality was evident in the 100 bp and 200 bp windows of the SMM (0.226). We also calculated the correlation between both programs. In agreement with our observation that most of the conservation signal was found in the immediate upstream region of about 300 bp, we also observed that both methods were most highly correlated at the smallest window size (Table 1).

We also correlated the data from the sliding window analysis with expression correlation (lower part of Table 1). The pattern in correlation mirrors that of what was observed for URS conservation in Figure 2; that is, increasing distance from the 5' TSS translates not only into a reduced level of sequence conservation but also into a reduced correlation with expression. Considering only the windows that had correlations significant at the 1% level, both programs reported high correlations within the three sliding windows most proximal to the 5' TSS. The same pattern was evident in terms of the inter-application correlations reported for the anchored window analysis: the closer to the 5' TSS, the more congruent the two methods were.

As a comparison, we examined the correlation between an updated version of the alignment software used in [9], DIALIGN-TX (dtx) [12, 13] and EC (Table 1). In both the anchored window and sliding window analyses, the IR and the SMM yielded higher correlation values with EC than dtx. Furthermore, the inter-program correlation values were higher between the IR and the SMM than either was with dtx.

We also measured properties of the coding sequences, such as sequence identity, synonymous (K_s) and non-synonymous (K_a) substitution rates and their ratio (K_a/K_s)

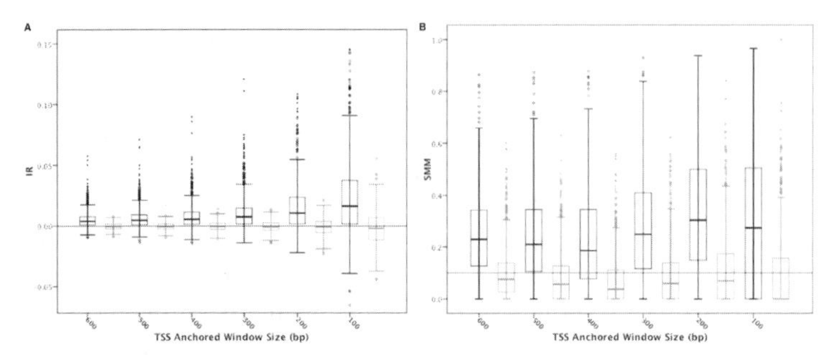

Fig. 2. Sliding window analysis. Observed and randomized data (black and red box plots, respectively) for the IR (Panel A) and the SMM (Panel B).

Table 1. Spearman rank correlations. Values in bold indicate significance at the 0.01 level, a single asterisk at the 0.05 level. See text for long form of acronyms.

Anchored	IR-EC	SMM-EC	dtx-EC	IR-SMM	IR-dtx	SMM-dtx
600	**0.222**	**0.159**	**0.158**	**0.451**	**0.251**	**0.391**
500	**0.228**	**0.169**	**0.153**	**0.471**	**0.273**	**0.401**
400	**0.247**	**0.192**	**0.191**	**0.510**	**0.269**	**0.437**
300	**0.277**	**0.207**	**0.175**	**0.525**	**0.295**	**0.388**
200	**0.252**	**0.226**	**0.178**	**0.582**	**0.313**	**0.412**
100	**0.214**	**0.226**	**0.097**	**0.619**	**0.292**	**0.380**
Sliding	IR-EC	SMM-EC	dtx-EC	IR-SMM	IR-dtx	SMM-dtx
400-600	0.059	0.059	0.021	**0.287**	0.016	**0.188**
350-550	0.049	0.074*	0.022	**0.261**	-0.016	**0.225**
300-500	0.064	0.042	0.034	**0.308**	-0.044	**0.165**
250-450	**0.094**	0.036	0.040	**0.265**	0.019	**0.190**
200-400	**0.121**	0.063	0.041	**0.329**	0.065	**0.209**
150-350	**0.190**	0.085*	0.042	**0.356**	**0.115**	**0.217**
100-300	**0.217**	**0.150**	**0.079**	**0.431**	**0.139**	**0.299**
50-250	**0.257**	**0.198**	**0.108**	**0.472**	**0.232**	**0.316**
0-200	**0.253**	**0.226**	**0.170**	**0.582**	**0314**	**0.404**

(Appendix Table 1[1]). A marginal but significant positive correlation was observed between protein sequence identity and expression correlation. In complete agreement with this result, we also observed a significant negative correlation between expression correlation and the rate of non-synonymous substitutions, K_a. On the other hand, no correlation was found between expression correlation and the rate of synonymous substitutions, K_s. Also, the transition to transversion ratio and the difference in length between duplicate pairs exhibited a significantly negative correlation. Taken together, there is concomitant functional constraint on the protein and its expression profile, as evidenced by the significantly negative correlation between expression correlation and K_a/K_s. Markedly, no significant correlation was observed between any of the URS windows and any evolutionary property of the coding sequences listed in Appendix Table 1 (data not shown).

[1] http://justus.genetik.uni-koeln.de:8200/people/Dale

3.3 Micro-chromosomal Rearrangements of Exact Matches

As similarity values alone do not encompass the entire range of sequence evolution, we analyzed the 300 bp URSs using the positional information and length of exact matches (7-19 bp) between these sequences. We analyzed the whole data set, gene pairs in the upper 25% and lower 25% quantiles based on EC (Fig. 3). We considered four mutually exclusive arrangement classes for the type of exact matches that can occur: proximal exact matches (pem), distal exact matches (dem), inverse proximal exact matches (ipem) and inverse distal exact matches (idem) (Fig. 3D). Pem and ipem are defined as an exact match in the query sequence that is located within the subject sequence constrained by ±30 bp boundaries (blue and green boxes in Fig. 3D), whereas dem and idem are exact matches that are located outside of these boundaries (red and purple boxes in Fig. 3D).

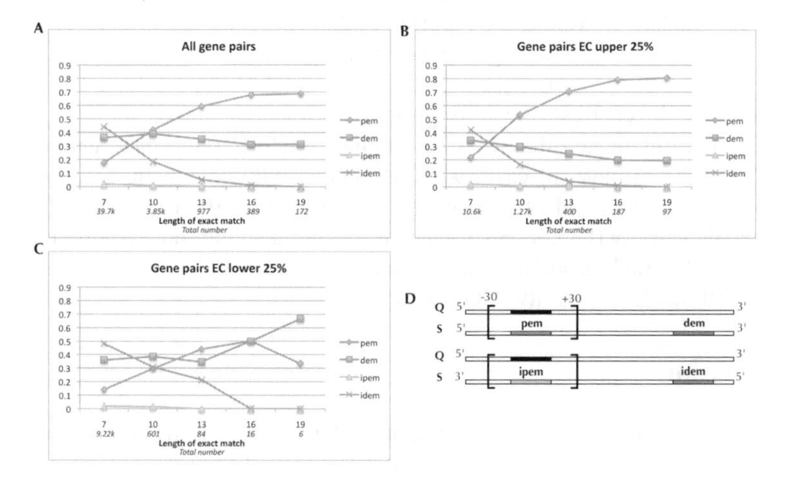

Fig. 3. Micro-chromosomal rearrangements of exact matches. Different lengths and types of exact matches and their frequencies are shown. In panels A-C, italicized numbers represent total counts for each type of exact match. Panel D depicts the possible configurations of exact matches. The black box shows the location of the exact match in the query sequence, whereas the colored boxes show where the match is located in the subject. Abbreviations: see text.

We observed that as the length of the exact match increases, there is concordant increase and decrease in the fraction of pem and idem (blue and purple lines, Fig. 3A). By contrast, the fraction of ipem and dem (red and green lines) remained relatively constant as length increased. A similar pattern was reflected in the upper 25% EC quantile gene pairs (Fig. 3B). However, a distinctly different pattern emerged in the lower 25% EC quantile gene pairs (Fig. 3C). Here, as length increased, the fraction of pem and dem decreased and increased, respectively. Also with increased length, the fraction of idem (purple line, Fig. 3C) was higher for the first three length categories.

3.4 Three-Way Rank Analysis

To understand what kind of gene functional biases result as a consequence of URS and CDS evolution within the context of expression correlation, we performed a joint rank analysis of three variables: the IR of the 300 bp anchored URS window, K_a of the CDS and the expression correlation of the duplicate pairs. For each variable there were three equally populated classes (n = 245): high conservation, middle conservation and low conservation (Appendix Table 2). Of the 27 possible combinations of each rank class, we focused on two specific categories: those gene pairs that fell into high rank categories for IR and expression correlation but had low K_a ranks (26 pairs) and a second grouping that included high rank categories for K_a and expression correlation but low IR ranks (36 pairs). We checked for overrepresented GO slim terms using the web interface at AmiGO [18] for each of these data sets. The enriched molecular function terms for the gene pairs with high ranked K_a and expression correlation but low ranked IR values are reported in Appendix Table 3. Interestingly, about half of the genes in this rank grouping are associated with transcription regulator activity (GO:0030528), transcription factor activity (GO:0003700) or DNA binding (GO:0003677). In contrast, the rank grouping with high ranked K_a and expression correlation but low ranked IR values have completely different functional biases (Appendix Table 4). Almost all cases were associated with enzymatic activity, such as various lyases, ligases and hydrolases, etc., with 42% of the genes enriched for the term catalytic activity (GO:0003824). There was not a single transcription factor term annotated in this rank grouping. Furthermore, 13.5% of the genes were enriched for the structural molecule activity (GO:0005198).

4 Discussion

It has been stated that expression divergence is the first step in the functional divergence between duplicated genes and is a determinant of their evolutionary fates [19, 20]. Here, we have profiled the characteristics of upstream regulatory sequence conservation between 815 duplicated gene pairs in Arabidopsis that originated in a single polyploidy event 20-60 million years ago using three distinct methodologies in order to evaluate the effects of URS evolution on expression correlation. Previously, it was reported in Arabidopsis that borderline significant similarity persists in URSs of WGD-derived duplicates and that this similarity was correlated to expression differences [9]. However, the use of a simple alignment measure likely underestimated the amount of conservation between duplicated URSs, and consequently, the correlations reported for URS similarity and expression correlation were probably also underestimated. The outperformance of DIALIGN-TX by specialized programs (IR and SMM) for measuring evolutionarily conserved regions in non-coding DNA sequences is not surprising (Table 1). This is partly explained by the nature of these specialized programs; both inherently check forward and reverse strands for similarity, whereas dtx considers only the forward strand in its alignments. Additionally, with the allowance for translocations and inversions of local regions of similarity, the specialized methods will be more sensitive at detecting conserved regions. These factors suggest that the failure to find stronger correlations in previous studies is due to the inferiority of traditional alignment-based approaches in measuring similarity in non-coding DNA.

Despite the high inter-program correlations observed between the SMM and the IR, their values for any individual sequence pair can be drastically different. For example, the sequence pair At3g49910 and At5g67510 (both code for 60S ribosomal protein) scored very differently with respect to the two programs. The IR value was 0.019 (median 300 bp anchored window = 0.007) and the SMM score was 0 (median 300 bp anchored window = 0.25). The number of exact matches shared between the two sequences reflects the difference in scores between the two programs. There are only 16 instances of exact matches of length greater than or equal to 8 bp, and no instances of length greater than or equal to 12 bp shared between At3g49910 and At5g67510. As mentioned in the methods, the SMM requires that an alignment score threshold be set (L=12 for the 300 bp anchored window); consequently, our chosen L value excluded all alignments that did not have an exact match of at least 12 bp in these two sequences. This scenario is likely repeated for other sequence pairs, giving rise to the observed differences in correlation values reported in Table 1.

Our analysis extended previous work by revealing the most relevant location of the URS harboring the majority of shared sequence elements. Given our results for the sliding window analysis and considering only congruent levels of significance between the three programs and expression correlation ($p \ll 0.01$) (last three rows of Table 1), the most relevant region in the URS which is most strongly associated with duplicate gene expression falls within the first 300 bp of the TSS. Furthermore, it is unlikely that basal promoter elements, which typically reside within the first 100 bp of the TSS [21] were solely responsible for the observed correlations, as significant correlations were observed well outside of this sequence range (Table 1). Therefore, based on the patterns of similarity (Fig. 1 and 2) and in lieu of the correlations reported in Table 1, duplicated URSs in Arabidopsis have diverged rapidly and have a very limited region of conservation restricted to the immediate vicinity of the TSS.

We observed strong correlations between URSs and expression correlations in our data. Although these correlations are relatively strong compared to previous studies [9], they can only explain a minority of the shared variance between what occurs on the sequence level in the URS and what is actualized at the expression level. This is to be expected since many factors can influence co- or divergent expression of duplicate genes, such as trans factors [22], possible stochastic epigenetic effects [23] and micro-chromosomal rearrangements in URSs. Therefore, by studying positional information on exact matches in paralogous URSs, we obtained a more comprehensive look at sequence evolution that was not previously seen on a genome-wide scale (Fig. 3). Interestingly, longer exact matches tend to occur in homologous positions between query and subject sequences (Fig. 3A and 3B), whereas translocations and inversions of exact matches are more prominent in divergently expressed gene pairs (Fig. 3C). Regardless of length, the least prominent type of exact match arrangement is always "homologous inversion" or matches in homologous positions on opposite strands (green lines in Fig. 3). This suggests that this type of evolutionary event is unlikely to occur and may be unfavored as an adaptation for transcriptional regulation of duplicate genes. Moreover, the pattern in the upper and lower 25% quantiles based on EC (Fig. 3B and 3C) illustrates how micro-chromosomal rearrangements are an important consideration in studying what contributes to the expression divergence of duplicate genes.

Intriguingly, while the evolution of CDSs is coupled to expression correlation, as is URS evolution (Table 1 and Appendix Table 1), neither of the two is themselves coupled. There is no evidence in the data to suggest that one is the consequence of the other, or that the order of events could be revealed. We hypothesize that this is because these regions face distinctly different selective constraints; that is, proteins have functions that are more constrained by form (implying function) and it is likely that they face heavier selective pressures against mutations, since most changes would probably be deleterious. On the other hand, URS regions are known to exhibit high plasticity, whereby due to their modular nature they are evolutionarily flexible [30]. Furthermore, plant URSs possess the facility to tolerate a variety of evolutionary changes, such as insertions, rearrangements and other forms of mutation driven novelty that could therefore explain the evolutionary "decoupling" between CDSs and URSs [31] (see above). Nevertheless, these results suggest that different selective forces have acted on different components of the genes throughout their post-duplication history, depending on the gene's functions. Our GO analysis reveals that duplicates exhibiting conserved URSs and expression correlation tend to be associated with transcription factor activity/regulation, whereas those pairs with conserved CDS and EC tend to be involved in metabolic pathways or are structural proteins or enzymes. The incongruence between our results and those reported in [27] can be attributed to our data set being more diverse, e.g., we did not only consider gene pairs involved in oxidative stress response. With respect to the lack of correlation between K_s and EC (Appendix Table 1), it is evident that many of the duplicated genes have evolved at different rates, as evidenced by the large range of K_s and K_a values -- a result in line with previous reports [9, 10].

Though the evolution of URSs and CDSs is uncoupled, both are intimately tied to the expression correlation of the paralogs. A three-way joint rank analysis on the IR, K_a and expression correlations revealed biases in molecular function for certain gene pair categories (Appendix Tables 3 and 4). The preponderance of terms associated with transcription factor activity for those gene pairs with relatively conserved URSs but diverged CDSs are in accord with previous reports on mammalian promoter sequences [32, 33]. In contrast to this, the multitude of terms associated with enzymatic activity (Appendix Table 4) and primary metabolic processes (data not shown) for those gene pairs with non-conserved URSs but conserved CDSs, agrees with a report on mammalian house-keeping genes [34]. This makes sense, because genes involved in primary metabolic processes are likely to be ubiquitously expressed in many tissues and consequently would not require the tight regulation of expression experienced by transcription factors. Taken together, the molecular functions of duplicate genes in Arabidopsis depend on the relative conservation profiles of either the URS or the CDS.

References

1. De Bodt, S., Maere, S., Van de Peer, Y.: Genome duplication and the origin of angiosperms. Trends Ecol. Evol. 20(11), 591–597 (2005)
2. Wendel, J.F.: Genome evolution in polyploids. Plant Mol. Biol. 42(1), 225–249 (2000)

3. Masterson, J.: Stomatal Size in Fossil Plants: Evidence for Polyploidy in Majority of Angiosperms. Science 264(5157), 421–424 (1994)
4. Vision, T.J., Brown, D.G., Tanksley, S.D.: The origins of genomic duplications in Arabidopsis. Science 290(5499), 2114–2117 (2000)
5. Simillion, C., et al.: The hidden duplication past of Arabidopsis thaliana. Proc. Natl. Acad. Sci. USA 99(21), 13627–13632 (2002)
6. Bowers, J.E., et al.: Unravelling angiosperm genome evolution by phylogenetic analysis of chromosomal duplication events. Nature 422(6930), 433–438 (2003)
7. Blanc, G., Hokamp, K., Wolfe, K.H.: A recent polyploidy superimposed on older large-scale duplications in the Arabidopsis genome. Genome Res. 13(2), 137–144 (2003)
8. Blanc, G., Wolfe, K.H.: Functional divergence of duplicated genes formed by polyploidy during Arabidopsis evolution. Plant Cell 16(7), 1679–1691 (2004)
9. Haberer, G., et al.: Transcriptional similarities, dissimilarities, and conservation of cis-elements in duplicated genes of Arabidopsis. Plant Physiol. 136(2), 3009–3022 (2004)
10. Ganko, E.W., Meyers, B.C., Vision, T.J.: Divergence in Expression Between Duplicated Genes in Arabidopsis. Mol. Biol. Evol. (2007)
11. Castillo-Davis, C.I., Hartl, D.L., Achaz, G.: cis-Regulatory and protein evolution in orthologous and duplicate genes. Genome Res. 14(8), 1530–1536 (2004)
12. Subramanian, A.R., et al.: DIALIGN-T: an improved algorithm for segment-based multiple sequence alignment. BMC Bioinformatics 6, 66 (2005)
13. Subramanian, A.R., Kaufmann, M., Morgenstern, B.: DIALIGN-TX: greedy and progressive approaches for segment-based multiple sequence alignment. Algorithms Mol. Biol. 3, 6 (2008)
14. Obayashi, T., et al.: ATTED-II: a database of co-expressed genes and cis elements for identifying co-regulated gene groups in Arabidopsis. Nucleic Acids Res. 35(Database issue), D863–D869 (2007)
15. Haubold, B., Domazet-Loso, M., Wiehe, T.: An Alignment-Free Distance Measure for Closely Related Genomes. In: Nelson, C.E., Vialette, S. (eds.) RECOMB-CG 2008. LNCS (LNBI), vol. 5267, pp. 87–99. Springer, Heidelberg (2008)
16. Haubold, B., Wiehe, T.: How repetitive are genomes? BMC Bioinformatics 7, 541 (2006)
17. Yang, Z., Nielsen, R.: Estimating synonymous and nonsynonymous substitution rates under realistic evolutionary models. Mol. Biol. Evol. 17(1), 32–43 (2000)
18. Carbon, S., et al.: AmiGO: online access to ontology and annotation data. Bioinformatics (2008)
19. Ohno, S.: Evolution by Gene Duplication. Springer, Heidelberg (1970)
20. Britten, R.J., Davidson, E.H.: Gene regulation for higher cells: a theory. Science 165(891), 349–357 (1969)
21. Molina, C., Grotewold, E.: Genome wide analysis of Arabidopsis core promoters. BMC Genomics 6(1), 25 (2005)
22. Zhang, Z., Gu, J., Gu, X.: How much expression divergence after yeast gene duplication could be explained by regulatory motif evolution? Trends Genet. 20(9), 403–407 (2004)
23. Wang, J., et al.: Stochastic and epigenetic changes of gene expression in Arabidopsis polyploids. Genetics 167(4), 1961–1973 (2004)
24. Wagner, A.: Decoupled evolution of coding region and mRNA expression patterns after gene duplication: implications for the neutralist-selectionist debate. Proc. Natl. Acad. Sci. USA 97(12), 6579–6584 (2000)
25. Conant, G.C., Wagner, A.: Asymmetric sequence divergence of duplicate genes. Genome Res. 13(9), 2052–2058 (2003)

26. Gu, Z., et al.: Rapid divergence in expression between duplicate genes inferred from microarray data. Trends Genet. 18(12), 609–613 (2002)
27. Stanley Kim, H., et al.: Transcriptional divergence of the duplicated oxidative stress-responsive genes in the Arabidopsis genome. Plant J. 41(2), 212–220 (2005)
28. Ingvarsson, P.K.: Gene expression and protein length influence codon usage and rates of sequence evolution in Populus tremula. Mol. Biol. Evol. 24(3), 836–844 (2007)
29. Duret, L., Mouchiroud, D.: Determinants of substitution rates in mammalian genes: expression pattern affects selection intensity but not mutation rate. Mol. Biol. Evol. 17(1), 68–74 (2000)
30. Kirchhamer, C.V., Yuh, C.H., Davidson, E.H.: Modular cis-regulatory organization of developmentally expressed genes: two genes transcribed territorially in the sea urchin embryo, and additional examples. Proc. Natl. Acad. Sci. USA 93(18), 9322–9328 (1996)
31. Wessler, S.R., Bureau, T.E., White, S.E.: LTR-retrotransposons and MITEs: important players in the evolution of plant genomes. Curr. Opin. Genet. Dev. 5(6), 814–821 (1995)
32. Iwama, H., Gojobori, T.: Highly conserved upstream sequences for transcription factor genes and implications for the regulatory network. Proc. Natl. Acad. Sci. USA 101(49), 17156–17161 (2004)
33. Lopez-Bigas, N., De, S., Teichmann, S.A.: Functional protein divergence in the evolution of Homo sapiens. Genome Biol. 9(2), R33 (2008)
34. Farre, D., et al.: Housekeeping genes tend to show reduced upstream sequence conservation. Genome Biol. 8(7), R140 (2007)

Transcription Factor Binding Probabilities in Orthologous Promoters: An Alignment-Free Approach to the Inference of Functional Regulatory Targets

Xiao Liu[1] and Neil D. Clarke[1,2]

[1] Department of Biophysics and Biophysical Chemistry, The Johns Hopkins University School of Medicine, Baltimore, MD, USA
[2] Computational and Systems Biology Group, Genome Institute of Singapore, Singapore
clarken@gis.a-star.edu.sg

Abstract. Using a physically principled method of scoring genomic sequences for the potential to be bound by transcription factors, we have developed an algorithm for assessing the conservation of predicted binding occupancy that does not rely on sequence alignment of promoters. The method, which we call ortholog-weighting, assesses the degree to which the predicted binding occupancy of a transcription factor in a reference gene is also predicted in the promoters of orthologous genes. The analysis was performed separately for over 100 different transcription factors in S. cerevisiae. Statistical significance was evaluated by simulation using permuted versions of the position weight matrices. Ortholog-weighting produced about twice as many significantly high scoring genes as were obtained from the S. cerevisiae genome alone. Gene Ontology analysis found a similar two-fold enrichment of genes. Both analyses suggest that ortholog-weighting improves the prediction of true regulatory targets. Interestingly, the method has only a marginal effect on the prediction of binding by chromatin immunoprecipitation (ChIP) assays. We suggest several possibilities for reconciling this result with the improved enrichment that we observe for functionally related promoters and for promoters that are under positive selection.

1 Introduction

Eukaryotic transcription factors bind motifs that are typically short and frequently degenerate. Consequently, the vast majority of potential binding sites in the genome may occur by chance, rather than as the result of functional selection. Because most motifs that closely resemble binding sites are likely to be there by chance, it is difficult to use consensus binding sites to predict reliably what the targets of a transcription factor are. We have shown previously that it is possible to achieve somewhat better results using a Position Weight Matrix (PWM) to predict the relative probability of binding for all genomic subsequences, and to then integrate the contributions of all such potential binding sites to obtain the probability of binding to a genomic region (e.g., a promoter). [1, 2] Scoring of this type also lends itself naturally to realistic modeling of cooperativity and competition in binding. [1] However, the accuracy of these predictions still leaves much to be desired. A major reason for this is the effect of chromatin on binding site selection. [3]

F.D. Ciccarelli and I. Miklós (Eds.): RECOMB-CG 2009, LNBI 5817, pp. 229–240, 2009.

One way to improve the prediction of target genes is to look for the conservation of binding sites in aligned promoter regions. A variety of these "phylogenetic footprinting" methods have been developed, and they can indeed be quite powerful. [4, 5] However, there are two intrinsic limitations to phylogenetic footprinting that allow functionally important sites to be missed. The first is that the regulatory regions must be correctly aligned in order to perform the analysis. Without a reliable alignment, it is not meaningful to ask whether a particular site is conserved. This is a particular challenge in mammalian systems, but alignment of non-coding sequences is a problem even among yeast species related to Saccharomyces cerevisiae. The second limitation of phylogenetic footprinting is that the binding sites that are involved in regulating orthologous genomes need not themselves be orthologous. That is, functional binding sites in the upstream regions of orthologous genes can have independent evolutionary origins. In principle, this can happen quite readily because eukaryotic binding motifs are small, and sequences that match or resemble the preferred binding site can be expected to arise and disappear in the course of evolution. An analysis of four Drosophila genomes suggests that about 5% of functional binding sites for the Zeste transcription factor, as defined by chromatin immunoprecipitation in D. melanogaster, were gained on the D. melanogaster lineage or lost in at least one of the other lineages. [6] In Saccharomyces, ChIP-chip analyses of Ste12 have been performed in three species, directly demonstrating that a large number of binding loci differ among these species. [7]. Similar results have been obtained for four tissue-specific transcription factors in mouse and human hepatocytes. Binding loci for these factors differ in 40-90% of the cases between the two species. Even when the promoters of orthologous genes are both bound by a transcription factor, the binding sites do not align in about two thirds of the cases. [8]

A number of groups have recognized that it can be helpful to score for binding sites within orthologous regions rather than to score for strict orthology of binding sites themselves (i.e., identically located sites witihin aligned regions). [9, 10] Here, we describe a new method that takes advantage of orthologous sequences, without requiring that the orthologous sequences be aligned. The method builds on an explicit thermodynamic model for DNA binding that we have described previously. [1] Briefly, the method scores every sub-sequence within a genomic region (e.g., a promoter) using a PWM representation of DNA binding specificity and an assumed (or parameterized) protein concentration. The program then integrates over all subsequence scores to produce a combined probability for occupancy of the promoter region. The program implements a view of gene regulation that considers it possible that even very low affinity sites can contribute to binding and regulation, and that they can be selected for evolutionarily. That this is a valid perspective was demonstrated by Tanay who showed that ranked lists of ChIP-enriched promoters show predicted binding scores that are higher than expected by chance even at p-value thresholds for binding that would ordinarily be considered insignificant. [11] Furthermore, using an evolutionary model for sequence divergence, it appears that many genes have orthologs with higher predicted binding scores than would be expected if only high affinity sites were under selection. [11]

Here, we have scored S. cerevisiae genes and their orthologs in six other Saccharomyces species for the predicted binding occupancy of 102 transcription factors, and we have combined these scores to obtain a measure of binding occupancy conservation. The method improves the sequence-based prediction of biologically relevant target genes.

2 Methods

2.1 Scoring Yeast Promoter Regions for the Binding Potential of 102 Transcription Factors

For each of the seven Saccharomyces species, files containing 1000bp sequences 5' to the all ORFs were downloaded in August, 2004 from the Saccharomyces Genome Database (SGD). [12] A file called fungalAlignHits.txt, obtained from SGD in November 2004, was used to define orthologs to S. cerevisiae genes. The small number of homolog pairs that are categorized as "unresolved" in this file were considered to be orthologs for the purposes of our analysis. Upstream sequences were truncated to the 600bp 5' to each ORF, and are defined as the promoter region for that gene. S. cerevisiae genes considered "Dubious" by SGD were excluded from the analysis. Promoter regions were scored using the program GOMER. [1] GOMER uses Position Weight Matrix (PWM) descriptions of DNA binding specificity to infer relative binding affinities and estimates the probability of binding somewhere within the genomic interval. All subsequences within the interval are scored and contribute to the overall probability of binding. An assumed protein concentration equal to the K_d for an optimal site was used in all GOMER calculations. PWMs for 102 transcription factors were obtained from the file Final_InTableS2_v24.motifs, downloaded from the Fraenkel laboratory at MIT (http://fraenkel.mit.edu/Harbison/release_v24/final_set/). Many of these were obtained by motif discovery from a large scale ChIP dataset [13]

2.2 Conservation of Binding Potential

For a given PWM and a given S. cerevisiae gene, GOMER was used to calculate the binding probability scores for the S. cerevisiae gene and its ortholog in each species (assuming the ortholog exists by the definition above). A cerevisiae-centric binding conservation score, B_{cons}, was then calculated as follows:

$$B_{cons} = \frac{B_{S.cer} + \sum \min(B_i, B_{S.cer})}{7} \quad (1)$$

where B_{Scer} is the GOMER binding probability score for the S. cerevisiae gene and B_i is the GOMER binding probability score for the ortholog from species. If an ortholog is lacking in a particular species, its binding score is taken to be zero. The denominator is 7 because that is the total number of species analyzed. This scoring scheme, which implicitly attaches equal weight to all non-cerevisiae species, was found to work reasonably well. However, we also explored the value of weighting species by their evolutionary distance from S. cerevisiae. For this purpose, we used a more general form of the scoring function:

$$B_{cons} = \frac{B_{S.cer} + \sum w_i \min(B_i, B_{S.cer})}{1 + \sum w_i} \quad (2)$$

where w_i is a species-specific weight for non-cerevisiae species i. The weights were based on the relative evolutionary distances, which in turn were inferred from measuring branch lengths in published phylogenetic analyses. [14, 15]. As the measure of evolutionary distance is arbitrary, we sought to identify an appropriate scaling factor, C, for the weights. We defined weights, w_i, for Equation 1 as follows:

$$w_i = 1 + (C-1)\left(\frac{d\{S_{cer}, S_i\} - d\{S_{cer}, S_{par}\}}{d\{S_{cer}, S_i\} - d\{S_{cer}, S_{klu}\}}\right) \quad (3)$$

where C is the scaling parameter and the terms d{...} denote the evolutionary distance between two species. S_i is the species whose weight, w_i, we wish to calculate, S_{par} is S. paradoxus (the species closest to S. cerevisiae) and S_{klu} is S. kluyveri (the species furthest from S. cerevisiae). For a given scale factor C, the predicted binding score for a S. kluyveri promoter gets a weight of C, for S. paradoxus a weight of 1, and for all other species the weights interpolated between these values based on their relative evolutionary distance to S. cerevisiae. Using permuted PWMs to estimate the number of genes with significantly high scores (see below), we found the best results with C=4. Nevertheless, the effect was modest for all values tested. Indeed, by the GO term enrichment criterion the simple unweighted form of the scoring function (Equation 1) was superior (see below). Therefore, except where stated otherwise, we used the simple form of the scoring function, which assigns equal weights to all species.

2.3 Use of Permuted PWMs to Estimate the Number of Genes with Significantly High Binding Potential

To assess the statistical significance of a given score, whether ortholog weighted or based on S. cerevisiae alone, we performed simulated analyses using randomly permuted PWMs as controls. For each PWM, fifty control PWMs were generated by permuting the order of the positions and, independently, the weights of the four bases at each position. The control PWMs were then used to calculate the predicted occupancy of all promoter regions in S. cerevisiae genes. Neglecting biases in the genome sequence, the information content of the permuted PWMs is identical to the real PWM. However, non-randomness in the genome makes the distribution of promoter binding probabilities quite different for different permuted PWMs. In particular, permuted PWMs with preferences for A or T yield substantially higher predicted binding because of the preponderance of AT basepairs in the yeast genome. We therefore scaled the predicted binding of permuted PWMs to the predicted binding for the corresponding real PWM. However, binding probabilities are saturable (that is, the maximum score asymptotically approaches 1) so shape of the score distribution is not quite the same for AT rich PWMs as it is for GC-rich PWMs. Therefore, scaling of scores for permuted PWMs to those of the real PWM was not performed directly in the space of GOMER binding probabilities. Instead, binding probabilities, B, were first transformed into relative free energies of promoter binding, which are not bounded:

$$\Delta G = -RT \ln\left(\frac{B}{1-B}\right) \quad (4)$$

The result of this transformation is similar to what would be obtained from a conventional PWM score summation, and in fact becomes equivalent in the limit of low scores. These transformed GOMER scores were then normalized to the median value of the promoter binding energies obtained with the real PWM, and scaled based on the interquartile differences. Finally, these normalized and scaled binding free energies were transformed back into predicted binding probabilities for the promoter through the reciprocal operation to equation 4. Results from the fifty permuted PWMs were combined and used to determine the number of genes expected by chance at any given value of the predicted binding probability. This distribution, in turn, was used to estimate the number of genes whose binding probabilities, predicted with the real PWM, are significantly high at a 5% false discovery rate. This method for calculating the expected score distribution was extended to the ortholog-weighted measure as well. For each permuted PWM, and for each genome, the distribution of promoter binding scores was normalized and scaled to the values for the real PWM in that genome. Then, for each promoter, the ortholog-weighted binding potential was calculated as described above for the real PWM. The score distribution for permuted PWMs was used to estimate the number of significantly high scoring genes, just as was done for the scoring based on the S. cerevisiae genome only.

2.4 Gene Ontology Analyses

To assess whether ortholog weighting helps distinguish promoters that are biologically functional from those that are not, we performed Gene Ontology analysis on the top 50 genes as ranked by binding probabilities. [16] Separate analyses were performed for the top-ranked genes as calculated using S. cerevisiae promoters only and for the top ranked genes calculated using ortholog-weighting. Genes were submitted to the SGD GeneOntology Term Finder service to find Process Ontology terms that are enriched at a threshold of 10^{-2}. The results were downloaded, and parsed to determine the total number of genes that have at least one enriched GO annotation, at p-value cutoffs of 10^{-2}, 10^{-3}, 10^{-4} or 10^{-5}.

2.5 Intersection between Predicted and Experimentally Measured Binding

Chromatin immunoprecipitation data for all 102 transcription factors studied here was obtained from the work of Harbison et al, and was downloaded from the associated web site. [13] If ChIP experiments for a particular factor were performed under more than one conditions, we used the lowest p-value associated with binding to that gene to represent its ability to be bound. We assessed the intersection between predicted binding and observed binding using a p-value of 10^{-3} or better to define observed binding. For predicted binding, the top 50 genes were used.

3 Results

3.1 Scoring Transcription Factor Binding Probabilities in the Promoters of Orthologous Genes

Calculation of an ortholog-weighted binding probability score starts with the predicted probability that a transcription factor can bind to the promoter region of an

ORF in the reference genome (here, S. cerevisiae). Upstream regions of orthologous ORFs are then scored in an identical manner. The most straightforward way to use these scores to assess conservation of binding probability would be to average the scores for a gene across all species. However, we found that averaging scores without regard to the score in S. cerevisiae resulted in a substantial background of moderately high scoring genes that owe their scores to high-affinity sites in one or two non-cerevisiae species. These may be examples of non-functional binding motifs that occur by chance, but in any case, since almost all of the data available with which to evaluate the method comes from S. cerevisiae, we chose to devise a scoring algorithm that is cerevisiae-centric. In particular, the algorithm asks whether the potential to be bound in S. cerevisiae is conserved in other species. For that reason, the contribution of a non-cerevisiae species to the ortholog-weighted score is based on the smaller of two numbers: the predicted binding probability for the gene in S. cerevisiae and the predicted binding probability for its ortholog. Because the contribution of each species is averaged, a species whose gene is predicted to be bound as well or better than the S. cerevisiae gene has the effect of leaving the ortholog-weighted score unchanged from the score in S. cerevisiae. A species whose gene is predicted to be bound less well than the S. cerevisiae ortholog decreases the ortholog-weighted score. In the simplest scheme, which is used throughout except where stated otherwise, the contribution of the non-cerevisiae species is equal (i.e., the scores are uniformly weighted). We also tested the effect of weighting species based on their evolutionary distance from S. cerevisiae. Details of the scoring function and the application of evolutionary distance weights are provided in the Methods section.

3.2 Criteria for Evaluating the Utility of Ortholog-Weighted Binding Probabilities

We previously showed that a PWM-based algorithm for scoring promoter regions yields gene rankings that are moderately well correlated with transcription factor binding and with gene regulation. [1, 2] Since orthologs tend to be bound and regulated by the same transcription factors, we can expect at least some orthologous promoter sequences to be about as predictive of binding and regulation the S. cerevisiae promoters. Consequently, we can expect that a method for combining predicted binding probabilities from different species should provide greater sensitivity and specificity in identifying true regulatory targets. Here, we have taken three different approaches to assessing our method: (i) determine whether there are more genes with unexpectedly high (statistically significant) scores when calculated with ortholog-weighting than when using S. cerevisiae alone; (ii) assess the enrichment of true targets among high-scoring genes using shared GeneOntology annotations as a proxy for targets under functional selection, and (iii) determine whether the ortholog-weighted scores are more predictive of observed binding, as defined by chromatin-IP experiments.

3.3 Ortholog-Weighting Yields More Genes with Significantly High Scores

We first asked whether the method can identify more genes with significantly high scores than can be obtained using the genome of a single species. To do this, we first needed to estimate the expected distribution of binding probability scores so that we could define a significantly high score. We did this by generating simulated data with randomized PWMs. This was performed separately for each PWM because the size, degeneracy, and base pair composition of the PWM has a substantial effect on the expected distribution of scores. For each of the 102 real PWMs, we generated 50 permuted versions, first by scrambling positions and then by scrambling the base-specific weights within each position. The 50 control PWMs that were derived from each real PWM were used to predict the probability of "binding" to the promoter regions of S. cerevisiae and to the promoter regions of the orthologous genes. These were then used to estimate the number of genes whose scores with the real PWM are significantly high at a 5% false discovery rate. This was done for S. cerevisiae promoters alone, and for the incorporation of orthologous promoter scores into an ortholog-weighted binding probability score. For about half of the transcription factors, there were no genes with scores significantly higher than expected, regardless whether multiple genomes were used or not. For the remaining PWMs, we found that ortholog-weighting almost always increased the number of significantly high scoring genes (Figure 1). This result was not sensitive to how the non-cerevisiae species were weighted: very similar results were obtained whether non-cerevisiae species were weighted uniformly, or in a way that depends on evolutionary distance from S. cerevisiae (data not shown). The greater number of genes that have significantly high scores is not a trivial consequence of conservation in general because if that were the case, the permuted PWMs would show the same level of "conservation" as the real PWMs.

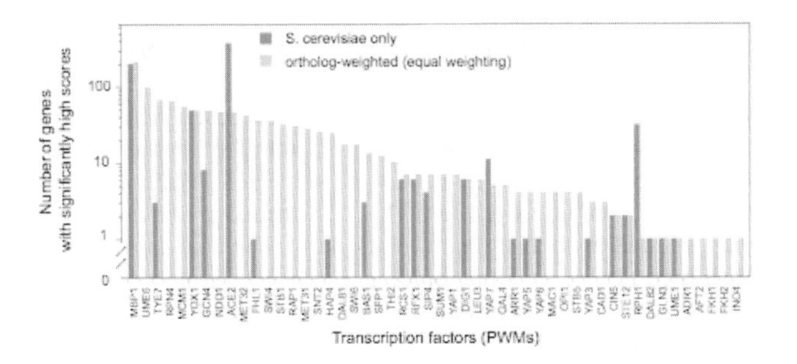

Fig. 1. Ortholog weighting increases about two-fold he number of genes with significantly high predicted-binding scores. For each transcription factor (PWM), all S. cerevisiae genes were scored for predicted binding occupancy, and the significance assessed (Methods). Similarly, promoters of orthologous genes were scored and used to determine an ortholog-weighted binding score, whose significance was assessed in a similar manner. For the subset of PWMs that yield significantly high scores, the number of genes identified is indicated. With three exceptions, more genes are found with ortholog weighting (gray) than without (black).

3.4 Ortholog Weighting Improves the Predicted Regulation of Genes with Shared Biological Functions

In the absence of a positive selection for binding, binding probability scores would be no more correlated across species than are the permuted PWMs. However, we find that ortholog-weighting does yield substantially more high-scoring genes, which means the algorithm is enriching for genes that are under positive selection for binding in more than one species. Selection for binding, of course, is really a consequence of selection for transcriptional control. Furthermore, since genes that are co-regulated are more likely to have shared biological functions, scoring genes for conserved binding probabilities should also enrich for genes that have Gene Ontology annotations in common.

To test this, we performed Gene Ontology analysis on the highest scoring genes as ranked by S. cerevisiae promoter scoring only and, separately, by the ortholog-weighted binding probability. For each of the 102 transcription factors (PWMs), the top 50 genes from each method were assessed for over-represented Gene Ontology annotations. As shown in Figure 2, ortholog-weighting identifies about twice as many genes with significantly enriched GO annotations in common as can be found using S. cerevisiae promoters alone.

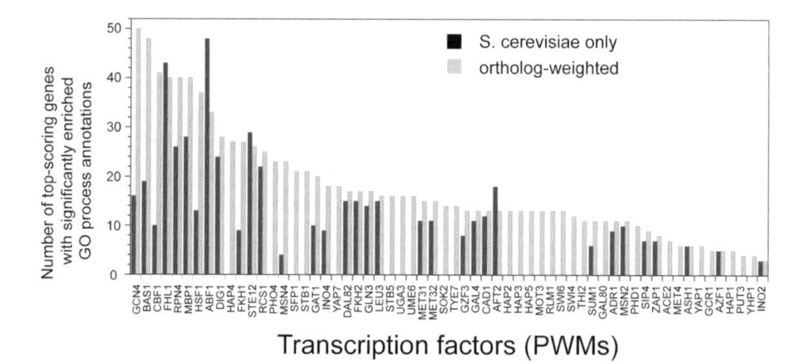

Fig. 2. Ortholog weighting increases the number of top-scoring genes that have significant Gene Ontology annotations. For each transcription factor (PWM) the top 50 genes based on predicted promoter occupancy in S. cerevisiae were subjected to GO analysis to find annotations that are over-represented at a p-value of 10^{-2}. The number of genes that had at least one such annotation is indicated in the histogram (black bars). The analysis was repeated for the top 50 genes as defined by ortholog-weighted binding occupancy (gray bars). Transcription factors that do not yield gene sets with GO enrichment using either method are not shown. The superior performance of the ortholog weighting method persists to a similar degree at higher stringencies of GO-term enrichment (e.g., $10^{-3}$and 10^{-5}; data not shown).

3.5 Ortholog-Weighted Binding Probabilities Are Only Marginally More Predictive of ChIP Data Than Are Predictions Based on a Single Genome

To test whether ortholog-weighted binding probabilities can improve the prediction of actual binding in S. cerevisiae, we took the same sets of 50 top-scoring genes used in the GO analysis and asked how much of an overlap there is between these genes and the genes that have been identified as being bound by Young and colleagues in their large scale ChIP analyses in S. cerevisiae. [13] There are substantial inaccuracies in these predictions due to inadequacies in the computational model, the effects of chromatin structure in modulating binding site selection and the availability of ChIP data for only a limited number of conditions. [1, 3] Nevertheless, as shown in Figure 3, most of the transcription factors (PWMs) that we studied predict binding to S. cerevisiae promoters that significantly overlaps with ChIP-defined binding. Furthermore, for many of the PWMs, ortholog-weighting of the predicted binding probability increases the number of genes in the intersection between predicted and experimentally bound. However, it is important to note that this improvement in predicted binding is considerably more modest than the improvements we observed in our other two criteria for evaluating ortholog-weighting. 44 of the PWMs find greater numbers of ChIP-bound genes using ortholog-weighting, but 21 find fewer. On average, a little less than one additional gene is correctly predicted to be bound for each transcription factor compared to the number obtained using S. cerevisiae promoters only.

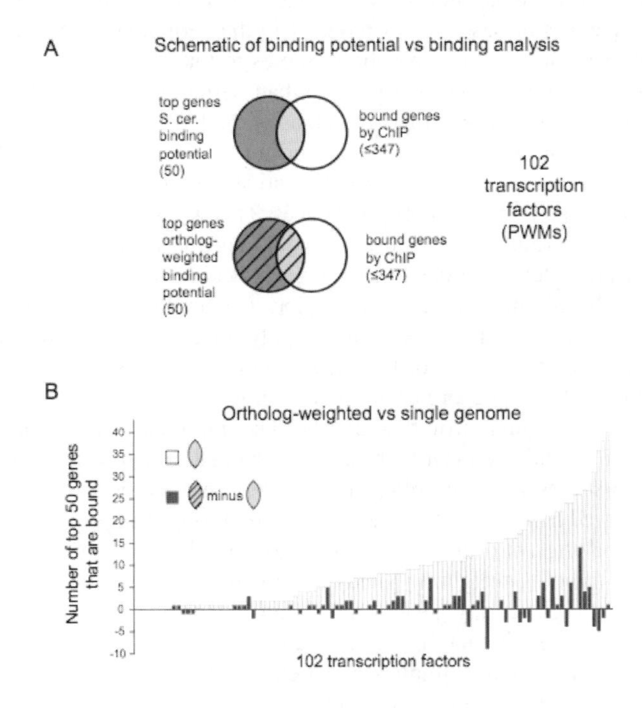

Fig. 3. Ortholog weighting has only a modest effect on the overlap between predicted and ChIP-defined binding. (A) For each transcription factor, the top 50 predicted targets were identified as defined using S.cervisiae alone (solid dark circle) or by ortholog-weighting (hatched dark circle). These lists were compared to bound promoter regions, (Harbison et al, p≤103). (B) For each transcription factor, the intersection between the top 50 predicted targets, as calculated using the S. cerevisiae genome alone, and the set of bound genes, as defined by ChIP, is indicated by the white bars. The change in this measure when ortholog-weighting is applied is shown in black.

4 Discussion

By two independent measures, we find that ortholog-weighting of binding probabilities increases the number of inferred regulatory targets by about two-fold. Interestingly, this did not translate into a substantial improvement in the prediction of binding. Before considering explanations for the poorer performance of the ChIP prediction criterion, it is worth reviewing what each of these criteria is actually measuring, how the assessment is done, and what the alternatives might have been.

The first measure of significance we assessed was the score of a gene when compared to the distribution of scores obtained with permuted PWMs. The method identifies genes with unexpectedly high score in a single species, and genes with unexpectedly high scores when orthologs are taken into account. The former implies functional selection for binding in S. cerevisiae, the latter implies conservation of that functional selection in other species. To construct the permuted PWMs, we permuted both the columns (positions) and rows (log-odds scores of the bases), We then dealt with the consequences of base bias differences in the permuted PWMs by scaling the distribution of promoter binding probability scores to that obtained for the real PWM. Although this is a more complicated process than permuting just the PWM columns, which would preserve the base composition bias of the PWM [9, 17], column permutation appears to be insufficiently random for our purposes, at least for PWMs that have only a few information-rich positions. An alternative method for assessing significance would have been to permute promoter sequences. This would not be quite as simple as the usual sequence permutation because we would need to preserve permutation order in orthologous promoters in order to assess conservation of binding probabilities. In retrospect, though, this may have been simpler. A completely different approach to assessing conservation would have been to apply an evolutionary model to the sequence divergence of promoters [6, 11, 18] An approach of this type has been used to show that promoter binding energies are more conserved than can be explained by the conservation of just the high-affinity sites. [11] The important thing is that the statistical criterion that we did use clearly shows a substantial enrichment in the number of genes with high scores when orthologs are included in the scoring. The two fold improvement observed here thus establishes a lower limit on the improvement that can be expected from inclusion of orthologous sequences from these genomes.

The fact that ortholog-weighting yields more genes with significant scores lead to the prediction that high scoring genes would show Gene Ontology term enrichment as well. The reasoning here is that high conservation scores imply selection for binding; a shared selection for binding implies a shared selection for gene regulation; and a shared selection for gene regulation implies a greater-than-random enrichment of shared biological functions. However, although there is a logical connection between an increase in high-scoring genes and an increase in shared Gene Ontology annotations, the methods for evaluating these criteria are independent. For the GO analysis, we did not require that genes be high scoring by the statistical criterion. Instead, we used the top 50 genes from each of the two scoring methods (single genome and ortholog-weighted). This was partly because we did not want the GO assessment to be dependent on the particular measure of statistical significance that we adopted, and partly because comparisons between the two methods methods can be made more readily if the same number of genes are used as input into the GO analysis. We found that orthology weighting increases the number of genes that have shared gene ontology terms by about two fold, and that roughly the same enrichment is observed at p-value stringencies for shared GO terms that range from 10^{-2} to 10^{-5}. We conclude that ortholog-weighting of binding probabilities enriches for genes with shared biological function. This would not be expected unless the genes with shared functions tended to be true targets of the transcription factor.

Given the evidence that ortholog-weighting helps identify genes that are true targets of the transcription factor, it is perhaps surprising that the overlap with ChIP data is so modest. There are at least three possible explanations for this. First, the thresholds for ChIP-enrichment ($p<10^{-3}$) and for high-scoring binding probabilities (top 50 genes) are arbitrary. A more dramatic effect might be revealed if we had chosen better criteria for evaluating the correlation between predicted and observed. We did try, of course, to find such criteria, but the fact that we did not succeed does not mean that they do not exist. Second, it may be that a substantial fraction of binding that can be observed in S. cerevisiae is not under functional selection for binding in other species, and perhaps not even in S. cerevisiae. To the extent that there is binding that is not under selection in other species, methods that rely on conversation will not show any improvement over methods that rely on a single genome. Third, the ChIP data are incomplete. For most factors, binding has only been assayed under normal laboratory growth conditions, but transcription factors themselves are expressed or activated differently in different media, and when cells are subject to various environmental stresses. Unless the ChIP experiment were performed under relevant conditions, we may not expect an improved prediction of binding to translate into an improved overlap with ChIP-enriched genes.

The method that we have described here may have special utility in helping to identify regulatory targets in mammals, where genomic alignments are often uncertain. Mammals, like other multicellular organisms, are also characterized by highly complex regulatory regions that involve multiple transcription factors. It has recently been shown that the conserved orientation and spacing of motif pairs in the regulatory regions of mouse and human orthologs allows regulatory interactions to be inferred in an alignment-free manner. [19] The algorithm described here can also be used to take into account cooperative binding, as the GOMER program that was used to calculate binding potentials has a flexible scoring function that readily accommodates multiple cooperative binding partners. A substantial difficulty in applying this or any other method, however, is knowing how to define a regulatory region in mammalian genes.

Acknowledgments. This work was funded by NIH grant GM065179 and by the Agency for Science, Technology and Research (Singapore).

References

1. Granek, J.A., Clarke, N.D.: Explicit equilibrium modeling of transcription-factor binding and gene regulation. Genome Biol. 6, R87 (2005)
2. Liu, X., Clarke, N.D.: Rationalization of gene regulation by a eukaryotic transcription factor: calculation of regulatory region occupancy from predicted binding affinities. J. Mol. Biol. 323, 1–8 (2002)
3. Liu, X., Lee, C.K., Granek, J.A., Clarke, N.D., Lieb, J.D.: Whole-genome comparison of Leu3 binding in vitro and in vivo reveals the importance of nucleosome occupancy in target site selection. Genome Res. 16, 1517–1528 (2006)

4. Cliften, P., Sudarsanam, P., Desikan, A., Fulton, L., Fulton, B., Majors, J., Waterston, R., Cohen, B.A., Johnston, M.: Finding functional features in Saccharomyces genomes by phylogenetic footprinting. Science 301, 71–76 (2003)
5. Wasserman, W.W., Sandelin, A.: Applied bioinformatics for the identification of regulatory elements. Nature reviews 5, 276–287 (2004)
6. Moses, A.M., Pollard, D.A., Nix, D.A., Iyer, V.N., Li, X.Y., Biggin, M.D., Eisen, M.B.: Large-scale turnover of functional transcription factor binding sites in Drosophila. PLoS computational biology 2, e130 (2006)
7. Borneman, A.R., Gianoulis, T.A., Zhang, Z.D., Yu, H., Rozowsky, J., Seringhaus, M.R., Wang, L.Y., Gerstein, M., Snyder, M.: Divergence of transcription factor binding sites across related yeast species. Science 317, 815–819 (2007)
8. Odom, D.T., Dowell, R.D., Jacobsen, E.S., Gordon, W., Danford, T.W., MacIsaac, K.D., Rolfe, P.A., Conboy, C.M., Gifford, D.K., Fraenkel, E.: Tissue-specific transcriptional regulation has diverged significantly between human and mouse. Nature genetics 39, 730–732 (2007)
9. Aerts, S., van Helden, J., Sand, O., Hassan, B.A.: Fine-Tuning Enhancer Models to Predict Transcriptional Targets across Multiple Genomes. PLoS ONE 2, e1115 (2007)
10. Pritsker, M., Liu, Y.C., Beer, M.A., Tavazoie, S.: Whole-genome discovery of transcription factor binding sites by network-level conservation. Genome Res. 14, 99–108 (2004)
11. Tanay, A.: Extensive low-affinity transcriptional interactions in the yeast genome. Genome Res. 16, 962–972 (2006)
12. Saccharomyces Genome Database
13. Harbison, C.T., Gordon, D.B., Lee, T.I., Rinaldi, N.J., Macisaac, K.D., Danford, T.W., Hannett, N.M., Tagne, J.B., Reynolds, D.B., Yoo, J., Jennings, E.G., Zeitlinger, J., Pokholok, D.K., Kellis, M., Rolfe, P.A., Takusagawa, K.T., Lander, E.S., Gifford, D.K., Fraenkel, E., Young, R.A.: Transcriptional regulatory code of a eukaryotic genome. Nature 431, 99–104 (2004)
14. Phillips, M.J., Delsuc, F., Penny, D.: Genome-scale phylogeny and the detection of systematic biases. Molecular biology and evolution 21, 1455–1458 (2004)
15. Rokas, A., Williams, B.L., King, N., Carroll, S.B.: Genome-scale approaches to resolving incongruence in molecular phylogenies. Nature 425, 798–804 (2003)
16. Ashburner, M., Ball, C.A., Blake, J.A., Botstein, D., Butler, H., Cherry, J.M., Davis, A.P., Dolinski, K., Dwight, S.S., Eppig, J.T., Harris, M.A., Hill, D.P., Issel-Tarver, L., Kasarskis, A., Lewis, S., Matese, J.C., Richardson, J.E., Ringwald, M., Rubin, G.M., Sherlock, G.: Gene ontology: tool for the unification of biology. The Gene Ontology Consortium. Nature Genetics 25, 25–29 (2000)
17. Vardhanabhuti, S., Wang, J., Hannenhalli, S.: Position and distance specificity are important determinants of cis-regulatory motifs in addition to evolutionary conservation. Nucleic acids research 35, 3203–3213 (2007)
18. Moses, A.M., Chiang, D.Y., Pollard, D.A., Iyer, V.N., Eisen, M.B.: MONKEY: identifying conserved transcription-factor binding sites in multiple alignments using a binding site-specific evolutionary model. Genome Biol. 5, R98 (2004)
19. Hu, Z., Hu, B., Collins, J.F.: Prediction of synergistic transcription factors by function conservation. Genome Biol. 8, R257 (2007)

Author Index